Nanotechnology Methods for Neurological Diseases and Brain Tumors

Nanotechnology Methods for Neurological Diseases and Brain Tumors

Drug Delivery across the Blood–Brain Barrier

Yasemin Gürsoy-Özdemir

Koç University School of Medicine
Research Center for Translational Medicine
Istanbul, Turkey

Sibel Bozdağ-Pehlivan

Hacettepe University
Faculty of Pharmacy, Department of Pharmaceutical Technology
Ankara, Turkey

Emine Sekerdag

Koç University
Research Center for Translational Medicine, Neuroscience Research Lab
Istanbul, Turkey

Academic Press is an imprint of Elsevier
125 London Wall, London EC2Y 5AS, United Kingdom
525 B Street, Suite 1800, San Diego, CA 92101-4495, United States
50 Hampshire Street, 5th Floor, Cambridge, MA 02139, United States
The Boulevard, Langford Lane, Kidlington, Oxford OX5 1GB, United Kingdom

Library of Congress Cataloging-in-Publication Data
A catalog record for this book is available from the Library of Congress

British Library Cataloguing-in-Publication Data
A catalogue record for this book is available from the British Library

ISBN: 978-0-12-803796-6

Publisher: Mara Conner
Acquisition Editor: Natalie Farra
Editorial Project Manager: Kathy Padilla
Production Project Manager: Lucía Pérez
Designer: Mark Rogers

Typeset by Thomson Digital

Contents

List of Contributors

Ayca Akgoz, MD Hacettepe University, Ankara, Turkey
Eren Aytekin, MD Hacettepe University, Ankara, Turkey
Sibel Bozdağ-Pehlivan, PhD Hacettepe University, Ankara, Turkey
Elif Bulut, MD Hacettepe University, Ankara, Turkey
Meltem Çetin, PhD Ataturk University, Erzurum, Turkey
Tugba Copur, BSc Hacettepe University, Ankara, Turkey
Ozgun F. Duzenli, BSc Hacettepe University, Ankara, Turkey
Melike Ekizoğlu, PhD Hacettepe University, Ankara, Turkey
Yasemin Gürsoy-Özdemir, MD, PhD Koç University School of Medicine; Research Center for Translational Medicine, Istanbul, Turkey
Tugba Gulsun, PhD Hacettepe University, Ankara, Turkey
Margareta Hammarlund-Udenaes, PhD Uppsala University, Uppsala, Sweden
Nihan Izat, BSc Hacettepe University, Ankara, Turkey
Asli Kara, MSc Hitit University, Corum, Turkey
Kader Karlı Oguz, MD Hacettepe University, Ankara, Turkey
Ayşe Filiz Oner, PhD Hacettepe University, Ankara, Turkey
Levent Oner, PhD Hacettepe University, Ankara, Turkey
Özgur Öztop-Çakmak, MD Koç University Hospital, Istanbul, Turkey
Naile Ozturk, BSc Hacettepe University, Ankara, Turkey
Muhammed Abdur Rauf, PhD Yeditepe University, Istanbul, Turkey
Selma Sahin, PhD Hacettepe University, Ankara, Turkey
Emine Sekerdag, MSc Koç University, Istanbul, Turkey
Erdem Tüzün, MD Institute for Experimental Medicine, Istanbul University, Istanbul, Turkey
Yagmur Cetin Tas, MD Koç University, Research Center for Translational Medicine, Istanbul, Turkey
Banu Cahide Tel, PhD Hacettepe University, Ankara, Turkey

Ebru N. Vanli-Yavuz, PhD Koç University Hospital, Istanbul, Turkey
Imran Vural, PhD Hacettepe University, Ankara, Turkey
Gul Yalçin Çakmakli, MD Hacettepe University, Ankara, Turkey
Burçin Yavuz, PhD Hacettepe University, Ankara, Turkey
Chi Zhang, MSc Renji Hospital, Shanghai Jiao Tong University School of Medicine, Shanghai, China
Qizhi Zhang, PhD Fudan University, Shanghai, China

About the Editors

Prof. Yasemin Gürsoy-Özdemir
E-mail: ygursoy@ku.edu.tr

Yasemin Gürsoy-Özdemir (MD, PhD) is a Professor of Neurology and Neuroscience at Koç University, School of Medicine in Istanbul, Turkey. Dr. Gürsoy-Özdemir's main research interests are basic pathophysiological aspects of neurological diseases, especially migraine, headache, stroke, neurodegenerative diseases, and their translation to clinical neurology. She conducted several experimental studies about the blood–brain barrier changes during neurological diseases and targeted drug delivery strategies. She is the author of several research articles published in peer-reviewed indexed journals. She did her postdoctoral studies at Mass General Hospital, Harvard University, Boston, MA, United States. She was the Associate Director of Hacettepe University Institute of Neurological Sciences and Psychiatry, Ankara, Turkey, between 2011 and 2014. She gave numerous lectures at both the Faculty of Medicine for Neurology Board examination and Institute of Neurological Sciences and Psychiatry for PhD programs. She received several awards, such as the Research Encouragement Award from the Brain Research Organization; Young Researcher Awards from Hacettepe University, Turkish Academy of Sciences; and Young Investigator Award from The Scientific and Technological Council of Turkey (TUBITAK).

Assoc. Prof. Sibel Bozdağ Pehlivan
E-mail: sbozdag@hacettepe.edu.tr

Sibel Bozdağ Pehlivan (PhD) is an Associate Professor at Hacettepe University, Faculty of Pharmacy Pharmaceutical Technology Department in Ankara, Turkey. Her research interests are designing and evaluating nanotechnology-based drug delivery systems in in vitro/in vivo settings, targeted drug delivery for brain tumors, and ocular drug delivery. Dr. Bozdağ Pehlivan acted as a supervisor and cosupervisor for several MSc and PhD research students under both the Pharmaceutical Technology and Nanotechnology Nanomedicine programs. She is the author of several research articles published in peer-reviewed indexed journals and is a member of the American Association of Pharmaceutical Scientists. She received several awards, such as the Hacettepe University Research Group Award and Hacettepe Technopolis Innovation Competition Award: Health Sciences and Technologies (2008 and 2009).

PhD Candidate, Emine Sekerdag
E-mail: esekerdag15@ku.edu.tr

Emine Sekerdag (MSc), earned both her Bachelor of Science degree and Master of Science degree in Biopharmaceutical Sciences at Leiden University, Leiden, The Netherlands. She pursued the first part of her Master's study at the Drug Delivery Technology Department of the Leiden Academic Center for Drug Research (LACDR) at Leiden University, where she worked on transdermal drug delivery with vaccine-coated solid microneedles, for which she developed a coating procedure. She also obtained in vitro/in vivo experience in this field, and drug delivery technologies became one of her passions. She completed the last part of her Master's study at the Pharmaceutical Technology Department of Hacettepe University in Ankara, Turkey, where she worked with hybrid nanoparticles for brain-targeted nasal drug delivery. Currently, she is a PhD candidate at the Neuroscience Department of Koç University, working with targeted treatment strategies for neurological disorders.

Preface

Neurological diseases are prevalent, they affect approximately a billion people worldwide, cause 12% of total deaths globally, and have led to an estimated economical cost of 139 billion euros in Europe alone in 2004.[1] According to the World Health Organization (WHO) this burden, a serious threat to the global public health, which generally starts with disability and rehabilitation and ends up in morbidity, will increase further to a global uncontrollable threat in 2030, unless immediate action is taken.

In addition to neurological diseases, brain tumors, especially malignant ones, are one of the world's leading causes of death. With 256,000 people affected worldwide in 2012,[2] the incidence of brain tumors is increasing every year. The expected number of brain and spinal cord cancer cases in the United States in 2014 was 23,380, and nearly 61% of the affected people (14,320) died from this burden in the same year.[3]

Neurological diseases, as well as brain tumors, with their complex pathogenesis are not yet fully understood and therefore are extremely difficult to treat. The biggest bottleneck in brain drug delivery is the blood–brain barrier (BBB), which allows the entry of only 1% of drugs, and hence requires special attention so that new treatment strategies can be developed. The currently available treatment approaches have led to many side effects and/or are not effective enough. In addition, invasive treatment approaches are no longer favored by patients, as well as by clinicians, hence they are considered to be the last step for treatment. To this end, new brain-targeted treatment strategies have become a need in the clinical and scientific fields of neurological diseases and neurooncology. The main challenge is for drugs to cross or bypass the BBB structure, which isolates the brain from the systemic circulation, so that most therapeutic compounds can enter into the brain.

In this volume, experts in the field have compiled the latest and potential treatment strategies of neurological diseases, in particular Alzheimer, Parkinson, and stroke, and brain tumors. Part 1 explains BBB structure in detail and will lead to a better understanding of how drugs can reach the brain through brain drug delivery. Part 2 focuses on the implementation of the nose-to-brain route in brain-targeted drug delivery. Furthermore, nanotechnology-based brain drug delivery, discussed in Part 3, can offer substantial improvement in the treatment of neurological diseases and brain tumors through the use of bio-engineered systems that interact with biological systems at a molecular level. Additional emphasis will be placed, in Part 4, on the need of brain-targeted experimental models that mimic disease conditions. Parts 5 and 6 will discuss the latest advances in targeted treatment strategies for neurological diseases and brain tumors. At last, a Future Outlook on this field is incorporated in this book.

References

1. World Health Organization. *Neurological Disorders: Public Health Challenges*. Geneva, Switzerland: World Health Organization; 2006.
2. International Agency for Research on Cancer, World Health Organization. GLOBOCAN 2012: estimated cancer incidence, mortality and prevalence worldwide in 2012. Available from: http://globocan.iarc.fr/Pages/fact_sheets_cancer.aspx
3. American Cancer Society. *Cancer Facts & Figures 2013*. Atlanta: American Cancer Society; 2013.

PART 1

The Rationale to Reach the Brain

CHAPTER 1

Anatomy and Physiology of the Blood–Brain Barrier

Yasemin Gürsoy-Özdemir, MD, PhD*,, Yagmur Cetin Tas, MD†**

*Koç University School of Medicine, Istanbul, Turkey;

**Research Center for Translational Medicine, Istanbul, Turkey;

†Koç University, Research Center for Translational Medicine, Istanbul, Turkey

CONTENTS

1 INTRODUCTION

It is crucial to maintain a proper hemostasis within the central nervous system (CNS) in order to establish undisturbed and proper brain functioning, since transmission of signals in the CNS occurs through combined action of both chemical and electrical signals. To sustain this signal transduction, it is mandatory to strictly regulate ionic hemostasis around the synapses, which are the main elements of signal transmission between neurons. While providing a proper balance for normal functioning, the required energy and materials must be carried from circulation into the brain tissue at the same time. For this reason, the relation of the brain tissue with systemic vasculature is quite different from other tissues. Hence, material transfer is provided by a special anatomical and physiological barrier, namely blood–brain barrier (BBB) for the CNS. It is quite specialized and tightly controlled. BBB endothelia are highly differentiated as "door keepers" to perform the complex control function of material transfer. Endothelia in other tissues allow free passage of substances into organs, whereas it is strickly regulated in highly specialized BBB-forming endothelial cells. In addition to this functional limited passage of necessary substances, brain is an immune-privileged tissue when compared to other organs in the body. As a result, there is a strict regulation of trafficking of cells, ions, and molecules from blood through the brain, as well as in the opposite direction, from the brain into circulating blood.

Presence of such a barrier was first described by Paul Ehrlich's research.[1,2] He injected water-soluble dyes into the systemic circulation of animals and found out that all the dyes he studied stayed in peripheral organs, but could not penetrate the brain tissue and spinal cord. Later, Ehrlich's student, Edwin Goldmann, demonstrated that a dye injected to brain and cerebral spinal fluid (CSF) remained in the CNS and could not pass through peripheral circulation, and hence to the

Nanotechnology Methods for Neurological Diseases and Brain Tumors. http://dx.doi.org/10.1016/B978-0-12-803796-6.00001-0

other organs.[3] The observations drawn from these dye studies brought about the concept of a barrier between blood and brain, as well as between blood and CSF.[4]

Later on studies have demonstrated that in addition to the BBB, CNS has other barriers. Three barrier layers contribute to the separation of the blood and neural tissues are listed below:

1. The BBB.
2. The blood–CSF barrier (BCSFB), with the choroid plexus epithelium, which secretes specialized CSF into the cerebral ventricles.
3. The arachnoid epithelium separating the blood from the subarachnoid CSF.[5]

However, these last two barriers do not have a big surface area compared to proper BBB, and are not very tightly regulated. That's why we will focus on the BBB in this chapter, as it is the main target for drug delivery across the brain tissue.

2 STRUCTURE OF THE BBB

BBB is a complex structure that is located at the interface of blood and the brain tissue. It consists of capillary endothelium connected through tight junctions (TJs), basal lamina, pericytes embedded in basal lamina and encircling the abluminal part of endothelium, astrocytic end-feet, as well as adjacent neurons (Fig. 1.1). It is necessary to understand the "neurovascular unit" concept to conceive the functions and importance of BBB for CNS. A neurovascular unit is the basic and structural functional unit of the CNS that enables transfer of materials from blood to CNS according to the needs of the brain tissue and transfers waste back to vasculature.[6,7] Furthermore, the neurovascular unit is mainly composed of cellular elements of the BBB, such as a capillary (feeding one neuron) and an astrocyte (providing the communication of both neuron and its surrounding with a capillary). Functional status of this one unique neuron is transferred via astrocyte to the capillary and in turn the capillary modifies the blood flow according to the needs of the neuron, hence necessary energy and nutrients are supplied. This process is called *neurovascular coupling*.[6] Via this coupling, microcirculation can sense the needs of functioning areas of the nervous system and increase or decrease the blood flow accordingly with the help of pericytes located at the abluminal side of the endothelium. The BBB has a big surface area so that it can establish this important transfer function throughout the whole brain tissue, which weighs around 1.3–1.4 kg.[8] A volume of 1 mm^3 of human cortex contains a surface area of microcirculation of about 10 cm^2.[9] This huge surface area is necessary for the normal functioning of CNS in physiological conditions, and it may serve as a potential surface area for drug delivery to brain if the selective drug transport systems could be established.

The innermost compartment of the neurovascular unit is a single-layered capillary endothelia (Fig. 1.1). This continuous and highly specialized endothelial

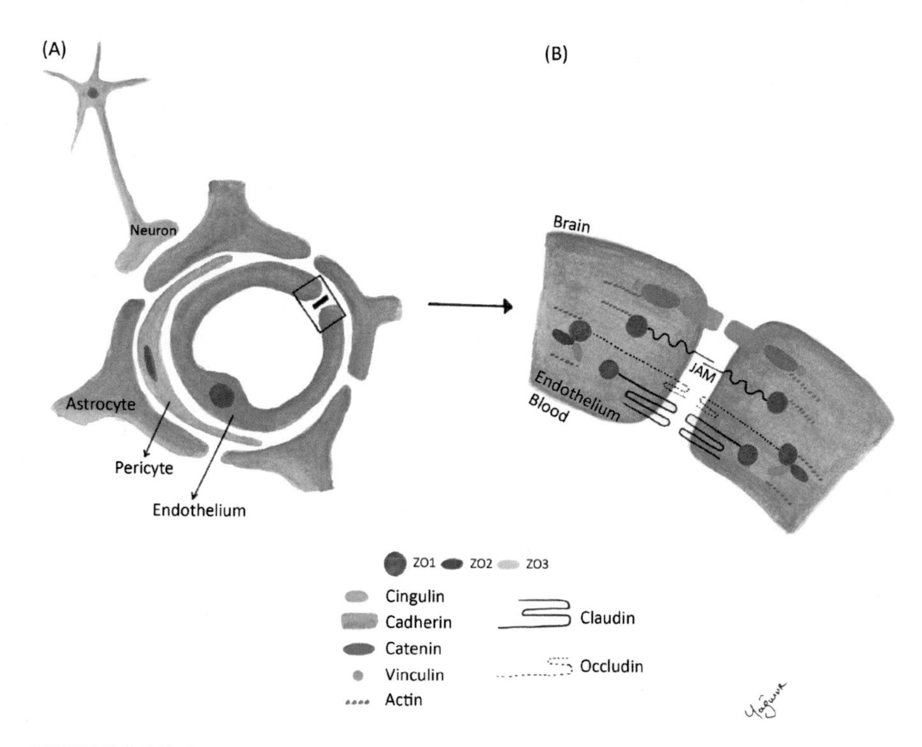

FIGURE 1.1 (A) Cellular elements of the blood–brain barrier (BBB) are displayed. There is a dynamic interaction of astrocyte end-feet, pericytes, and endothelial cells. (B) Structure of tight (TJ) and adherence junctions (AJ) are schematized.

cell layer is the most important part for the formation of tight regulation across vasculature to establish a controlled pathway. CNS endothelia have significantly different properties compared to endothelial cells in other tissues. They have TJs and adherence junctions (AJ), very few pinocytic vesicles, contain more mitochondria,[2,5,10] and they lack fenestrations, but have specific transport systems. These transport systems mediate the directed and controlled transport of nutrients from blood into CNS and the removal of toxic metabolites out of CNS.[11]

Junctional complexes between endothelia are important for the formation of barrier properties. They are formed from TJs and AJs (Fig. 1.1). TJs are located more apical than AJs and limit the passage of polar solutes through paracellular pathways.[12] They are formed by occludin, claudins, and junctional adhesion molecules.[12] Occludin and claudins are linked to cytoplasmic zonula occludens proteins. Presence of these proteins is important for the proper functioning of BBB, as knocking out claudin proteins can lead to BBB disruption.[13] On the other hand, AJs are formed by cadherin family proteins and they are mainly responsible for structural support. Their presence in the endothelia produces tightness of BBB, which can be measured via the electrical resistance through endothelia (transendothelial resistance or TEER), which is high compared to endothelial lining of vasculature in other tissues.

Basement membrane located beneath the endothelia and embedding pericytes and astrocytic end-feet is composed of collagen type 4, laminin, fibronectin, proteoglycans, and extracellular matrix proteins.[2,12] Multiple basal lamina proteins, matrix metalloproteases (MMPs) and their inhibitors, the tissue inhibitor of metalloproteases (TIMPs), are involved in the dynamic regulation of the BBB in physiological, as well as inflammatory conditions.[14] For example, during disease processes, such as stroke or migraine, MMP-9 is released from cells located in BBB and leads to the breakdown of BBB and plasma leakage from vasculature into the brain tissue.[15,16]

Pericytes are contractile cells located around the abluminal side of the endothelium at precapillary arterioles, capillaries, and postcapillary venules. They can be considered as a continuation of arterial smooth muscle cells. They can regulate microcirculatory blood flow through constriction and relaxation as responce to signals drived from neural parencyhma according to neuronal needs. They are located in close relation to astrocytes and neurons.[17] Astrocytes with their end-feet cover nearly 90% of capillary abluminal area in CNS. Both astrocytes and pericytes have important functions for the formation and maintenance of a functional BBB.

3 BBB DEVELOPMENTAL STEPS

Formation of BBB occurs during embryonic life and is completed before birth.[2,18] Most of the properties and complete BBB function, including blockage of systemic dye entrance to brain tissue, is established around the 15th day of embryonic life.[19] Significant amount of studies have demonstrated that Wnt/beta catenin signaling pathway is necessary for both the formation and maintenance of a proper BBB, as genetic disruption of the pathway leads to loss of BBB properties, together with severe CNS-specific angiogenesis defects.[20–22]

Other than endothelial signaling pathways, astrocytes and pericytes have roles in BBB formation and maintenance. Astrocytes play an important role in the generation of BBB characteristics of endothelial cells. In vitro cocultures of endothelial cells with astrocytes or astrocyte-conditioned media have been shown to induce more complex TJs, elevated expression of transporters, and increased transendothelial electrical resistance.[23,24]

Similarly, pericyte recruitment is crucial for the establishment of BBB characteristics. Complete loss of pericytes in platelet-derived growth factor beta (Pdgfb) or Pdgfrb knockout mice results in CNS microhemorrhages, dysfunctional TJs, increased vascular permeability, and embryonic lethality.[25,26] Pericytes are also vital to BBB integrity during adulthood because Pdgfrb knockout mice exhibit age-dependent BBB dysfunction as a result of reduced TJ protein expression.[27]

4 TRANSPORT ACROSS THE BBB

Endothelial cells are the main location for the transfer of nutrients to the tissues. As it is discussed previously in this chapter, BBB-forming endothelial cells are specialized in this aspect. These endothelial transport pathways are the main lines for entrance to CNS and they constitute important tools for novel drug passage strategies and targeted drug delivery systems. CNS endothelial cells are highly polarized with different expression and localization patterns of proteins, such as TJs and carrier systems at either luminal or abluminal compartments (Fig. 1.2). This polarization helps endothelial cells to regulate influx and efflux transport.[28,29] Material and nutrient transfer is especially important for proper nervous system functioning, but contrary to endothelial cells in the periphery, BBB endothelia have a very limited transcytosis capacity (also known as vesicle-mediated transport), giving rise to an important obstacle for material transfer. Actually, the transfer of materials to and out of the brain under normal physiological conditions is controlled via four main routes:

1. passive transfer;
2. solute carrier proteins;
3. efflux transporters, also known as ATP-binding cassette (ABC) transporters; and
4. transport systems for macromolecules.

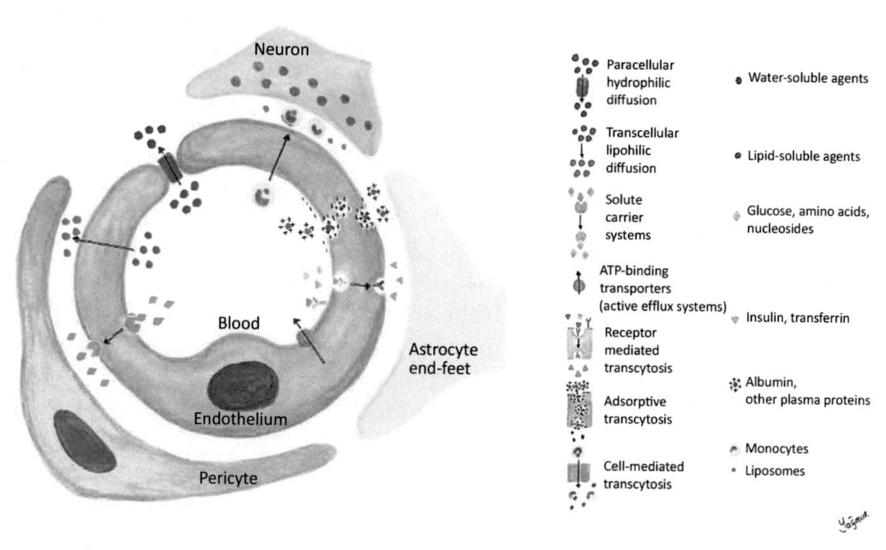

FIGURE 1.2 Schematic representation of transport systems located on BBB-forming endothelial cells.

These transport pathways may be good targets for drug delivery to the CNS. We will briefly describe these routes in more detail.

4.1 Passive Transfer or Diffusion

Passive diffusion in BBB either occurs through paracellular hydrophilic diffusion or transcellular lipophilic diffusion (Fig. 1.2). Most important factors determining passive transfer are lipophilicity, amount of hydrogen bonds, and molecular weight. In general, Lipinski's "rule of five," as well as the Abraham's equation, can be used to predict the passive transport of a drug molecule across the BBB.[30–32] Lipophilic drugs smaller than 400–600 Da can pass through endothelia freely and molecules with fewer than 10 hydrogen bonds may enter the brain via the transcellular route.[33] Bases carrying a positive charge have better penetration due to their cationic nature and ability to interact with charged heparin sulfate proteoglycans.[12] Presence of TJs and AJs (Figs. 1.1 and 1.2) is the main point for limited transfer of materials to the brain tissue. There are several lines of studies trying to open TJs and hence allowing material transfer, especially hydrophilic substance transfer, across this pathway.[34–36] Recently in a study focused on the TJ proteins, researchers have developed and tested several modulators of TJ molecules. Those TJ modulator proteins were especially successful when they targeted claudins, as detected by TEER measurement in a cell culture system. They have exposed endothelial cells to these designed peptides and produced BBB permeability changes lasting till 24 h.[34] The transient opening of BBB through TJs, with focused ultrasound application from outside of the skull and the targeting brain tissue, is also being pursued.[35] Another novel approach is the activation of A2A adenosine receptors. Researchers were able to transiently open TJs for 0.5–2.0 h with adenosine receptor–activating ligands.[36] It looks like this line of transfer is attractive and will be in focus for a long time in future studies.

4.2 Solute Carrier System

Presence of TJs, as well as junctional adhesion molecules, strictly regulates paracellular diffusion; hence many essential polar nutrients, such as glucose and amino acids, necessary for brain metabolism can't pass through. Solute carriers (transporters) located on BBB endothelia overcome this limitation (Fig. 1.2). BBB endothelia contain several important specified carriers to supply the CNS with the substances, such as glucose, amino acids, monocarboxylic acids, hormones, fatty acids, nucleotides, organic anions, amines, choline, vitamins, and hormones. Some of the well-known ones are listed in Table 1.1.[2,12,37,38] One of the well-studied transporter systems is GLUT1, which transports glucose from the circulation into the brain. Other than glucose transport, it has important roles for normal brain functioning, as GLUT-1 deficiencies in humans causes infantile seizures and mental motor retardation, and experimental studies demonstrate its important role for BBB integrity and brain glucose transport.[39]

Table 1.1 Several Examples of Significant Solute Carrier Transporter Systems

Name of Transporters	Target Molecules
Energy transport system	
GLUT1	Glucose
MCT	Monocarboxylic acids (lactate, pyruvate, and ketone bodies)
CRT	Creatine
Amino acid transport system	
LAT1	Large neutral amino acids
CAT1	Cationic amino acids
EAAT	Anionic amino acids
Neurotransmitter transport system	
GAT	Aminobutyric acid
SERT	Serotonin
NET	Norepinephrine
GLYT	Glycine
Organic anion transport system	
OAT2	Dicarboxylate
OAT3	Homovanillic acid
Oatp	Digoxin and organic anions
Oatp14	Thyroid hormones
Nucleic acid transport system	
CNT	Nucleosides, nucleotides, and nucleobases
ENT	Nucleosides, nucleotides, and nucleobases

Its endothelial expression patterns also show variations depending on disease conditions, species, and interindividual differences.[40]

Another example of an important solute carrier–mediated transporter is LAT1 for neutral large amino acids. Some of the amino acid–mimetic drugs use this pathway through the brain tissue. However, LAT1 has its own binding kinetics that are saturated with its endogenous binding proteins, such as dopamine.[38] Although solute carrier systems seem to be good targets for drug delivery to the brain, their substance specificity, together with their binding kinetics, limit their use for this purpose.

4.3 ATP-Binding Transporters of Efflux Transporters

Other than solute carrier transport systems, there are active efflux systems located in the BBB, which are members of the ABC transporter family (Fig. 1.2). They are mainly known through their efflux patterns, where the most important ones are P-glycoproteins (Pgps; multidrug resistance protein, ABCB1), the multidrug resistance–associated proteins (MRPs; ABCC1, 2,4,5, and possibly 3 and 6), and breast cancer resistance protein (BRCP; ABCG2).[12,41] Expression of Pgps is found to be significantly high in tumors and epileptic brains, which limits efficient drug transport to the brain, leading to insufficient therapeutic drug concentration in the extracellular environment due to efflux of drugs back into the circulation.[42,43] Their use for drug delivery is limited due to their intrinsic property of carrying out efflux, rather than influx. On the other hand inhibition of these efflux transporters may provide better penetration of some of the drugs that are cleaned from CNS by this route.

4.4 Transport of Macromolecules

For large molecules, such as growth hormones and most of the other proteins, pinocytosis and transcytosis are the usual way of carriage of substances across the endothelia.[2,12] Pinocytosis is a kind of (fluid-phase) endocytosis, which is a common way of substance uptake into cells in the body. However, endocytosis has three different forms:

1. Fluid-phase endocytosis
2. Adsorptive endocytosis (AMT)
3. Receptor-mediated endocytosis (RMT)
4. Cell-mediated transcytosis

Negative surface charge of endothelia may interact with positively charged proteins or molecules in the blood, leading to AMT, which is a nonselective way of transport across BBB. Albumin transport mainly occurs via this pathway, and cationized albumin was used as brain-targeted drug delivery strategy in experimental models of neurodegeneration.[44] On the other hand, most of the macromolecules, such as proteins and peptides, can be transported into the brain tissue through receptor-mediated transcytosis. In normal conditions large peptides and proteins cannot enter the brain via either passive transfer or carrier-mediated transport and AMT. Receptor-mediated transport systems (transcytosis) are the specialized transport pathways for this kind of material transfer. The exact meaning of transcytosis is transfer of large molecules from the apical or luminal side of endothelia to the basolateral or abluminal side via membrane-bound vesicles. BBB-forming endothelial cells have a high expression of several receptors for receptor-mediated transcytosis, such as insulin receptor, transferrin receptor, and low-density lipoprotein receptor-related protein 1, and they are used as targets for CNS drug delivery sites.[45–47] New studies point out that in the future more receptors will be defined and may be targeted for RMT.[48,49]

Other than proteins and peptides, cellular passage is very limited to the brain tissue, making it immune privileged. Under normal conditions, inflammatory cells, such as neutrophils pass to CNS if there is any kind of injury, such as ischemia, trauma, infection, or BBB breakdown. However very limited amount of mononuclear cellular passage occurs directly from endothelium via diapedesis,[50] as well as through the paracellular pathway, such as TJs and AJs, under normal physiological conditions.[51]

5 CONCLUSIONS

Transport of most of the drugs to the brain tissue is a challenge due to the presence of BBB. To overcome this obstacle, physiological transport mechanisms that are present in the CNS vasculature may introduce useful information for new targets. Further studies about the details of anatomy and physiology of BBB are fundamental for generation of such novel drug transport systems and for a better understanding of BBB characteristics in physiology, as well as in pathological conditions.

Abbreviations

ABC ATP-binding cassette
AJ Adherence junction
AMT Adsorptive endocytosis
BBB Blood–brain barrier
BCSFB Blood–CSF barrier
BRCP Breast cancer resistance protein
CNS Central nervous system
CSF Cerebrospinal fluid
MMP Matrix metalloprotease
MRP Multidrug resistance–associated protein
Pdgfb Platelet-derived growth factor beta
Pgp P-glycoproteins
RMT Receptor-mediated endocytosis
TEER Transendothelial resistance
TIMP Tissue inhibitor of matrix metalloprotease
TJ Tight junction

References

1. Ehrlich P. *Das Sauerstoff-Bediirfniss des Organismus*. Berlin: A Hirschwald; 1885.
2. Serlin Y, et al. Anatomy and physiology of the blood-brain barrier. *Sem Cell Dev Biol*. 2015;38:2–6.
3. Goldmann E. Vitalfärbung am zentralnervensystem: beitrag zur physiopathologie des plexus choriodeus und der hirnhäute. *Königl. Akademie der Wissenschafte*. 1913;1:1–60.
4. Thomas PD. History of blood-brain barrier. Available from: http://davislab.med.arizona.edu/content/history-blood-brain-barrier
5. Abbott NJ, Rönnbäck L, Hansson E. Astrocyte-endothelial interactions at the blood-brain barrier. *Nat Rev Neurosci*. 2006;7(1):41–53.
6. Öztop-Çakmak O, Solaroglu I, Gursoy-Ozdemir Y. The role of pericytes in neurovascular unit: emphasis on stroke. *Curr Drug Targets*. 2016 (Epub ahead of print).

7. Gursoy-Ozdemir Y, Yemisci M, Dalkara T. Microvascular protection is essential for successful neuroprotection in stroke. *J Neurochem*. 2012;123(suppl 2):2–11.
8. Chudler EH. Brain Facts and Figures. Available from: https://faculty.washington.edu/chudler/facts.html
9. Lauwers F, et al. Morphometry of the human cerebral cortex microcirculation: general characteristics and space-related profiles. *NeuroImage*. 2008;39(3):936–948.
10. Oldendorf WH, Cornford ME, Brown WJ. The large apparent work capability of the blood-brain barrier: a study of the mitochondrial content of capillary endothelial cells in brain and other tissues of the rat. *Ann Neurol*. 1977;1(5):409–417.
11. Dermietzel R, Spray D, Needergard M. *Blood–Brain Interfaces: From Ontogeny to Artificial Barriers*. VCH, Weinheim: Wiley; 2006.
12. Abbott NJ, et al. Structure and function of the blood-brain barrier. *Neurobiol Dis*. 2010;37(1):13–25.
13. Nitta T, et al. Size-selective loosening of the blood-brain barrier in claudin-5-deficient mice. *J Cell Biol*. 2003;161(3):653–660.
14. Weiss N, et al. The blood-brain barrier in brain homeostasis and neurological diseases. *Biochim Biophys Acta*. 2009;1788(4):842–857.
15. Gursoy-Ozdemir Y, Qiu J, et al. Cortical spreading depression activates and upregulates MMP-9. *J Clin Invest*. 2004;113(10):1447–1455.
16. Gursoy-Ozdemir Y, Can A, Dalkara T. Reperfusion-induced oxidative/nitrative injury to neurovascular unit after focal cerebral ischemia. *Stroke*. 2004;35(6):1449–1453.
17. Dalkara T, Gursoy-Ozdemir Y, Yemisci M. Brain microvascular pericytes in health and disease. *Acta Neuropathol*. 2011;122(1):1–9.
18. Saunders NR, Knott GW, Dziegielewska KM. Barriers in the immature brain. *Cell Mol Neurobiol*. 2000;20(1):29–40.
19. Hallmann R, et al. Novel mouse endothelial cell surface marker is suppressed during differentiation of the blood brain barrier. *Dev Dyn*. 1995;202(4):325–332.
20. Blanchette M, Daneman R. Formation and maintenance of the BBB. *Mech Dev*. 2015;138 (pt 1):8–16.
21. Daneman R, et al. Wnt/beta-catenin signaling is required for CNS, but not non-CNS, angiogenesis. *Proc Natl Acad Sci USA*. 2009;106(2):641–646.
22. Liebner S, et al. Wnt/beta-catenin signaling controls development of the blood-brain barrier. *J Cell Biol*. 2008;183(3):409–417.
23. Dehouck M-P, et al. An easier, reproducible, and mass-production method to study the blood? Brain barrier in vitro. *J Neurochem*. 1990;54(5):1798–1801.
24. Hayashi Y, et al. Induction of various blood-brain barrier properties in non-neural endothelial cells by close apposition to co-cultured astrocytes. *Glia*. 1997;19(1):13–26.
25. Daneman R, et al. Pericytes are required for blood-brain barrier integrity during embryogenesis. *Nature*. 2010;468(7323):562–566.
26. Lindahl P. Pericyte loss and microaneurysm formation in PDGF-B-deficient mice. *Science*. 1997;277(5323):242–245.
27. Bell RD, et al. Pericytes control key neurovascular functions and neuronal phenotype in the adult brain and during brain aging. *Neuron*. 2010;68(3):409–427.
28. Betz A, Goldstein G. Polarity of the blood-brain barrier: neutral amino acid transport into isolated brain capillaries. *Science*. 1978;202(4364):225–227.
29. Chow BW, Gu C. The molecular constituents of the blood–brain barrier. *Trends Neurosci*. 2015;38(10):598–608.

30. Abraham MH, Chadha HS, Mitchell RC. Hydrogen bonding. 33. Factors that influence the distribution of solutes between blood and brain. *J Pharm Sci*. 1994;83(9):1257–1268.
31. Lipinski CA, et al. Experimental and computational approaches to estimate solubility and permeability in drug discovery and development settings. *Adv Drug Deliv Rev*. 2012;64:4–17.
32. Pardridge WM. Transport of small molecules through the blood-brain barrier: biology and methodology. *Adv Drug Deliv Rev*. 1995;15(1–3):5–36.
33. Pardridge WM. Alzheimer's disease drug development and the problem of the blood-brain barrier. *Alzheimers Dement*. 2009;5(5):427–432.
34. Bocsik A, et al. Reversible opening of intercellular junctions of intestinal epithelial and brain endothelial cells with tight junction modulator peptides. *J Pharm Sci*. 2016;105(2):754–765.
35. Alonso A. Ultrasound-induced blood-brain barrier opening for drug delivery. *Front Neurol Neurosci*. 2015;36:106–115.
36. Gao X, et al. Overcoming the blood-brain barrier for delivering drugs into the brain by using adenosine receptor nanoagonist. *ACS Nano*. 2014;8(4):3678–3689.
37. Ohtsuki S, Terasaki T. Contribution of carrier-mediated transport systems to the blood-brain barrier as a supporting and protecting interface for the brain; importance for CNS drug discovery and development. *Pharm Res*. 2007;24(9):1745–1758.
38. Pardridge WM. Drug targeting to the brain. *Pharm Res*. 2007;24(9):1733–1744.
39. Keaney J, Campbell M. The dynamic blood-brain barrier. *FEBS J*. 2015;282(21):4067–4079.
40. Cornford EM, Hyman S. Localization of brain endothelial luminal and abluminal transporters with immunogold electron microscopy. *NeuroRx*. 2005;2(1):27–43.
41. Begley D. ABC transporters and the blood-brain barrier. *Curr Pharm Design*. 2004;10(12): 1295–1312.
42. Cornford EM, Diep CP, Pardridge WM. Blood-brain barrier transport of valproic acid. *J Neurochem*. 1985;44(5):1541–1550.
43. Yu X-Y, et al. Role of P-glycoprotein in limiting the brain penetration of glabridin, an active isoflavan from the root of *Glycyrrhiza glabra*. *Pharm Res*. 2007;24(9):1668–1690.
44. Kamalinia G, et al. Cationic albumin-conjugated chelating agent as a novel brain drug delivery system in neurodegeneration. *Chem Biol Drug Design*. 2015;86(5):1203–1214.
45. Tian X, et al. LRP-1-mediated intracellular antibody delivery to the central nervous system. *Sci Rep*. 2015;5:11990.
46. Witt KA, et al. Insulin enhancement of opioid peptide transport across the blood-brain barrier and assessment of analgesic effect. *J Pharmacol Exp Ther*. 2000;295(3):972–978.
47. Yemisci M, et al. Systemically administered brain-targeted nanoparticles transport peptides across the blood–brain barrier and provide neuroprotection. *J Cereb Blood Flow Metab*. 2015;35(3):469–475.
48. Kim SY, et al. Transthyretin as a new transporter of nanoparticles for receptor-mediated transcytosis in rat brain microvessels. *Colloids Surf B*. 2015;136:989–996.
49. Zuchero YJY, et al. Discovery of novel blood-brain barrier targets to enhance brain uptake of therapeutic antibodies. *Neuron*. 2016;89(1):70–82.
50. Wolburg H, Wolburg-Buchholz K, Engelhardt B. Diapedesis of mononuclear cells across cerebral venules during experimental autoimmune encephalomyelitis leaves tight junctions intact. *Acta Neuropathol*. 2005;109(2):181–190.
51. Greenwood J, et al. Review: leucocyte-endothelial cell crosstalk at the blood-brain barrier: a prerequisite for successful immune cell entry to the brain. *Neuropathol Appl Neurobiol*. 2011;37(1):24–39.

CHAPTER 2

Blood–Brain Barrier: Genomics, Proteomics, Disease Targets, and Drug Delivery

Ozgun F. Duzenli, BSc, Ayşe Filiz Oner, PhD
Hacettepe University, Ankara, Turkey

CONTENTS

1 INTRODUCTION

Understanding molecular and cellular events of the brain microvasculature is an essential issue for designing novel neuropharmaceuticals that will be used in neurological diseases, such as Alzheimer's disease (AD), stroke, Parkinson's disease (PD), multiple sclerosis (MS), brain tumors, and autoimmune and infectious diseases of the brain. Despite the importance of this subject, little is known about the molecular mechanisms of the blood–brain barrier (BBB), which limits the entry of molecules into the brain.

The BBB is formed essentially from brain endothelial cells (BECs) linked together by intracellular tight junctions (TJs).[1–3] Although, brain microvessels, so called BBB, protect the brain from toxic/harmful molecules, inflammation, and disease, this protective structure also provides a barrier for delivering drugs to the central nervous system (CNS).[3,4] Selectively permeable microvasculature of the brain (BBB) consists of different cell types that have interdependent gene expression.[5] The different cell types and cell–cell interactions in the perivascular space regulate molecular transport in the brain. Endotheial cells (ECs) are the main cell types of the brain microvessels, but properties of the BBB are affected by interactions with other cell types in the neurovascular space, such as mural cells, immune cells, glial cells, and neural cells.[2,6,7] Expressed efflux transporters provide a barrier to small lipophilic molecules, whereas nutrient transporters facilitate transport of some nutrients from the blood to the brain and help in the removal of the waste products from the brain into the circulation.[8,9] Pericytes (PCs), the mural cells of blood microvessels, have unique properties in the CNS compared to other cell types. These cells play important roles in angiogenesis, formation of BBB, and maintenance of its function.[8,10] Astrocytes are glial cells, which secrete factors to regulate BBB function to modulate and maintain the barrier after it is formed.[6,7] Perivascular macrophages and microglial cells are the main types of immune cells that can interact with the CNS vessels. Both CNS cell types are assumed to regulate BBB

Nanotechnology Methods for Neurological Diseases and Brain Tumors. http://dx.doi.org/10.1016/B978-0-12-803796-6.00002-2

by responding to infection, injury, and diseases via increasing vascular permeability by activating T-cells and macrophages.[11]

Endothelial cells are connected by TJs, which are high-resistance barriers for entering molecules and ions into the brain.[12] The combination of transmembrane molecules and transmembrane adhesion complex proteins are involved in the TJ structure to form a barrier-functioning seal. Membrane proteins, such as occludin, claudins, and junctional adhesion molecules (JAMs) are involved in intercellular contacts and interactions with transmembrane adhesion proteins and other cytoplasmic accessory proteins, such as zonula occludens (ZO), cingulin, protein kinases, and heterotrimeric G-proteins.[13–15] Of these membrane proteins, occludin is one of the first identified molecule, suggesting key functions for the resistance of the barrier and regulation of calcification.[13] Claudins are a family of 25 proteins that are responsible for paracellular barrier formation with different functions, depending on the tissue they are expressed in.[14] It has been shown that deficiency of claudin-5 in mice cause a leakage through the BBB, depending on the size of the molecule.[15,16] JAMs belong to the immunoglobulin superfamily, and participate in TJ formation and regulation of paracellular permeability.[12,17]

Other molecular components of the BBB, efflux transporters, and nutrient transporters control movement of molecules and ions between the blood and the brain. In addition to these molecules, the expression of leukocyte adhesion molecules, including selectins, differs in CNS during inflammation and disease states of the BBB.[18] Signaling pathways and molecules of the BBB are other important aspects that regulate the formation and barrier function of the CNS vasculature. Understanding the unique molecular elements of the BBB and impacts of these molecules in disease states allow the development of pharmaceuticals that can deliver drugs to the brain and can restore pathological conditions. These molecules will be discussed in the following parts of this chapter from different point of views.

2 BLOOD–BRAIN BARRIER GENOMICS

2.1 Genome Products Selectively Expressed at the BBB

Developments in molecular biology and genomics have expanded the knowledge of the BBB function and dysfunction. Brain genomics and BBB genomics differ from each other due to differences in cell types and gene expression.[19] BBB genomic research in based on cells and molecules of the BBB microvasculature, and continuous development of new technologies in gene expression profiling allows the analysis of a huge number of genes and proteins. Application of genomics methodology to BBB microvasculature provides novel findings for brain physiology and pathology, as well as for brain drug delivery and targeting.

Distinctions between specific genes of brain microvessel endothelial cells and other ECs can be identified by genomic techniques, but the data do not correlate directly with the protein expression levels in the related cells. Proteomic studies are required to complement these genomic datasets to analyze BBB function and disease status. In the genomic studies, important genes for BBB function can be summarized as: TJ (claudin-5, occludin, Zo-1, and Zo-2), efflux transporters (P-glycoprotein and Bcrp), nutrient transporters (Glut-1, Mct-1, and Lat-1), and others (basigin and carbonic anhydrase). In addition, Wnt/β-catenin signaling components are enriched in BECs and are necessary for CNS-specific angiogenesis and BBB regulation.[9,20–22] New molecules for the targeted drug delivery to the brain may be generated from a large number of molecular transporters highly expressed in BECs. Transferrin receptor (TfR) is a well-known target for the BBB-specific receptor–mediated delivery.[9] Genes of known functions, such as myelin-related genes, amyloid-related genes, and tissue plasminogen activator (TPA)–encoding genes that are expressed at the BBB have been identified; however, a number of BBB-enriched genes have also been identified with unknown functions. Summary of the identified genes from genomics studies for the formation and functions of the BBB are presented in Table 2.1.

2.2 Methods Used in BBB Genomics

Isolation of brain capillaries is the initial step in a BBB genomic research program. Either mechanical homogenization or enzymatic methods can be used to isolate vessels properly.[40,41] In BBB genomic studies, determining the accurate cells and tissues to be analyzed and the method of purification are very important parameters.[9] The accurate cellular material is necessary for isolating mRNA from brain vasculature. Vascular enrichment and laser capture microdissection (LCM) methods are earlier techniques, which do not yield specifically purified BECs.[9,42] To isolate BECs specifically from tissues, gradient centrifugation,[43] fluorescence-activated cell sorting (FACS), or antibody-based purification methods can be used as well.[10,43] In these studies some of the conditions, such as mechanical shearing, enzymatic digestion, and temperature, may affect expression profiles even if the isolation and genomic profiling of BECs are carefully performed. Besides, the expression of proteins and their levels vary depending upon the species from which they are isolated, isolation method, and BBB culture model.[44]

Various methods are used to identify BBB-specific genes expressed by BECs. Gene microarray technologies, serial analysis of gene expression (SAGE), and suppression subtractive hybridization (SSH) methods are used extensively in BBB genomics research.[44] SSH is a polymerase chain reaction (PCR)–based method for identifying tissue-specific gene transcripts. Subtractive hybridization techniques are powerful methods, which detect BBB-derived complementary

Table 2.1 Identified Genes From Genomic Studies for the Formation and Functions of the Blood–Brain Barrier (BBB)

Genes[a]	Proteins	BBB Formation Functions	References
LSR	Lsr	Formation of tTJs	[23]
MARVELD2 (TRIC)	Tricellulin	Sealing of TJs	[10,24]
OCLN	Occludin	BBB permeability and integrity	[25–27]
F11R	Jam1	TJs integrity	[28]
CLDN1	Claudin-1	BBB permeability	[29]
CLDN3	Claudin-3	BBB permeability and development	[30]
CLDN5 (BEC1, CPETRL1)	Claudin-5	BBB integrity and permeability	[26,31,32]
CLDN12	Claudin-12	Supporting BBB integrity	[26]
TJP1	ZO-1	Supporting BBB permeability	[25]
FZD6, LEF1, AXIN2, APCDD1, TNFRSF21, TNFRSF19	Fzd6, Lef1, Axin2, Apcdd1, DR6, Troy	Wnt/β-catenin signaling for CNS-specific angiogenesis	[33,34]
NOS2, PRKCβ, NFKB1	iNOS, PKCβII, PKCβI, NF-KB	Proinflammatory signaling pathway induced by TNF-α, TLR4, TNFα-R1; changes on P-glycoprotein transport function	[35]
ABCB1	P-glycoprotein	Transporter; organic cations, weak organic bases	[35]
ABCG2	BCRP	Transporter; organic cations, some anionic drugs	[35]
ABCC1, ABCC2, ABCC4, ABCC5	Mrp1, Mrp2, Mrp4, Mrp5	Transporter; organic anions, cyclic nucleotides	[35]
SLC2A1	Glut-1	Transporter; glucose, galactose, mannose, glucosamine	[10,36]
SLC7A5	Lat-1	Transporter; large neutral L-amino acids, T3, T4, L-DOPA, BCH	[10,36]
SLC16A1	Mct-1	Transporter; lactate, pyruvate, ketone	[10,36]
SLC19A3	THTR2	Transporter; thiamin	[10]
SLCO1A2, SLC22A1	OATP1A2, OCT1, OCT2	Transporter; organic anions and cations	[35]
SLC25A20	Slc25a20	Transporter; carnitine	[10]
SLC27A1, SLC27A4	FATP1, FATP4	Transporter; LCFA	[37]
TFRC	Tf receptor	Receptor (drug delivery target)	[38]
INSR	Insulin receptor	Receptor (drug delivery target)	[39]

Apcdd1, Adenomatosis polyposis coli downregulated 1; BCRP, breast cancer–related protein; FAT, fatty acid transporter; Fzd, Frizzled receptors; Glut-1, glucose transporter-1; iNOS, inducible nitric oxide synthase; JAM, junctional adhesion molecule; Lat-1, L-type amino acid transporter-1; LCFA, long-chain fatty acids; Lef1, lymphoid enhancer-binding factor 1; Lsr, lipolysis-stimulated receptor; Mct-1, monocarboxylate transporter 1; MRP, multidrug resistance–associated protein; NF-KB, nuclear factor kappa B; OATP1A2, organic anion transporting polypeptide 1A2; OCT, organic cation transporter; PKCβ, protein kinase C β; SLC, solute carrier; Tf, transferrin; THTR2, thiamine transporter 2; TJP, tight junction protein; TLR, Toll-like receptor; TNFα, tumor necrosis factor-α; TNFα-R1, tumor necrosis factor-α receptor-1; TNFRSF, tumor necrosis factor receptor superfamily; tTJs, tricellular tight junctions; ZO, zonula occludens.

[a] *Gene symbols were obtained from http://www.genecards.org/database*

DNA (cDNA) relative to obtained cDNA subtracted from other tissues. Tester cDNA from rat brain capillary and driver cDNA from rat liver or rat kidney were used to generate a BBB library. Clones showing a strong hybridization signal are selected and subjected to DNA sequence analysis.[19] SAGE gives information about the entire gene expression profile by using small messenger RNA (mRNA) fragments (tags), which allow the quantitative cataloging of all expressed genes in cells or tissues.[45] Tissue-specific gene products can be detected by microarray screening, but due to the low ratio of brain capillaries over the whole brain and low sensitivity of the microarray gene detection methods, only highly expressed gene transcripts can be detected from the whole-brain samples.[46] MicroRNA (miRNA) arrays and high-throughput quantitative transcriptomic RNA sequencing techniques will provide detailed knowledge with respect to BBB development and dysfunction.[9]

3 BLOOD–BRAIN BARRIER PROTEOMICS

Interacting cells and molecules in the BBB affect both functional pathways and functions of the brain. Although genomics and proteomics provide different types of data, they are complementary techniques, which can be used to solve problems due to genome- or proteome-related molecular dysfunctions and diseases. If only genomics data are used, misinterpretation of the findings is possible because of the dynamic regulation of mRNA and posttranslational modifications in the cell.[5,47]

Membrane proteins are of significant importance in proteomics as they are recognized as targets for transporting molecules across the BBB. Highly purified BECs obtained by magnetic cell sorting with protease treatment were not found suitable for membrane proteomic analysis.[48] Findings from proteomic analysis had a great impact on BBB studies. By using the LCM method, expression of hundreds of proteins from mouse brain capillaries can be identified at the BBB in different regions.[49] Obtaining pure and sufficient amount of capillary material is a critical limiting factor in the proteomic research, as in the case of genomic studies. Contamination of other neural cells in BBB cells causes expression of proteins from these cells that do not belong to BBB. Due to this concern human pluripotent stem cells may be a useful material in BBB proteomic studies.[50] In addition, species differences are of critical importance in drug development studies, and comparison of protein levels can better reflect functional differences than mRNA levels.[48] Quantities of organic anion transporters (OAT) can be measured in mouse models, however, these molecules were below quantification levels in humans as compared to rodents, as an example, MDR1/mdr1a/ABCB1 is a major efflux pump, which pumps hydrophobic drugs out of BECs into blood circulation; similar to anion transporters, MDR1 expression is lower in humans than mice.[51,52] In this context, quantitative targeted proteomics is an essential method that can provide detailed information about the permeability and transport functions of BBB, protein expression levels in different species, and differences in vivo–in vitro models.[48] However, further improvement is necessary to increase the number of proteins that can be analyzed simultaneously.

Other technologies, including sodium dodecyl sulfate polyacrylamide gel electrophoresis (SDS-PAGE), two-dimensional electrophoresis (2DE), two-dimensional fluorescence difference gel electrophoresis (2D DIGE), stable isotope labeling with amino acids in cell culture (SILAC), multidimensional protein identification technology (MUDPIT), protein array, and mass spectrometry are also used in proteomics to analyze BBB membrane proteins. 2DE and SDS-PAGE are standard techniques for the separation and isolation of proteins from cells; however, these methods do not provide high-throughput protein identification for the proteomic analysis of the cell. Mass spectrometry revolutionized proteomics studies with high sensitivity and resolving power by rapidly fragmenting and identifying peptides generally in conjunction with other methods. In the label-free approach relative protein quantification from different experimental samples is possible.[53,54] Another precise proteomic technique based on relative identification of isotope-labeled protein samples is known as SILAC, where matrix-assisted laser desorption/ionization time-of-flight (MALDI-TOF) mass spectrometry analysis performed after fragmentation of the proteins,[55] and MUDPIT is used to identify a larger number of proteins in complex cellular mixtures.[56] By using these techniques in proteomic research, massive data are obtained for understanding molecules of BBB, which can be used for diagnostic and therapeutic purposes in brain disorders. With the help of bioinformatics this vast amount of proteomic datasets can be processed for functional analysis and data-mining.[57] In BBB-focused bioinformatics it is difficult to organize proteomic data into a useful database format and to mine data to extract accurate information for understanding mechanisms and pathways underlying brain diseases.[58]

4 DISEASE TARGETS

Developments in molecular biology, genomics, and proteomics provide a strong basis for drug discovery in BBB-linked diseases. Molecular changes can be determined by comparing molecules in healthy and diseased states of the BBB for identifying therapeutic targets in drug delivery.[59] Knowledge of the disease targets is essential in biopharmaceutical industry to develop new drugs that can cross BBB and treat brain disorders, such as amyloid diseases, stroke, brain tumors, and others. For example, one of the most studied BBB dysfunction is observed in amyloid diseases due to the altered gene expression levels in response to the amyloid β peptides (Aβ), which provides clues for understanding pathological effects of Aβ as a disease target on brain microvasculature.[59,60] Different Aβ-targeted therapeutic strategies are developed, including modulation of Aβ production, inhibition of Aβ aggregation, and enhancement of Aβ degradation.[61]

Some neurodegenerative conditions, termed as tauopathies, are mediated by the intracellular protein tau. Tau is a soluble microtubule-binding and -stabilizing protein under normal conditions, but hyperphosphorylated tau is insoluble and accumulation of insoluble aggregates indicates various neurodegenerative disorders, including AD. Anti-tau antibody research studies are seen as promising therapeutic approaches in animal models.[62] Stroke is another example of the most frequent neurological disease condition resulting from cerebral ischemia and hypoxia. Protein expression profiles showed that genes were differentially expressed by downregulation or upregulation of protein functions in stroke conditions.[63,64] Identified tissue-specific key proteins or genes for aforementioned neurological disease states or other pathological conditions would be useful targets for the treatment of BBB dysfunction and related diseases. Examples for these molecules and conditions are summarized in Table 2.2.

5 DRUG DELIVERY

In drug delivery research, BBB is a major obstacle for the treatment of neurological disorders, as it prevents penetration of drugs into the brain. Developments in pharmaceutical biotechnology assisted the drug development efforts through the use of biologic molecules, recombinant peptides–proteins, monoclonal antibodies (mAbs), and nucleic acids (DNA and siRNA) as active drug substances, drug targeting molecules, or drug delivery vectors.[4] Since the last decade biomolecules have been considered as effective neuroprotective agents: preventing the progression of neurodegenerative disorders and affecting healing in damaged cells.[98–101] Biological drug development efforts to stop or to cure brain-related disorders and diseases, mainly for AD, stroke, PD, and brain tumors, are under research in a large number of university and industry projects.

In a BBB drug delivery system design, multiple functional characteristics of BBB as a physical (TJs), transport, metabolic, or enzymatic barrier and as an immunologic barrier in healthy and pathological conditions should be taken into account.[102] Also drug transport routes across the BBB should be evaluated for the designing of drugs. These routes are classified as paracellular and transcellular diffusion, carrier-mediated transport, efflux pumps (transporters), receptor-mediated transcytosis, adsorptive transcytosis, and cell-mediated transcytosis.[31,102]

Table 2.2 Specific Disease Targets for Alzheimer's Disease (AD), Stroke, Ischemia, and Neuroinflammatory Conditions

Molecules (Proteins)	Localization (secretion or specificity)	Study Models	Target Diseases	References
IL-1α, IL-1β, IL-6 (precursor of CRP)	Microglia, astrocytes, neurons	Mouse	Neuroinflammation AD, ischemic brain injury	[65–68]
IL-22, IL-17	IL-22, IL-17 receptors on human BBB ECs	Human BBB ECs, mouse	Neuroinflammation	[69]
IL-34	Neurons in CNS	Mouse	Neuroinflammation	[70]
TPA	Microglia, astrocytes	Mouse, glial cell culture	Neuroinflammation AD, ischemia	[67,71,72]
VEGF	Astrocytes	Mouse	CNS inflammation	[65]
Aβ protein	Microglia, astrocytes, BECs	BECs of mouse	Neuroinflammation AD	[67,73]
Apolipoprotein E	Microglia, astrocytes	Mouse	Neuroinflammation AD	[67,74]
PPAR-γ	Microglia, astrocytes	Rat	AD	[75]
TNFα	Microglia	Rat PCs	Neuroinflammation AD	[67,76]
CCL2 (MCP-1)	Activated astrocytes	Mouse	AD	[77]
CD40L	Microglia in AD patient brain	Mouse	AD	[78,79]
CRP and AP	Hippocampus	Human	AD, inflammation	[80]
TLR2, TLR4, TLR5, TLR6, TLR7, TLR9	Microglia, astrocyte	Mouse astrocytes	AD, ICH, neuroinflammation	[81–85]
Endophilin-1	BECs	hCMEC/D3	Brain disorders	[25]
BCRP	BECs	Rat, mouse	Brain tumors	[86]
RAGE	BECs, microglia	BMECs culture, mouse, rat	Neuroinflammation AD	[73,87,88]
MMP-2	BECs, astrocytes	Mouse	Ischemia	[73,89,90]
MMP-3	Microglia, astrocytes, neurons	Mouse	Ischemic stroke	[91]
MMP-8	Microglia	Mouse	Neuroinflammation ischemia	[92,93]
MMP-9	PCs	Rat, mouse	Ischemia and stroke	[76,94]
PDGFR-α	Astrocytes	Mouse	ICH	[95]
CXCR4	Astrocytes	Mouse	Ischemia	[96,97]

Aβ, Amyloid beta; AP, amyloid P; BBB, blood–brain barrier;BCRP, breast cancer-related protein; BECs, brain endothelial cells; CCL2 (MCP-1), monocyte chemotactic protein-1; CD40L, cluster of differentiation 40 ligand; CNS; central nervous system, CRP, C reactive protein; CXCR4, chemokine (C-X-C) receptor type 4; ECs, endothelial cells; hCMEC/D3, human BEC cell line; ICH, intracerebral hemorrhage; IL, interleukin; MMP, matrix metalloproteinase; PC, pericyte; PDGFR-α, platelet-derived growth factor receptor α; PPAR-γ, peroxisome proliferator–activated receptor-γ; RAGE, receptor for advanced glycation end products; TLR, Toll-like receptor; TNF-α, tumor necrosis factor-α; TPA, tissue plasminogen activator; VEGF, vascular endothelial growth factor.

5.1 Recombinant Proteins and Monoclonal Antibodies

Recombinant proteins and mAbs are too large to cross BBB, but due to their promising therapeutic effects in CNS there is a great demand for novel biopharmaceutical dosage forms prepared with these large molecules. Different strategies are studied to formulate these peptides, proteins, and mAbs against brain disorders. Due to the problems in BBB delivery, anti-Aβ immunotherapies for AD and other brain diseases failed in previous clinical trials.[103] Developments in the drug delivery and nanotechnology provided new possibilities for delivering large biologic molecules to the brain. Recent studies demonstrate that biomolecules can be delivered by fusion protein technology or Aβ-binding antibody fragments.[104,105] Antigens localized at the BBB are suitable targets for biological drugs that will be used in the treatment of BBB dysfunctions. The genetically engineered recombinant fusion proteins, which have both transport and receptor-binding functions, are promising protein conjugates in brain drug delivery research. Proteins that bind certain BBB receptors, such as TfR and human insulin receptors (HIR), are used for delivering drugs to the brain with a process known as receptor-mediated trancystosis.[106] A therapeutic peptide–protein drug is fused to a molecular Trojan horse, which is a second peptide or a nucleotide that binds a specific receptor on the BBB to ferry the fused protein to the brain. Selective binding affinities of Trojan horses enable the receptor-mediated delivery of the fusion protein across the BBB, so that they can be effective pharmacologically.[107] For example, TfRmAb–glial-derived neurotrophic factor (GDNF), TfR–tumor necrosis factor receptor (TNFR), or HIRmAb–GDNF fusion conjugates can cross the BBB in a mouse model against stroke, causing a reduction in the stroke volume.[108,109] Peptide, protein, mAb, and protein-mAb fusion protein examples are summarized in Table 2.3.

5.2 Gene-Based Systems

Genes and gene fragments, similar to proteins, are also too large as to cross BBB without properly designed delivery systems. Gene-based delivery approaches for BBB include formulating exogenous nucleic acids, such as DNA, mRNA, siRNA, miRNA, or antisense oligonucleotides, in nanodelivery dosage forms.[129] Tumor sites can be actively targeted by coupling gene delivery systems with targeting peptides, antibodies, or chemical molecules against specifically expressed receptors on tumor cells.[130] Current gene therapy approaches to expressneurotrophic factors, such as brain-derived neurotrophic factor (BDNF) and fibroblast growth factor 2 (FGF2), in AD and PD for neuroprotection in the brain include use of viral vectors, such as adenoassociated virus serotype-2 (AAV-2).[126] By using an AAV-2 vector carrying human granulocyte colony-stimulating factor (hG-CSF)–cDNA, neuroprotection can be monitored in a mouse model of AD.[131] Transcytosis, such as receptor-mediated transcytosis, is a process for viruses to overcome BBB and to shuttle molecules to the other site of

Table 2.3 Effects of Biologic Molecules on the Blood–Brain Barrier (BBB)–Related Disorders in Animal Models

Biologic Molecules	Study Models	Target Diseases	Observed Effects in Study	References
Peptides–proteins–mAbs				
TGF-β1	Rat	Ischemia	Reduced basement membrane degradation, BBB disruption, hemorrhagic transformation	[110]
ASN12	Mouse	AD	Reduced intraneuronal Aβ, improved memory	[111]
DRα1–MOG-35-55	Mouse	Stroke	Reduced infarct size	[112]
FAM19A3	Mouse	Stroke	Attenuated cerebral ischemia	[113]
6D11	Mouse	AD	Recovery in cognitive learning	[114]
HJ8.5	Mouse	AD	Decreased brain atrophy and improved motor function	[62]
TOMA	Mouse	AD	Improved cognitive performance	[115]
mAb47	Mouse	PD	Reduced motor dysfunction	[116]
9E4	Mouse	PD and DLB	Ameliorated motor and learning deficits and synaptic pathology	[117]
Anti-HMGB1 mAb	Rat	Ischemia	Reduced brain edema	[118]
Clone MM17F3	Mouse	Stroke	Reduced infarct size and improved clinical outcome	[119]
11C7	Rat	Stroke	Significant improvement in a chronic lesion–induced deficit of skilled forelimb reaching	[120]
Peptide–mAb fusions				
ANG4043	Mouse	Tumor	Improved survival	[121]
cTfRmAb–TNFR fusion protein	Mouse	Ischemia	Reduced hemispheric, cortical, and subcortical stroke volumes and neuronal deficit	[107]
The HIRmAb–GDNF fusion protein	Rat	Stroke	Reduced hemispheric stroke volume	[108]

Nucleic acids				
AAV-2–shCDK5miR	Mouse	AD	Prevented intracellular β-amyloidosis	[122]
AAV–IL-10	Mouse	AD	Improved spatial learning	[123]
AAV–7ND	Mouse	AD	Reduced Aβ accumulation, improved spatial learning	[124]
AAV–IL-4	Mouse	AD	Suppressed gliosis and β-amyloidosis, enhanced neurogenesis	[125]
AAV-2/1–FGF2	Mouse	AD	Enhanced spatial learning and neurogenesis, suppressed β-amyloidosis	[126]
AAV-5–hGFAP–GDNF	Mouse and rat	PD	Protected DA neurons from degeneration	[127]
α-Synuclein siRNA	Primate	PD	Suppressed both α-synuclein mRNA and protein, 40%–50% reduction of nigral α-synuclein	[128]

Aβ, Amyloid β; AAV-2, adenoassociated virus serotype 2; AAV2/1–FGF2, AAV2/1 expressing fibroblast growth factor-2; AAV–IL-10, AAV serotype 2/1 hybrid expressing IL-10; AAV–IL-4, recombinant AAV expressing IL-4; AAV–7ND, AAV expressing N-terminal deletion of formally monocyte chemotactic protein-1 (MCP-1); AD, Alzheimer's disease; ANG4043, conjugation between angiopep-2 and anti-HER2 mAb; ASN12, brain-targeted recombinant protein; 11C7, Anti-Nogo-A antibody; CDK5, cyclin-dependent kinase 5; clone MM17F3, mouse monoclonal anti-murine IL-17A antibody; cTfR, chimeric transferrin receptor; cTfRmAb–TNFR, fusion of the type II human TNFR to engineered chimeric mAb against TfR; 6D11, anti-PrPC monoclonal antibody (mAb); DA, dopamine small interfering RNA (siRNA); DLB, dementia with Lewy bodies; DRα1–MOG-35-55, recombinant protein comprised of the HLA-DRα1 domain linked to MOG-35-55 peptide; 9E4, mAb against C-terminus of α-synuclein; FAM19A3, secreted protein; GDNF, glial-derived neurotrophic factor; hGFAP, human glial fibrillary acidic protein; HIRmAb, monoclonal antibody to the human insulin receptor; HJ8.5, anti-tau antibody; HMGB1, high-mobility group box-1; IL, interleukin; mAb47, α-synuclein protofibril selective mAb; PD, Parkinson's disease; PrPC, cellular prion protein; shCDK5miR, short hairpin microRNA (sh-miR) sequences used to silence CDK5; TGF-β1, transforming growth factor beta 1; TNFR, tumor necrosis factor receptor; TOMA, tau oligomer–specfic mAb.

the barrier.[132] Recombinant AAV virus expressing IL-4 (AAV–IL-4) delivery system suppressed beta amyloidosis in another mouse study, but safety concerns with viral vectors still remain. For this reason besides viral delivery systems, nonviral nanosized vectors, such as high-branching dendrimers, liposomes, nanoparticles, and other nanosized carriers, are being designed for delivering DNA into BECs. High transfection efficiency can be achieved by applying a polyamidoamine (PAMAM) polyethylene glycol (PEG)–Tf/DNA complex in mouse.[133] In another example, plasmid DNA, which can encode short interfering RNA (shRNA), can be delivered across the BBB with nonviral immunoliposomes.[134] Targeting methods that allow the efficient delivery of chemically synthesized siRNA duplex molecules to BECs must be used, so that the silencing of functionally relevant genes can be achieved.[134,135] The application of stem cells is another promising approach for delivering genes to tumor sites in the brain.[136] More examples related to BBB gene delivery are presented in Table 2.3.

6 CONCLUSIONS

To sum up, recent rapid developments in genomics, proteomics, molecular biology, pharmaceutical biotechnology, and nanotechnology for the identification, slowing down, stopping, healing, or curing of neurodegenerative pathological conditions have accelerated research on BBB formation and functions. Massive amounts of datasets are collected from CNS research studies, but further studies and improved methods are still required to extract and combine promising results from these datasets, specifically for BBB functions and dysfunctions.

Abbreviations

AAV-2 Adenoassociated virus serotype-2
Aβ Amyloid beta peptide
AD Alzheimer's disease
BBB Blood–brain barrier
BDNF Brain-derived neurotrophic factor
BECs Brain endothelial cells
cDNA Complementary DNA
CNS Central nervous system
2D DIGE Two-dimensional fluorescence difference gel electrophoresis
2DE Two-dimensional electrophoresis
ECs Endothelial cells
FACS Fluorescence-activated cell sorting
FGF2 Fibroblast growth factor 2
GDNF Glial-derivedneurotrophic factor
HIR Human insulin receptor

hG-CSF	Human granulocyte colony-stimulating factor
JAMs	Junctional adhesion molecules
LCM	Laser capture microdissection
mAbs	Monoclonal antibodies
miRNA	MicroRNA
mRNA	Messenger RNA
MS	Multiple sclerosis
MUDPIT	Multidimensional protein identification technology
OATs	Organic anion transporters
PAMAM	Polyamidoamine
PCR	Polymerase chain reaction
PCs	Pericytes
PD	Parkinson's disease
PEG	Polyethylene glycol
SAGE	Serial analysis of gene expression
SDS-PAGE	Sodium dodecyl sulfate polyacrylamide gel electrophoresis
shRNA	Short interfering RNA
SILAC	Stable isotope labeling with amino acids in cell culture
siRNA	Small interfering RNA
SSH	Suppression subtractive hybridization
TfR	Transferrin receptor
TJs	Tight junctions
TNFR	Tumor necrosis factor receptor
TPA	Tissue plasminogen activator
ZO	Zonula occludens

References

1. Reese TS, Karnovsky MJ. Fine structural localization of a blood-brain barrier to exogenous peroxidase. *J Cell Biol.* 1967;34(1):207–217.
2. Daneman R, Prat A. The blood–brain barrier. *Cold Spring Harb Perspect Biol.* 2015;7(1):a020412.
3. Abbott NJ, Patabendige AAK, Dolman DEM, Yusof SR, Begley DJ. Structure and function of the blood–brain barrier. *Neurobiol Dis.* 2010;37(1):13–25.
4. Pardridge WM. Blood–brain barrier delivery. *Drug Discov Today.* 2007;12(1–2):54–61.
5. Shusta EV. Blood-brain barrier genomics, proteomics, and new transporter discovery. *Neurotherapeutics.* 2005;2(1):151–161.
6. Abbott NJ, Ronnback L, Hansson E. Astrocyte-endothelial interactions at the blood-brain barrier. *Nat Rev Neurosci.* 2006;7(1):41–53.
7. Armulik A, Genove G, Mae M, Nisancioglu MH, Wallgard E, Niaudet C, et al. Pericytes regulate the blood-brain barrier. *Nature.* 2010;468(7323):557–561.
8. Armulik A, Genové G, Betsholtz C. Pericytes: developmental, physiological, and pathological perspectives, problems, and promises. *Dev Cell.* 2011;21(2):193–215.
9. Huntley MA, Bien-Ly N, Daneman R, Watts RJ. Dissecting gene expression at the blood-brain barrier. *Front Neurosci.* 2014;8:355.
10. Daneman R, Zhou L, Agalliu D, Cahoy JD, Kaushal A, Barres BA. The mouse blood-brain barrier transcriptome: a new resource for understanding the development and function of brain endothelial cells. *PloS One.* 2010;5(10):e13741.

11. Hudson LC, Bragg DC, Tompkins MB, Meeker RB. Astrocytes and microglia differentially regulate trafficking of lymphocyte subsets across brain endothelial cells. *Brain Res*. 2005;1058(1–2):148–160.
12. Liu WY, Wang ZB, Zhang LC, Wei X, Li L. Tight junction in blood-brain barrier: an overview of structure, regulation, and regulator substances. *CNS Neurosci Ther*. 2012;18(8):609–615.
13. Saitou M, Furuse M, Sasaki H, Schulzke J-D, Fromm M, Takano H, et al. Complex phenotype of mice lacking occludin, a component of tight junction strands. *Mol Biol Cell*. 2000;11(12):4131–4142.
14. Zhang J, Piontek J, Wolburg H, Piehl C, Liss M, Otten C, et al. Establishment of a neuroepithelial barrier by Claudin5a is essential for zebrafish brain ventricular lumen expansion. *Proc Natl Acad Sci USA*. 2010;107(4):1425–1430.
15. Mandell KJ, Parkos CA. The JAM family of proteins. *Adv Drug Deliv Rev*. 2005;57(6):857–867.
16. Nitta T, Hata M, Gotoh S, Seo Y, Sasaki H, Hashimoto N, et al. Size-selective loosening of the blood-brain barrier in claudin-5–deficient mice. *J Cell Biol*. 2003;161(3):653–660.
17. González-Mariscal L, Betanzos A, Nava P, Jaramillo BE. Tight junction proteins. *Prog Biophys Mol Biol*. 2003;81(1):1–44.
18. Huang J, Upadhyay UM, Tamargo RJ. Inflammation in stroke and focal cerebral ischemia. *Surg Neurol*. 2006;66(3):232–245.
19. Pardridge WM. Blood-brain barrier genomics. *Stroke*. 2007;38(2):686–690.
20. Liebner S, Fischmann A, Rascher G, Duffner F, Grote E-H, Kalbacher H, et al. Claudin-1 and claudin-5 expression and tight junction morphology are altered in blood vessels of human glioblastoma multiforme. *Acta Neuropathol*. 2000;100(3):323–331.
21. Liebner S, Corada M, Bangsow T, Babbage J, Taddei A, Czupalla CJ, et al. Wnt/β-catenin signaling controls development of the blood–brain barrier. *J Cell Biol*. 2008;183(3):409–417.
22. Wang Y, Rattner A, Zhou Y, Williams J, Smallwood PM, Nathans J. Norrin/Frizzled4 signaling in retinal vascular development and blood brain barrier plasticity. *Cell*. 2012;151(6):1332–1344.
23. Masuda S, Oda Y, Sasaki H, Ikenouchi J, Higashi T, Akashi M, et al. LSR defines cell corners for tricellular tight junction formation in epithelial cells. *J Cell Sci*. 2011;124(4):548–555.
24. Iwamoto N, Higashi T, Furuse M. Localization of angulin-1/LSR and tricellulin at tricellular contacts of brain and retinal endothelial cells in vivo. *Cell Struct Funct*. 2014;39(1):1–8.
25. Liu W, Wang P, Shang C, Chen L, Cai H, Ma J, et al. Endophilin-1 regulates blood–brain barrier permeability by controlling ZO-1 and occludin expression via the EGFR–ERK1/2 pathway. *Brain Res*. 2014;1573:17–26.
26. Kanoski SE, Zhang Y, Zheng W, Davidson TL. The effects of a high-energy diet on hippocampal function and blood-brain barrier integrity in the rat. *J Alzheimer Dis*. 2010;21(1):207–219.
27. Xu R, Feng X, Xie X, Zhang J, Wu D, Xu L. HIV-1 Tat protein increases the permeability of brain endothelial cells by both inhibiting occludin expression and cleaving occludin via matrix metalloproteinase-9. *Brain Res*. 2012;1436:13–19.
28. Yeung D, Manias JL, Stewart DJ, Nag S. Decreased junctional adhesion molecule-A expression during blood–brain barrier breakdown. *Acta Neuropathol*. 2008;115(6):635–642.
29. Pfeiffer F, Schäfer J, Lyck R, Makrides V, Brunner S, Schaeren-Wiemers N, et al. Claudin-1 induced sealing of blood–brain barrier tight junctions ameliorates chronic experimental autoimmune encephalomyelitis. *Acta Neuropathol*. 2011;122(5):601–614.
30. Wolburg H, Wolburg-Buchholz K, Kraus J, Rascher-Eggstein G, Liebner S, Hamm S, et al. Localization of claudin-3 in tight junctions of the blood-brain barrier is selectively lost during experimental autoimmune encephalomyelitis and human glioblastoma multiforme. *Acta Neuropathol*. 2003;105(6):586–592.

31. Abbott NJ, Romero IA. Transporting therapeutics across the blood-brain barrier. *Mol Med Today*. 1996;2(3):106–113.
32. Honda M, Nakagawa S, Hayashi K, Kitagawa N, Tsutsumi K, Nagata I, et al. Adrenomedullin improves the blood–brain barrier function through the expression of claudin-5. *Cell Mol Neurobiol*. 2006;26(2):109–118.
33. Tam Stephen J, Richmond David L, Kaminker Joshua S, Modrusan Z, Martin-McNulty B, Cao Tim C, et al. Death receptors DR6 and TROY regulate brain vascular development. *Dev Cell*. 2012;22(2):403–417.
34. Obermeier B, Daneman R, Ransohoff RM. Development, maintenance and disruption of the blood-brain barrier. *Nat Med*. 2013;19(12):1584–1596.
35. Miller DS. Regulation of P-glycoprotein and other ABC drug transporters at the blood–brain barrier. *Trends Pharmacol Sci*. 2010;31(6):246–254.
36. Lyck R, Ruderisch N, Moll AG, Steiner O, Cohen CD, Engelhardt B, et al. Culture-induced changes in blood-brain barrier transcriptome: implications for amino-acid transporters in vivo. *J Cereb Blood Flow Metab*. 2009;29(9):1491–1502.
37. Mitchell RW, On NH, Del Bigio MR, Miller DW, Hatch GM. Fatty acid transport protein expression in human brain and potential role in fatty acid transport across human brain microvessel endothelial cells. *J Neurochem*. 2011;117(4):735–746.
38. Zhou Q-H, Boado RJ, Hui EK-W, Lu JZ, Pardridge WM. Brain-penetrating tumor necrosis factor decoy receptor in the mouse. *Drug Metab Dispos*. 2011;39(1):71–76.
39. Boado RJ, Lu JZ, Hui EKW, Sumbria RK, Pardridge WM. Pharmacokinetics and brain uptake in the rhesus monkey of a fusion protein of arylsulfatase a and a monoclonal antibody against the human insulin receptor. *Biotechnol Bioeng*. 2013;110(5):1456–1465.
40. Bowman P, Betz AL, Ar D, Wolinsky J, Penney J, Shivers R, et al. Primary culture of capillary endothelium from rat brain. *In Vitro*. 1981;17(4):353–362.
41. Deli MA, Ábrahám CS, Niwa M, Falus A. *N,N*-Diethyl-2-[4-(phenylmethyl)phenoxy]ethanamine increases the permeability of primary mouse cerebral endothelial cell monolayers. *Inflamm Res*. 2003;52(1):s39–s40.
42. Ball HJ, McParland B, Driussi C, Hunt NH. Isolating vessels from the mouse brain for gene expression analysis using laser capture microdissection. *Brain Res Protoc*. 2002;9(3):206–213.
43. Wang S, Qaisar U, Yin X, Grammas P. Gene expression profiling in Alzheimer's disease brain microvessels. *J Alzheimer Dis*. 2011;31(1):193–205.
44. Calabria AR, Shusta EV. A genomic comparison of in vivo and in vitro brain microvascular endothelial cells. *J Cereb Blood Flow Metab*. 2008;28(1):135–148.
45. Diatchenko L, Lau YF, Campbell AP, Chenchik A, Moqadam F, Huang B, et al. Suppression subtractive hybridization: a method for generating differentially regulated or tissue-specific cDNA probes and libraries. *Proc Natl Acad Sci USA*. 1996;93(12):6025–6030.
46. Pardridge WM. Molecular biology of the blood-brain barrier. *Mol Biotechnol*. 2005;30(1): 57–69.
47. Anderson L, Seilhamer J. A comparison of selected mRNA and protein abundances in human liver. *Electrophoresis*. 1997;18(3–4):533–537.
48. Ohtsuki S, Hirayama M, Ito S, Uchida Y, Tachikawa M, Terasaki T. Quantitative targeted proteomics for understanding the blood–brain barrier: towards pharmacoproteomics. *Exp Rev Proteomics*. 2014;11(3):303–313.
49. Lu Q, Murugesan N, Macdonald JA, Wu SL, Pachter JS, Hancock WS. Analysis of mouse brain microvascular endothelium using immuno-laser capture microdissection coupled to a hybrid linear ion trap with Fourier transform-mass spectrometry proteomics platform. *Electrophoresis*. 2008;29(12):2689–2695.

50. Lippmann ES, Azarin SM, Kay JE, Nessler RA, Wilson HK, Al-Ahmad A, et al. Derivation of blood-brain barrier endothelial cells from human pluripotent stem cells. *Nat Biotechnol.* 2012;30(8):783–791.

51. Kamiie J, Ohtsuki S, Iwase R, Ohmine K, Katsukura Y, Yanai K, et al. Quantitative atlas of membrane transporter proteins: development and application of a highly sensitive simultaneous LC/MS/MS method combined with novel in-silico peptide selection criteria. *Pharm Res.* 2008;25(6):1469–1483.

52. Uchida Y, Ohtsuki S, Katsukura Y, Ikeda C, Suzuki T, Kamiie J, et al. Quantitative targeted absolute proteomics of human blood–brain barrier transporters and receptors. *J Neurochem.* 2011;117(2):333–345.

53. Torbett BE, Baird A, Eliceiri BP. Understanding the rules of the road: proteomic approaches to interrogate the blood brain barrier. *Front Neurosci.* 2015;9:70.

54. Mueller LN, Brusniak M-Y, Mani D, Aebersold R. An assessment of software solutions for the analysis of mass spectrometry based quantitative proteomics data. *J Proteome Res.* 2008;7(1):51–61.

55. Karamanos Y. Studying molecular aspects of the blood-brain barrier using an in vitro model: contribution of a global proteomics strategy. *J Biosci Med.* 2014;2(1):18–25.

56. Washburn MP, Wolters D, Yates JR. Large-scale analysis of the yeast proteome by multidimensional protein identification technology. *Nat Biotechnol.* 2001;19(3):242–247.

57. Chen S-S, Haskins WE, Ottens AK, Hayes RL, Denslow N, Wang KK. Bioinformatics for traumatic brain injury: proteomic data mining. Data Mining in Biomedicine. New York: Springer; 2007:(pp. 363–387).

58. Guingab-Cagmat J, Cagmat E, Hayes RL, Anagli J. Integration of proteomics, bioinformatics, and systems biology in traumatic brain injury biomarker discovery. *Front Neurol.* 2013;4:61.

59. Calabria AR, Shusta EV. Blood–brain barrier genomics and proteomics: elucidating phenotype, identifying disease targets and enabling brain drug delivery. *Drug Discov Today.* 2006;11(17):792–799.

60. Paris D, Ait-Ghezala G, Mathura VS, Patel N, Quadros A, Laporte V, et al. Anti-angiogenic activity of the mutant Dutch Aβ peptide on human brain microvascular endothelial cells. *Brain Red Mol Brain Res.* 2005;136(1–2):212–230.

61. Citron M. Alzheimer's disease: strategies for disease modification. *Nat Rev Drug Discov.* 2010;9(5):387–398.

62. Yanamandra K, Jiang H, Mahan TE, Maloney SE, Wozniak DF, Diamond MI, et al. Anti-tau antibody reduces insoluble tau and decreases brain atrophy. *Ann Clin Transl Neurol.* 2015;2(3):278–288.

63. Kirsch T, Wellner M, Luft FC, Haller H, Lippoldt A. Altered gene expression in cerebral capillaries of stroke-prone spontaneously hypertensive rats. *Brain Res.* 2001;910(1–2):106–115.

64. Haqqani AS, Nesic M, Preston E, Baumann E, Kelly J, Stanimirovic D. Characterization of vascular protein expression patterns in cerebral ischemia/reperfusion using laser capture microdissection and ICAT-nanoLC-MS/MS. *FASEB J.* 2005;19(13):1809–1821.

65. Argaw AT, Asp L, Zhang J, Navrazhina K, Pham T, Mariani JN, et al. Astrocyte-derived VEGF-A drives blood-brain barrier disruption in CNS inflammatory disease. *J Clin Invest.* 2012;122(7):2454–2468.

66. Luheshi N, Kovacs K, Lopez-Castejon G, Brough D, Denes A. Interleukin-1alpha expression precedes IL-1beta after ischemic brain injury and is localised to areas of focal neuronal loss and penumbral tissues. *J Neuroinflam.* 2011;8(1):186.

67. McGeer PL, McGeer EG. Inflammation, autotoxicity and Alzheimer disease. *Neurobiol Aging.* 2001;22(6):799–809.

68. Yang SH, Gangidine M, Pritts TA, Goodman MD, Lentsch AB. Interleukin 6 mediates neuroinflammation and motor coordination deficits after mild traumatic brain injury and brief hypoxia in mice. *Shock*. 2013;40(6):471–475.

69. Kebir H, Kreymborg K, Ifergan I, Dodelet-Devillers A, Cayrol R, Bernard M, et al. Human TH17 lymphocytes promote blood-brain barrier disruption and central nervous system inflammation. *Nat Med*. 2007;13(10):1173–1175.

70. Jin S, Sonobe Y, Kawanokuchi J, Horiuchi H, Cheng Y, Wang Y, et al. Interleukin-34 restores blood–brain barrier integrity by upregulating tight junction proteins in endothelial cells. *PLoS One*. 2014;9(12):e115981.

71. Pineda D, Ampurdanes C, Medina MG, Serratosa J, Tusell JM, Saura J, et al. Tissue plasminogen activator induces microglial inflammation via a noncatalytic molecular mechanism involving activation of mitogen-activated protein kinases and Akt signaling pathways and AnnexinA2 and Galectin-1 receptors. *Glia*. 2012;60(4):526–540.

72. Tsuji K, Aoki T, Tejima E, Arai K, Lee S-R, Atochin DN, et al. Tissue plasminogen activator promotes matrix metalloproteinase-9 upregulation after focal cerebral ischemia. *Stroke*. 2005;36(9):1954–1959.

73. Du H, Li P, Wang J, Qing X, Li W. The interaction of amyloid β and the receptor for advanced glycation endproducts induces matrix metalloproteinase-2 expression in brain endothelial cells. *Cell Mol Neurobiol*. 2012;32(1):141–147.

74. Pankiewicz JE, Guridi M, Kim J, Asuni AA, Sanchez S, Sullivan PM, et al. Blocking the apoE/Aβ interaction ameliorates Aβ-related pathology in APOE ε2 and ε4 targeted replacement Alzheimer model mice. *Acta Neuropathol Commun*. 2014;2(1):75.

75. Prakash A, Kumar A. Role of nuclear receptor on regulation of BDNF and neuroinflammation in hippocampus of β-amyloid animal model of Alzheimer's disease. *Neurotox Res*. 2014;25(4):335–347.

76. Takata F, Dohgu S, Matsumoto J, Takahashi H, Machida T, Wakigawa T, et al. Brain pericytes among cells constituting the blood-brain barrier are highly sensitive to tumor necrosis factor-α, releasing matrix metalloproteinase-9 and migrating in vitro. *J Neuroinflam*. 2011;8(106):1–12.

77. Yamamoto M, Horiba M, Buescher JL, Huang D, Gendelman HE, Ransohoff RM, et al. Overexpression of monocyte chemotactic protein-1/CCL2 in β-Amyloid precursor protein transgenic mice show accelerated diffuse β-amyloid deposition. *Am J Pathol*. 2005;166(5):1475–1485.

78. Togo T, Akiyama H, Kondo H, Ikeda K, Kato M, Iseki E, et al. Expression of CD40 in the brain of Alzheimer's disease and other neurological diseases. *Brain Res*. 2000;885(1):117–121.

79. Tan J, Town T, Crawford F, Mori T, DelleDonne A, Crescentini R, et al. Role of CD40 ligand in amyloidosis in transgenic Alzheimer's mice. *Nat Neurosci*. 2002;5(12):1288–1293.

80. Yasojima K, Schwab C, McGeer EG, McGeer PL. Human neurons generate C-reactive protein and amyloid P: upregulation in Alzheimer's disease. *Brain Res*. 2000;887(1):80–89.

81. Gorina R, Font-Nieves M, Márquez-Kisinousky L, Santalucia T, Planas AM. Astrocyte TLR4 activation induces a proinflammatory environment through the interplay between MyD88-dependent NFκB signaling, MAPK, and Jak1/Stat1 pathways. *Glia*. 2011;59(2):242–2455.

82. Ma D, Jin S, Li E, Doi Y, Parajuli B, Noda M, et al. The neurotoxic effect of astrocytes activated with toll-like receptor ligands. *J Neuroimmunol*. 2013;254(1):10–18.

83. Min H, Hong J, Cho I-H, Jang YH, Lee H, Kim D, et al. TLR2-induced astrocyte MMP9 activation compromises the blood brain barrier and exacerbates intracerebral hemorrhage in animal models. *Mol Brain*. 2015;8(1):23.

84. Butchi NB, Woods T, Du M, Morgan TW, Peterson KE. TLR7 and TLR9 trigger distinct neuroinflammatory responses in the CNS. *Am J Pathol*. 2011;179(2):783–794.

85. Jin J-J, Kim H-D, Maxwell JA, Li L, Fukuchi K-i. Toll-like receptor 4-dependent upregulation of cytokines in a transgenic mouse model of Alzheimer's disease. *J Neuroinflam.* 2008;5(23):2094–2095.

86. Hartz AM, Mahringer A, Miller DS, Bauer B. 17-β-Estradiol: a powerful modulator of blood–brain barrier BCRP activity. *J Cereb Blood Flow Metab.* 2010;30(10):1742–1755.

87. Li X-H, Lv B-L, Xie J-Z, Liu J, Zhou X-W, Wang J-Z. AGEs induce Alzheimer-like tau pathology and memory deficit via RAGE-mediated GSK-3 activation. *Neurobiol Aging.* 2012;33(7):1400–1410.

88. Fang F, Lue L-F, Yan S, Xu H, Luddy JS, Chen D, et al. RAGE-dependent signaling in microglia contributes to neuroinflammation, Aβ accumulation, and impaired learning/memory in a mouse model of Alzheimer's disease. *FASEB J.* 2010;24(4):1043–1055.

89. Lu A, Suofu Y, Guan F, Broderick JP, Wagner KR, Clark JF. Matrix metalloproteinase-2 deletions protect against hemorrhagic transformation after 1h of cerebral ischemia and 23h of reperfusion. *Neuroscience.* 2013;253:361–367.

90. Suofu Y, Clark JF, Broderick JP, Kurosawa Y, Wagner KR, Lu A. Matrix metalloproteinase-2 or-9 deletions protect against hemorrhagic transformation during early stage of cerebral ischemia and reperfusion. *Neuroscience.* 2012;212:180–189.

91. Suzuki Y, Nagai N, Umemura K, Collen D, Lijnen H. Stromelysin-1 (MMP-3) is critical for intracranial bleeding after t-PA treatment of stroke in mice. *J Thromb Haemost.* 2007;5(8):1732–1739.

92. Lee E-J, Han JE, Woo M-S, Shin JA, Park E-M, Kang JL, et al. Matrix metalloproteinase-8 plays a pivotal role in neuroinflammation by modulating TNF-α activation. *J Immunol.* 2014;193(5):2384–2393.

93. Han J, Lee E-J, Moon E, Ryu J, Choi J, Kim H-S. Matrix metalloproteinase-8 is a novel pathogenetic factor in focal cerebral ischemia. *Mol Neurobiol.* 2014;:1–9.

94. Chaudhry K, Rogers R, Guo M, Lai Q, Goel G, Liebelt B, et al. Matrix metalloproteinase-9 (MMP-9) expression and extracellular signal-regulated kinase 1 and 2 (ERK1/2) activation in exercise-reduced neuronal apoptosis after stroke. *Neurosci Lett.* 2010;474(2):109–114.

95. Ma Q, Huang B, Khatibi N, Rolland W, Suzuki H, Zhang JH, et al. PDGFR-α inhibition preserves blood-brain barrier after intracerebral hemorrhage. *Ann Neurol.* 2011;70(6):920–931.

96. Li G-H, Anderson C, Jaeger L, Do T, Major EO, Nath A. Cell-to-cell contact facilitates HIV transmission from lymphocytes to astrocytes via CXCR4. *AIDS.* 2015;29(7):755–766.

97. Huang J, Li Y, Tang Y, Tang G, Yang G-Y, Wang Y. CXCR4 antagonist AMD3100 protects blood–brain barrier integrity and reduces inflammatory response after focal ischemia in mice. *Stroke.* 2013;44(1):190–197.

98. Mizee M, Nijland P, van der Pol SA, Drexhage JR, van het Hof B, Mebius R, et al. Astrocyte-derived retinoic acid: a novel regulator of blood–brain barrier function in multiple sclerosis. *Acta Neuropathol.* 2014;128(5):691–703.

99. Zhang Y, Pardridge WM. Blood–brain barrier targeting of BDNF improves motor function in rats with middle cerebral artery occlusion. *Brain Res.* 2006;1111(1):227–229.

100. Zhang Y, Zhang Y-f, Bryant J, Charles A, Boado RJ, Pardridge WM. Intravenous RNA interference gene therapy targeting the human epidermal growth factor receptor prolongs survival in intracranial brain cancer. *Clin Cancer Res.* 2004;10(11):3667–3677.

101. Coloma MJ, Lee H, Kurihara A, Landaw E, Boado R, Morrison S, et al. Transport across the primate blood-brain barrier of a genetically engineered chimeric monoclonal antibody to the human insulin receptor. *Pharm Res.* 2000;17(3):266–274.

102. Chen Y, Liu L. Modern methods for delivery of drugs across the blood–brain barrier. *Adv Drug Deliv Rev.* 2012;64(7):640–665.

103. Tayeb HO, Murray ED, Price BH, Tarazi FI. Bapineuzumab and solanezumab for Alzheimer's disease: is the 'amyloid cascade hypothesis' still alive?. *Exp Opin Biol Ther*. 2013;13(7): 1075–1084.

104. Pardridge WM. Transport of protein and antibody therapeutics across the blood–brain barrier. Blood-Brain Barrier in Drug Discovery. New Jersey: John Wiley and Sons, Inc.; 2015:(pp. 146–166).

105. Zhou Q-H, Boado RJ, Pardridge WM. Selective plasma pharmacokinetics and brain uptake in the mouse of enzyme fusion proteins derived from species-specific receptor-targeted antibodies. *J Drug Target*. 2012;20(8):715–719.

106. Jones A, Shusta E. Blood–brain barrier transport of therapeutics via receptor-mediation. *Pharm Res*. 2007;24(9):1759–1771.

107. Sumbria RK, Boado RJ, Pardridge WM. Brain protection from stroke with intravenous TNFα decoy receptor-Trojan horse fusion protein. *J Cereb Blood Flow Metab*. 2012;32(10): 1933–1938.

108. Boado RJ, Zhang Y, Zhang Y, Wang Y, Pardridge WM. GDNF fusion protein for targeted-drug delivery across the human blood–brain barrier. *Biotechnol Bioeng*. 2008;100(2):387–396.

109. Sumbria RK, Boado RJ, Pardridge WM. Combination stroke therapy in the mouse with blood–brain barrier penetrating IgG–GDNF and IgG–TNF decoy receptor fusion proteins. *Brain Res*. 2013;1507:91–96.

110. Cai Y, Liu X, Chen W, Wang Z, Xu G, Zeng Y, et al. TGF-β1 prevents blood–brain barrier damage and hemorrhagic transformation after thrombolysis in rats. *Exp Neurol*. 2015;266:120–126.

111. Spencer B, Verma I, Desplats P, Morvinski D, Rockenstein E, Adame A, et al. A neuroprotective brain-penetrating endopeptidase fusion protein ameliorates Alzheimer disease pathology and restores neurogenesis. *J Biol Chem*. 2014;289(25):17917–21731.

112. Benedek G, Zhu W, Libal N, Casper A, Yu X, Meza-Romero R, et al. A novel HLA-DRα1-MOG-35-55 construct treats experimental stroke. *Metab Brain Dis*. 2014;29(1):37–45.

113. Shao Y, Deng T, Zhang T, Li P, Wang Y. FAM19A3, a novel secreted protein, modulates the microglia/macrophage polarization dynamics and ameliorates cerebral ischemia. *FEBS Lett*. 2015;589(4):467–475.

114. Chung E, Ji Y, Sun Y, Kascsak RJ, Kascsak RB, Mehta PD, et al. Anti-PrPC monoclonal antibody infusion as a novel treatment for cognitive deficits in an Alzheimer's disease model mouse. *BMC Neurosci*. 2010;11(1):130.

115. Castillo-Carranza DL, Guerrero-Muñoz MJ, Sengupta U, Hernandez C, Barrett AD, Dineley K, et al. Tau immunotherapy modulates both pathological tau and upstream amyloid pathology in an Alzheimer's disease mouse model. *J Neurosci*. 2015;35(12):4857–4868.

116. Lindström V, Fagerqvist T, Nordström E, Eriksson F, Lord A, Tucker S, et al. Immunotherapy targeting α-synuclein protofibrils reduced pathology in (Thy-1)-h [A30P] α-synuclein mice. *Neurobiol Dis*. 2014;69:134–143.

117. Masliah E, Rockenstein E, Mante M, Crews L, Spencer B, Adame A, et al. Passive immunization reduces behavioral and neuropathological deficits in an alpha-synuclein transgenic model of Lewy body disease. *PloS One*. 2011;6(4):e19338.

118. Zhang J, Takahashi HK, Liu K, Wake H, Liu R, Maruo T, et al. Anti-high mobility group box-1 monoclonal antibody protects the blood–brain barrier from ischemia-induced disruption in rats. *Stroke*. 2011;42(5):1420–1428.

119. Gelderblom M, Weymar A, Bernreuther C, Velden J, Arunachalam P, Steinbach K, et al. Neutralization of the IL-17 axis diminishes neutrophil invasion and protects from ischemic stroke. *Blood*. 2012;120(18):3793–3802.

120. Tsai S-Y, Papadopoulos CM, Schwab ME, Kartje GL. Delayed anti-nogo-a therapy improves function after chronic stroke in adult rats. *Stroke*. 2011;42(1):186–190.

121. Regina A, Demeule M, Tripathy S, Lord-Dufour S, Currie J-C, Iddir M, et al. ANG4043, a novel brain-penetrant peptide–mAb conjugate, is efficacious against HER2-positive intracranial tumors in mice. *Mol Cancer Ther*. 2015;14(1):129–140.

122. Castro-Alvarez JF, Uribe-Arias A, Cardona-Gómez GP. Cyclin-Dependent kinase 5 targeting prevents β-amyloid aggregation involving glycogen synthase kinase 3β and phosphatases. *J Neurosci Res*. 2015;93(8):1258–1266.

123. Kiyota T, Ingraham KL, Swan RJ, Jacobsen MT, Andrews SJ, Ikezu T. AAV serotype 2/1-mediated gene delivery of anti-inflammatory interleukin-10 enhances neurogenesis and cognitive function in APP+PS1 mice. *Gene Ther*. 2012;19(7):724–733.

124. Kiyota T, Yamamoto M, Schroder B, Jacobsen MT, Swan RJ, Lambert MP, et al. AAV1/2-mediated CNS gene delivery of dominant-negative CCL2 mutant suppresses gliosis, β-amyloidosis, and learning impairment of APP/PS1 mice. *Mol Ther*. 2009;17(5):803–809.

125. Kiyota T, Okuyama S, Swan RJ, Jacobsen MT, Gendelman HE, Ikezu T. CNS expression of anti-inflammatory cytokine interleukin-4 attenuates Alzheimer's disease-like pathogenesis in APP+PS1 bigenic mice. *FASEB J*. 2010;24(8):3093–3102.

126. Kiyota T, Ingraham KL, Jacobsen MT, Xiong H, Ikezu T. FGF2 gene transfer restores hippocampal functions in mouse models of Alzheimer's disease and has therapeutic implications for neurocognitive disorders. *Proc Natl Acad Sci USA*. 2011;108(49):E1339–E1348.

127. Drinkut A, Tereshchenko Y, Schulz JB, Bähr M, Kügler S. Efficient gene therapy for Parkinson's disease using astrocytes as hosts for localized neurotrophic factor delivery. *Mol Ther*. 2012;20(3):534–543.

128. McCormack AL, Mak SK, Henderson JM, Bumcrot D, Farrer MJ, Di Monte DA. Alpha-synuclein suppression by targeted small interfering RNA in the primate substantia nigra. *PLoS One*. 2010;5(8):e12122.

129. Pardridge WM. Blood–brain barrier delivery of protein and non-viral gene therapeutics with molecular Trojan horses. *J Control Release*. 2007;122(3):345–348.

130. Yue P-J, He L, Qiu S-W, Li Y, Liao Y-J, Li X-P, et al. OX26/CTX-conjugated PEGylated liposome as a dual-targeting gene delivery system for brain glioma. *Mol Cancer*. 2014;13(1):1–13.

131. Ren J, Chen YI, Liu CH, Chen PC, Prentice H, Wu JY, et al. Noninvasive tracking of gene transcript and neuroprotection after gene therapy. *Gene Ther*. 2015;23(1):1–9.

132. Di Pasquale G, Chiorini JA. AAV transcytosis through barrier epithelia and endothelium. *Mol Ther*. 2006;13(3):506–516.

133. Huang R-Q, Qu Y-H, Ke W-L, Zhu J-H, Pei Y-Y, Jiang C. Efficient gene delivery targeted to the brain using a transferrin-conjugated polyethyleneglycol-modified polyamidoamine dendrimer. *FASEB J*. 2007;21(4):1117–1125.

134. Pardridge WM. shRNA and siRNA delivery to the brain. *Adv Drug Deliv Rev*. 2007;59(2–3):141–152.

135. Slanina H, Schmutzler M, Christodoulides M, Kim KS, Schubert-Unkmeir A. Effective plasmid DNA and small interfering RNA delivery to diseased human brain microvascular endothelial cells. *J Mol Microbiol Biotechnol*. 2012;22(4):245–257.

136. Mariotti V, Greco SJ, Mohan RD, Nahas GR, Rameshwar P. Stem cell in alternative treatments for brain tumors: potential for gene delivery. *Mol Cell Ther*. 2014;2(1):24.

CHAPTER 3

Brain and the Drug Transporters

Tugba Gulsun, PhD, Nihan Izat, BSc, Selma Sahin, PhD
Hacettepe University, Ankara, Turkey

1 INTRODUCTION

The central nervous system (CNS) consists of the brain and spinal cord. The brain is one of the largest, most vital, delicate, and complicated organ in the human body and is protected from potentially harmful endogenous and exogenous substances by two important barriers, namely blood–brain barrier (BBB) and blood–cerebrospinal fluid barrier (BCSFB).[1] The BBB provides a physical and metabolic barrier between the brain and the systemic circulation. The main function of the BBB is to preserve the stable internal environment in the CNS, and to provide essential nutrients from the blood. Furthermore, the BBB is composed of a monolayer of brain capillary endothelial cells, which are connected by tight junctions. These endothelial cells are surrounded by pericytes, which help maintain the BBB and several other homeostatic and hemostatic functions of the brain,[2] and perivascular astrocytes. Under physiological conditions, tight junctions form a continuous, almost impermeable cellular barrier and limit paracellular flux and transport, as well as the influx of endogenous and exogenous substances (except very small lipid-soluble molecules).[3] Several receptors, ion channels, and influx–efflux transport proteins are expressed prominently at the BBB.[4] The major interface between the systemic circulation and the CSF is the BCSFB, which is formed by the choroid plexus (CP). This BCSFB is composed of fenestrated capillaries that are surrounded by a monolayer of epithelial cells joined together by tight junctions.[5] These tight junctions form the structural basis of the BCSFB and seal together adjacent polarized epithelial cells (ependymal cells). Once a solute has crossed the capillary wall, it must also permeate these ependymal cells before entering the CSF. The primary function of the CP is to produce CSF continuously, for the maintenance of its composition. The CP displays polarized expression of various receptors, ion channels, and transport systems that regulate the CSF composition via secretion and reabsorption.[6] The brain parenchyma consists of neurons and the surrounding glial cells. Neurons form the basic structural and functional component of the CNS.

CONTENTS

Nanotechnology Methods for Neurological Diseases and Brain Tumors. http://dx.doi.org/10.1016/B978-0-12-803796-6.00003-4

The primary function of neurons is to respond to stimuli by conducting electrical signals along conductive processes. Glial cells play an instrumental role in the regulation and maintenance of CNS homeostasis, and can be classified into two groups: the macroglia (astrocytes and oligodendrocytes) and the microglia.[7] The most abundant cell type in the brain is the astrocyte. Astrocytes form a frame to protect the biochemical environment in which neurons can function. Astrocytes branches or legs are in close contact with the blood vessels, so that they can maintain function and integrity of the BBB.[2,8] The primary function of oligodendrocytes is to form the insulating myelin sheath in the CNS. Myelin is a lipid-rich biological membrane that surrounds neuronal axons, and increases the resistance for electrical impulses during an action potential.[9] Microglia are mononuclear macrophages derived from blood, which protect the CNS from changes in physiological and pathological conditions.[10] Microglia are distributed widely within the CNS, with the basal ganglia and cerebellum having considerably greater numbers than the cerebral cortex.[11]

Therapeutic compounds may cross BBB and BCSFB by various uptake processes, including passive diffusion, carrier-mediated (facilitated) transport, active transport, transcytosis, and receptor-mediated endocytosis.[12] After crossing these initial barriers, brain drug accumulation can be restricted further by passive efflux within the CSF, metabolic degradation, and/or active efflux transport. Brain parenchymal cells (i.e., astrocytes, microglia, oligodendrocytes, and neurons) also play an important role in regulating CNS drug distribution. These cells express several drug transport proteins, which underscore the complexity of xenobiotic disposition within the brain. The objective of this chapter is to summarize cellular localization and functional activity of influx (solute lipid carriers or SLC) and efflux (ATP-binding cassette transporters or ABC) drug transporters in the brain barriers and brain parenchyma.

2 BRAIN TRANSPORTERS

For the treatment of various CNS diseases, sufficient amount of drugs has to cross the brain barriers (i.e., BBB and BCSFB). In some cases, they also have to permeate through the cellular compartments of the brain parenchyma (i.e., astrocytes, microglia, oligodendrocytes, and neurons). Although, passive diffusion is the route of entry to the brain for small, nonionic, lipid-soluble compounds, this transport mechanism is not possible for CNS permeation of larger, water-soluble, and/or ionic substances. For such substances, uptake into the brain and efflux from the brain may be regulated by various drug transport proteins. These drug transporters can be classified in different ways, such as efflux versus influx transporters and ABC transporters versus SLC transporters. Although ABC transporters require ATP hydrolysis for the transport of substrates across biological membranes, SLC transporters do not have ATP-binding sites. Some SLC transporters use an electrochemical potential difference in the substrate transported, where others use an ion gradient for the transport.[13]

Syvänen et al.[14] investigated how the nature, location, and capacity of the efflux processes in relation to the permeability properties influence brain concentrations. Based on simulation studies, it was concluded that hindrance of the influx process is more effective for keeping brain concentrations low. The relationship between the influx and efflux of the drug across the BBB determines the steady-state ratio of brain-to-plasma concentrations of the unbound drug.[14]

In the following section, we will briefly summarize the CNS localization and functional expression of the ABC and SLC membrane drug transporters. Localization of various transporters at the human brain barriers and in cellular compartments of the brain parenchyma is depicted in Fig. 3.1.

2.1 ABC Drug Transporters

With 50 known human ABC members belonging to seven different subfamilies (ABCA–G), the ABC superfamily is one of the largest efflux transporter families. ABC transporters are multidomain integral membrane proteins, which use

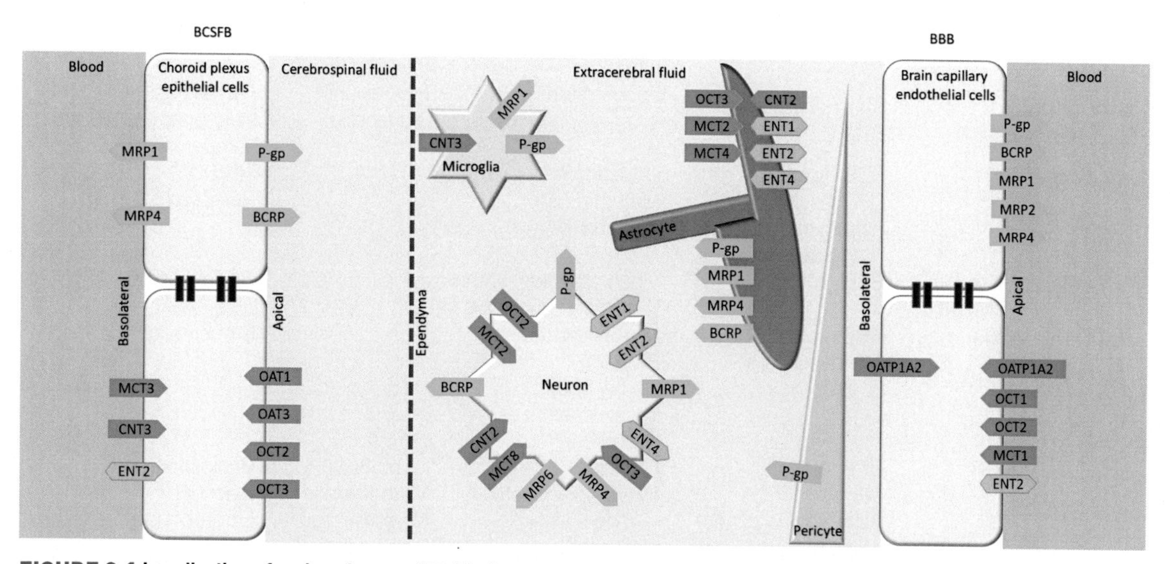

FIGURE 3.1 Localization of various human ATP-binding cassette (ABC)/solute lipid carrier (SLC) transporters at blood–brain barrier (BBB)/blood–cerebrospinal fluid barrier (BCSFB) and in brain parenchyma (astrocytes, microglia, neurons, and pericytes).

BCRP, Breast cancer resistance protein; *CNT*, Na^+-dependent concentrative nucleoside transporter; *ENT*, Na^+-independent equilibrative nucleoside transporter; *MCT*, monocarboxylate transporter; *MRP*, multidrug resistance–associated protein; *OAT*, organic anion transporter; *OCT*, organic cation transporter; *P-gp*, P-glycoprotein.

the energy of ATP hydrolysis to transport endogenous (e.g., inorganic anions, metal ions, peptides, amino acids, and sugars) and exogenous (e.g., antiviral agents, chemotherapeutics, antidepressants, and antihypertensive drugs) (Table 3.1) substances across plasma and intracellular membranes in all mammalian species. ABC transporters are generally composed of four domains: two

Table 3.1 Localization, Substrate, and Inhibitor Examples of Main ABC Transporters Located in the Central Nervous System (CNS)

Genes	Aliases	Localizations	Substrates		Inhibitors
ABCB1	P-gp, MDR1	BBB: apical[16] BBB: basolateral[17] CP: apical[18] Astrocytes, pericytes, and neurons[17]	Aliskiren[19] Ambrisentan[20] Colchicine[21] Cortisone[22] Dabigatran etexilate[23] Digoxine[24] Everolimus[25] Fexofenadine[26] Flesinoxan[27] Gefitinib[28] Glabridin[29] Imatinib mesylate[30] Indinavir[31] Ivermectin[32] Lapatinib[33] Loperamide[21]	Loratadine[34] Maraviroc[35] Nelfinavir[31] Nilotinib[36] Ondansetron[37] Paclitaxel[38] Prednisone[22] Ranolazine[39] Saquinavir[31] Salinomycin[40] Sirolimus[41] Sitagliptin[42] Sparfloxacin[43] Talinolol[44] Tolvaptan[45] Topotecan[46] Quinacrine[47] Vinblastine[48]	Amiodarone[49] Carvedilol[50] Clarithromycin[51] Conivaptan[52] Cyclosporin[53] Diltiazem[54] Dronedarone[49] Elacridar[30] Erythromycin[51] Itraconazole[52] Ketoconazole[52] Lopinavir[52] Ritonavir[52] Quercetin[55] Quinidine[49] Ranolazine[39] Valspodar (PSC 833)[38] Verapamil[49]
ABCC1	MRP1	BBB: apical[56] CP: basolateral[18] Astrocytes, microglia, and neurons[17]	Difloxacin[57] Methotrexate[57] Saquinavir[57]		MK571[58] Reversan[59]
ABCC2	MRP2, CMOAT	BBB: apical[60–62]	Vinca alkaloids, anthracyclines, taxanes, cisplatin, methotrexate[63]		Cyclosporin[64]
ABCC4	MRP4	BBB: apical[65] CP: basolateral[65]	Methotrexate[63] PMEA[66] Ro 64-0802[67] Thiopurines[63] Topotecan[65]		Probenecid[65]
ABCG2	BCRP	BBB: apical[68] CP: apical[17] Astrocytes and neurons[17]	Daidzein[69] Dihyroepiandosterone sulfate[70] Gefitinib[28] Genistein[69] Imatinib[71] Irinotecan (SN-38)[72] Lapatinib[73]	LY2228820[72,74] Methotrexate[30] Mitoxantrone[63] Rosuvastatin[75] Sulfasalazine[75] Topotecan[76]	Cyclosporin A[77] Elacridar[76] Eltrombopag[78] Gefitinib[79] Lapatinib[80] Sirolimus[77] Tacrolimus[77]

CMOAT, Canalicular multispecific organic anion transporter 1; CP, choroid plexus; MDR, multidrug resistance; PMEA, 9-(2-phosphonylmethoxyethyl)adenine.

nucleotide-binding domains (NMDs) that bind and hydrolyze ATP, and two transmembrane domains (TMDs) that are responsible for substrate recognition and transfer across the lipid membranes. ABC transporters are classified based on the two ATP-binding motifs (Walker A and Walker B) and the ABC signature C motif.[15]

ABC transporters are widely expressed in the body, including liver, intestine, kidneys, placenta, CNS barriers (BBB and BCSFB), and blood–testis barrier. These membrane-bound proteins function as gatekeepers, influence pharmacokinetic profiles of the substrate compounds, and frequently cooperate with drug-metabolizing enzymes.[17,81–84]

Among all ABC transporters, ABCB1 (P-glycoprotein, P-gp/multidrug resistance1, MDR1), ABCG2 (breast cancer resistance protein or BCRP), and ABCC members (multidrug resistance–associated protein or MRPs) are the most studied transporters. They are located at the apical (luminal) membranes of brain capillary and CP endothelial cells. They can limit the brain penetration of their substrates by unidirectional transport, regardless of the concentration gradient.[85,86] The presence of ABC transporters in other nervous system cells (pericytes, perivascular astrocytes, microglia, and neurons) was also demonstrated.[87,88] Although P-gp, Mrp1, and Mrp4 localizations at the basolateral (abluminal) membrane of brain capillary endothelial cells were shown, their contribution to barrier function was not clear.[89–93]

Daood et al. showed expressions of P-gp and BCRP at BBB, and MRP1 at BCSFB in the developing human CNS, and suggested that these transporters may restrict the CNS uptake of substrate compounds in neonates.[16] Localization, substrate, and inhibitor examples of main ABC transporters on the CNS are summarized in Table 3.1.

Expression levels and functions of ABC transporters may differ in CNS diseases. For instance, overexpression of P-gp in brain tumors (e.g., glioblastomas) and regional overexpression of P-gp and MRPs in the therapy of refractory epilepsy were demonstrated, and attributed to insufficient pharmacological treatment.[94–97] Zhang et al. investigated *ABCG2* expression in normal and glioblastoma tumors by RT-PCR and showed higher expression of *ABCG2*.[98] Recently, an inverse correlation was reported between vascular P-gp expression and deposition of β-amyloid protein (a P-gp substrate)[99] in the brain parenchyma, indicating that P-gp may be involved in the brain clearance of β-amyloid protein.[100] In the case of Parkinson's disease, it has been suggested that genetic polymorphisms in the MDR1 gene may represent a risk factor, as P-gp is involved in the active efflux of toxins from the brain.[101–103] In AIDS patients with HIV encephalitis, expression levels of P-gp were variable in astrocytes and microglia; however, a total increase was demonstrated.[104] Furthermore, overexpression of P-gp has been reported in mice with amyotrophic lateral sclerosis disease when compared to presymptomatic ones.[105]

2.1.1 ABCB1: P-Glycoprotein

P-gp is a well-characterized ABC transporter discovered initially in Chinese hamster ovary cells by Juliano and Ling in 1976.[106] It is a 170-kDa integral membrane protein encoded by the MDR gene. Two isoforms of the MDR gene (MDR1 and MDR2) have been cloned and sequenced in humans. P-gp is encoded by the mdr1a, mdr1b, and mdr2 genes in rodents.[107] MDR2/mdr2 is expressed primarily in the liver and involved in the translocation of phosphatidylcholine into the bile. In humans, the MDR1 comprises 1280 amino acids and consists of two homologous halves, each made up of six TMDs and one ATP-binding site. Mammalian P-gp may possess between two and four oligosaccharide side chains on the first extracellular loop.[108] It was reported that the nucleotide-bound form of catalytically active P-gp shows an asymmetry between two ATP-binding sites having high and low affinities. It was proposed that this asymmetric state represents the next intermediate on the path toward ATP hydrolysis after nucleotide binding, and an alternating sites mode of action is achieved by simultaneous switching of the two active sites between high- and low-affinity states.[109]

P-gp is an efflux transporter that hinders substrate drugs from penetrating tissues (i.e., brain and gut) and is also involved in biliary excretion and renal excretion of drugs. Schinkel et al. investigated the efflux activity of P-gp in mdr1a (−/−) knockout mice, and found that permeation of ivermectin and vinblastine was increased across BBB[32] when compared with the wildtype mice as the control group. Two main models are used to explain the efflux action of P-gp namely the hydrophobic vacuum cleaner and flippase models. For many years P-gp has been known as a "hydrophobic vacuum cleaner" as its substrates are mostly hydrophobic compounds. In this model, drugs (both substrates and modulators) partition into the lipid bilayer, interact with P-gp within the membrane, and then are subsequently effluxed into the aqueous phase on the extracellular side of the membrane. It was later found that P-gp plays a role in translocation of fluorescently labeled phospholipids, and shows similarity with ABCB4 and a mouse P-gp (mdr2). In the flippase model, drugs partition into the membrane, interact with the drug-binding pocket in P-gp (which is located within the cytoplasmic leaflet), and are then translocated, or flipped, to the outer membrane leaflet.[110–113]

P-gp is widely distributed throughout the body, and transports a large variety of compounds, including chemotherapeutic agents, antiviral agents, antihypertensive drugs, immunosuppressive agents, cardiac glycosides, antibiotics, antiallergenics, β-blockers, antiepileptics, steroid hormones, and antimycotic agents. P-gp transport is inhibited by specific (e.g., cyclosporin A analog valspodar) or nonspecific (verapamil) inhibitors. Many of these compounds act as both P-gp substrates and inhibitors, and may interact with P-gp at more than one binding site.

In the CNS, P-gp is both expressed primarily at the brain vascular barriers (i.e., BBB and BCSFB). P-gp has been localized on the apical side of the CP epithelia. Localization of P-gp on the luminal surface of the mammalian brain endothelium, neighboring astrocyte foot processes, microglia, pericytes, and neurons has been identified. It was also reported that P-gp is localized on both the luminal and abluminal sides of the brain microvascular endothelium.[18,93,114,115] The role of P-gp on caspase-dependent apoptosis was studied. Application of anti–P-gp monoclonal antibodies reversed the decreased level of caspase-dependent cell death (mediated by cytotoxic drugs or Fas ligation) in specific tumor cell lines that express high levels of P-gp.[18,115]

Recently, Broccatelli et al. debated the value of using biopharmaceutics drug disposition classification system (BDDCS) to improve BBB predictions of orally administered drugs. They combined BDDCS class membership with in vitro P-gp efflux and in silico permeability data to create a simple three-step classification tree for accurate prediction of CNS disposition of compounds. It was reported that about 98% of BDDCS class 1 drugs were found to markedly distribute throughout the brain, including a number of BDDCS class 1 drugs shown to be P-gp substrates.[116]

2.1.2 ABCC: Multidrug Resistance–Associated Protein

ABCC transporters, known as MRPs, are transporters involved in CNS transport of compounds. The human ABCC subfamily consists of 12 members: the 9 MRPs (MRP1–9), as well as the cystic fibrosis transmembrane conductance regulator (CFTR), and the 2 sulfonylurea receptors (SUR1 and SUR2).[56] MRP1–3, 6, and 7 contain three TMDs (TMD0, TMD1, and TMD2), two NBDs (NBD1 and NBD2), and a cytoplasmic linker (L0). MRP4, MRP5, MRP8, and possibly MRP9 do not contain TMD0, but do retain the cytoplasmic linker. Several anticancer drugs (e.g., vincristine, methotrexate, and cisplatin), physiological substrates (e.g., glutathione), and therapeutic compounds have been identified to be MRP/Mrp substrates (Table 3.1). Expression of MRP/Mrp has been reported at the BBB and more importantly, it has been found to be concentrated in the CP epithelium at the BCSFB.[18]

MRP1 is localized to the luminal side of BBB in humans.[117] It also contributes to the basolateral drug permeation in CP.[18] At the mammalian BBB, there is also evidence for the expression of Mrp2, MRP4/Mrp4, Mrp5, and Mrp6.[13] Besides, it was found that cMRP/MRP2 was upregulated in human epileptic tissue.[60] In rodents, abluminal expression of Mrp1 was reported in the CP epithelial cells of the BCSFB.[18] Also, MRP4/Mrp4 is located at both luminal and abluminal plasma membranes of CP epithelial cells.[118]

In the brain parenchyma, MRP1/Mrp1 and Mrp3–Mrp5 have been reported in astrocytes, microglia, oligodendrocytes, and neurons.[117,119] Mrp6 expression has not been observed in astrocytes, microglia, and oligodendrocytes.[119] MRP6 is highly expressed in human neurons.[120]

2.1.3 *ABCG2: Breast Cancer Resistance Protein*

BCRP (also known as ABCG2) is the third member of ABC superfamily. It was discovered in a human MCF-breast cancer cell line showing efflux activity without the contribution of P-gp.[121] In 1999, Miyake et al. isolated an efflux transporter in mitoxantrone-resistant human colon carcinoma cell line (S1-M1-80) and named it as MXR.[122] Due to the high expression of ABCG2 in placenta, it is also called as ABCP. In fact, BCRP, MXR, and ABCP are all homologous proteins having small differences in one or two amino acid sequences.[123]

BCRP is uniquely different from the members of other ABC subfamilies. This 72-kDa protein is composed of 663 amino acids having only six transmembrane helices (TMHs) and only one NBD.[124] It is considered as "half transporter," meaning that two subunits bind as homodimers to form an active transporter. Subfamily G comprises also "reverse half transporters" forming the second half as heterodimers.[15] A wide range of molecules, including chemotherapeutic drugs, antiviral agents, hypolipidemics, and flavonoids, have been reported as substrates and/or inhibitors of BCRP transporter system (Table 3.1).

BCRP is expressed in the small intestine, colon, liver, lungs, testis, ovary, placenta, and CNS. In CNS, it is located at the luminal side of the microvessel endothelium of normal human brain, gliomas, and meningiomas.[68,125] The role of BCRP on the restriction of drug transport across blood–testis/fetal/brain barriers has been revealed in the Abcg2 knockout mice. BCRP shows efflux activity concertedly with P-gp and MRPs. It was demonstrated that Bcrp1 expression was lower than that of P-gp in the murine BBB. On the other hand, mRNA levels of BCRP was found to be eightfold higher than P-gp mRNA levels in human brain microvessels.[126] In regard to brain parenchyma, low expression of Abcg2 has been shown in primary cultures of rat astrocytes and microglia, as well as in the MLS-9 cell line[127] and in human fetal astrocytes.[98]

2.2 SLC Transporters

The SLC transporters in human are grouped into 52 families based on their sequences, number of TMHs (typically 10–14 TMHs), and biological functions. They transport solutes, such as ions, metabolites, peptides, and drugs, across biological membranes.[128] Unlike ABC transporters, SLC transporters do not possess ATP-binding sites. Where some SLC transporters are classified as *facilitated transporters* because they use an electrochemical potential difference in the substrate transported; other SLC transporters are classified as *secondary active transporters* because they use an ion gradient produced by primary active transporters and transport substrates against an electrochemical difference.[13]

The main SLC members are organic anion–transporting polypeptides (OATPs), organic anion transporters (OATs), organic cation transporters (OCTs), nucleoside transporters (Na^+-dependent concentrative nucleoside transporters or CNTs, and Na^+-independent equilibrative nucleoside transporters or ENTs), monocarboxylate transporters (MCTs), and peptide transporters (PEPTs). Among many SLC members, OATP1A2, OATP3A1, OAT1, OAT3, OCT1–3, organic carnitine transporter 2 (OCTN2), ENT4, MCTs (MCT1, 2, 4, 8), and PEPT2 were identified in the CNS, including BBB, BCSFB, and/or brain parenchyma. Localization, substrate, and inhibitor examples of main SLC transporters in the CNS are summarized in Table 3.2.

To date, little is known about the expression levels of SLC transporters in CNS diseases, including various types of brain cancer. The mRNA expression of various OATP isoforms (OATP-A, OATP-B, and OATP-C) has been described in isolated human gliomas.[203] The functional expression of a Na^+-dependent nucleoside transporter,[193] and also an OCT, such as uptake system for zidovudine,[204] have been described in microglia. In addition, the low-affinity uptake of zidovudine and stavudine has been attributed to nucleoside transporters, such as hCNT1.[205]

2.2.1 SLCO1: Organic Anion Transporting Polypeptides

OATPs are members of SLCO/Slco family encoded by SLCO1 genes. Eleven OATPs were isolated in various human tissues.[206] It was predicted that OATP/Oatp membrane topology consists of 12 TMDs. All OATPs/Oatps share many structural features, including a large extracellular loop between TMDs 9 and 10[206] and N-glycosylation sites in extracellular loops 2 and 5. All OATP/Oatp family members have a conserved amino acid sequence, known as the *OATP superfamily signature.*[207]

OATP isoforms detected in human brain are OATP1A2,[208] OATP2B1 (OATP-B),[206] OATP3A1 (OATP-D),[209] OATP4A1 (OATP-E),[210] and OATP-F.[211] Two variants of OATP3A1 (OATP3A1-v1 and OATP3A1-v2) were found mainly in the basolateral membrane of CP epithelial cells and apical side of CP epithelial cells, respectively. They were found to be related with prostaglandins (E1 and E2), thyroxin, and vasopressin transport.[212] Gao et al. used immunofluorescence to investigate the localization of OATP1A2, and results were positive for both apical and basolateral sides of brain endothelium. However, the results were negative in astrocytes and neurons, suggesting that these cells do not express OATP-A.[129]

Although gene and protein expression of Oatp1 have been observed intracellulary in neonatal rat CP, it is localized primarily at the apical surface of CP

Table 3.2 Localization, Substrate, and Inhibitor Examples of Main SLC Transporters Located in the CNS

Genes	Aliases	Localizations	Substrates	Inhibitors
SLCO1A2	OATP1A2	BBB: apical, basolateral[129]	Deltorphin II[129] Fexofenadine[130] Levofloxacin[131] Methotrexate[132] Microcystins[133] D-Penicillamine(2,5)enkephalin[129]	Fluoroquinolones (ciprofloxacin, enoxacin, grepafloxacin, moxifloxacin)[131]
SLC22A6	OAT1	CP: apical[134]	Acyclovir[135] Adefovir[136] Furosemide[137] Lamivudine[138] Methotrexate[139] Oseltamivir[140] Tenofovir[141] Zalcitabine[142] Zidovudine[142]	Probenecid[135] Cefadroxil[143] Cefamandole[143] Cefazolin[143]
SLC22A8	OAT3	BMECs,[17] CP: apical[134]	17β-Estradiol-D-17β-glucuronide[144] 6-Mercaptopurine[145] 6-Thioguanune[71] Acyclovir[146] Baclofen[147] Benzoic acid[147] Benzylpenicillin[148] Furosemide[137] Indoxyl sulfate[147] Ketoprofen[147] Methotrexate[149] Olmesartan[148] *p*-Aminohippurat[150] Pravastatin[148] Rosuvastatin[148] Trichlormethiazide[148] Valsartan[148] Zidovudine[151]	Cimetidine[147] Diclofenac[152] Probenecid[144]
SLC22A1	OCT1	BBB: apical[153]	Acyclovir[154] Cimetidine[154] Metformin[154] Oxaplatin[155] Terbutaline[154] Tetraethylammonium[155]	Disopyramide[155]
SLC22A2	OCT2	BBB: apical, CP: apical, neurons[153,156,157]	Amantadine[158] Amiloride[159] Cimetidine[154] Dopamine[160] Memantine[161] Metformin[162] Procainamide[163] Varenicline[164] Oxaliplatin[155]	Cimetidine[155] Desipramine[165] Phenoxybenzamine[161] Quinidine[161]
SLC22A3	OCT3	CP: apical, neurons, astrocytes [17]	4-(4-(Dimethylamino)-styryl)-*N*-methylpyridinium iodide [166] Dopamine[167] Epinephrine[167] Histamine[167] Metformin[168] Norepinephrine[167]	Desipramine[169] Sertraline[169]

SLC16A1	MCT1	BBB: apical,[170] oligodendrocytes,[171] CP[172]	γ-Hydroxybutyrate[173] Lactate[174] Pyruvate[175]		2-Oxo-4-methyl-pentanoate, phenyl-pyruvate α-Cyano-4-hydroxycinnamate[176] Phloretin[177]
SLC16A7	MCT2	Astrocytes, neurons[178,179]	Acetoacetate, β-hydroxybutyrate, lactate, pyruvate[172]		α-Cyano-4-hydroxycinnamate[180] Phloretin[177,180]
SLC16A3	MCT4	Astrocytes[181]	Acetoacetate, β-hydroxybutyrate, lactate, pyruvate[172]		α-Cyano-4-hydroxycinnamate[176]
Slc15a2	Pept2	CP: apical, astrocytes, neurons[17]	Ampicillin[182] Amoxicillin[182,183] Cefadroxil[184] Enalapril[185] Glycylsarcosine[184] Valacyclovir[186]		Ala–Ala[187]
SLC28A2	CNT2	BBB: apical, CP: apical, neurons, astrocytes[17,188–190]	Adenosine[145] Inosine[145] Guanosine[191] Ribavirin[145] Mizoribine[192]		—
SLC28A3	CNT3	CP: basolateral, microglia[17,193,194]	Adenosine[195] Benzamide riboside[196] Cytarabine[197] Cytidine[198] Fludarabine[197]	Gemcitabine[197] Guanosine[198] Ribavirin[195] Uridine[198]	Phloridzin[199]
SLC29A1	ENT1	Astrocytes, neurons[17]	Adenosine[200] Cytarabine[200] Gemcitabine[200] Uridine[200]		Dipyridamole[200] Draflazine[200] Nitrobenzylthioinosine[200]
SLC29A2	ENT2	BBB: apical, CP: basolateral, astrocytes, neurons[17]	Adenosine[200,201] Cytarabine[200] Gemcitabine[200] Hypoxanthine[200] Uracil[200]	Uridine[200,201] Zalcitabine[200] Zidovudine[200]	Dipyridamole[202] Nitrobenzylthioinosine[202]
SLC29A4	ENT4	Astrocytes, neurons[17]	Adenosine[200]		

BMECs, Brain microvessel endothelial cells; PEPT, peptide transporter.

epithelial cells in adult rats.[213] The expression of other Oatp isoforms (e.g., Oatp2, Oatp3,and Oatp14) has been reported in brain capillary–enriched fractions, brain capillary endothelial cells, and in CP epithelial cells.[107,214,215] A few studies have reported the neuronal expression of both Oatp3 and Oatp9 in the brain parenchyma.[216,217]

Substrates sustaining sodium-independent uptake of OATPs include antibiotics (e.g., levofloxacin), β-blockers (e.g., atenolol, labetalol, and talinolol), 3-hydroxy-3-methylglutaryl-coenzyme A (HMG-CoA) reductase inhibitors (e.g., pitavastatin and rosuvastatin), synthetic peptides [e.g., deltorphin II and D-penicillamine (2,5)-encephalin], fexofenadine, and glyburide (detailed in Table 3.2). Polymorphism in OATPs may influence uptake of substrates into the brain. For example, methotrexate is a dihydrofolate reductase inhibitor used in the treatment of rheumatoid arthritis, Crohn's disease, and cancer. During high-dose methotrexate therapies, several CNS side effects occur, especially in populations having OATP1A2 polymorphism: Ile13Thr induces a twofold increased methotrexate uptake into the CNS.[218]

2.2.2 SLC22A6–8: Organic Anion Transporters

OATs (OAT1–10) are also members of the SLC22 family having 12 domains with intracellular carboxyl and amino terminals. They contribute to cellular uptake of organic anionic compounds with Na^+-dependent, Na^+-independent, or ATP-dependent mechanisms.[17] Sekine et al. first cloned OAT1 in 1997,[219] and OAT2 in 1998.[220] OAT3 was identified in 1999 by Kusuhara et al.[221] OATs are mostly known as kidney transporters due to their higher expression levels compared to other tissues. Among 10 OAT members, OAT1 (encoded by SLC22A6) and OAT3 (encoded by SLC22A8) are the most studied transporters, which have been identified in multiple compartments of human brain.[222,223] In humans, OAT1 is highly expressed in kidney and shows weaker expression in brain, CP, spinal cord, and iris cilliary bodies. OAT1, located at the apical side of CP, transports its anionic ligands in the CSF-to-blood direction.[17,134] Murine Oat1 has also been detected in neurons of the cerebral cortex and hippocampus, as well as in the ependymal cell layer of the CP.[224] OAT2 protein is not expressed in the brain, but Oat2 was detected in rat CP at the mRNA level.[107] On the other hand, OAT3/Oat3 is highly expressed in the brain. Immunohistochemical analysis revealed abluminal and faint luminal localizations of Oat3 in rat brain capillary endothelial cells,[223] apical membrane localization in rat,[225] and human choroid epithelial cells.[134] Expression of OAT3 has been also reported in a human brain microvessel endothelial cell (BMEC) line (BB19). OAT4 and OAT10 were determined in human BB19 brain microvessel endothelial cell line and brain tissue, respectively. OAT4 mRNA in the CP epithelial cells and brain microvessel endothelial cells has been shown by RT-PCR studies.[226,227] On the other hand OAT5–7 is not expressed in the human brain.

Also, OAT expression in glial cells has not been detected yet.[222,228] The mouse homolog protein (RST, SLC22A12) is highly expressed in the CP and brain capillary-enriched fraction. Mouse RST, in coordination with OAT3, could potentially play a role in the extrusion of anionic metabolites, generated by neurotransmitters, from the CNS.[229,230]

OATs substrates, including antiretrovirals, nonsteroidal antiinflammatory drugs, neurotransmitter metabolites, and HMG-CoA reductase inhibitors, are shown in Table 3.2.

2.2.3 SLC22A: Organic Cation Transporters

Two classes of OCT (SLC22A) systems have been defined, namely potential-sensitive OCTs (OCT1–3/Oct1–3) and H^+ gradient–dependent novel OCTs (OCTN1–3/Octn1–3). OCTs, localized to the basolateral membrane of various cells, are usually involved in the influx of organic cations into a cell. OCTNs mediate the cellular efflux of cationic substrates.[231–235] OCT1–3 have 554, 555, and 556 amino acids, respectively, with 50% amino acid identity.[236] They consist of 12 α-helical TMDs having extracellular and intracellular loops in between domain 1–2 and domain 6–7, respectively. Moreover, they facilitate an ion gradient–dependent transport across the membranes primarily into cells.[156] Also, they have an important role in the facilitated diffusion of organic cations, including some endogenous (1-methyl-4-phenylpyridinium, acetylcholine, dopamine, and epinephrine) and exogenous compounds (e.g., metformin, acyclovir, famotidine, cisplatin, and quinidine).[156] OCTs are located in various components of CNS, including the BBB, glial cells, neurons, and BCSFB (Table 3.2). In 1999, OCT1 was first identified in the cultured human glioma cell line SK-MG-1.[237] Since then, isolations of OCT1, OCT2, and OCT3 were confirmed in different glioma cell lines.[238] The OCT1/Oct1 mRNA expression in the brain (CP and astrocytes) was reported to be very little.[107,239] In the CNS, OCT2 may be involved in the uptake of choline into neurons, and choline reabsorption from the CSF. In the rat brain, Oct2 was initiated with serotonin and norepinephrine clearance.[157] OCT3 (SLC22A3) is expressed weakly in the brain compared to the other organs, such as liver, skeletal muscle, kidney, and heart.[236,240,241] The expression of Oct3 mRNA in the brain was reported to be greater than that of Oct1 and Oct2.[242] Expression of Oct3 in rat was detected in dopaminergic neurons of the substantia nigra, nonaminergic neurons of the ventral segmental area, substantia nigra reticulata, locus coeruleus, hippocampus, and cortex.[243] Functionally, in Oct3-deficient mice, the monoamine neurotransmission was degenerated accompanied with increased levels of stress and anxiety.[243,244] In this respect, Oct3 was suggested as a target for the treatment of anxiety disorders. OCT3 also plays a significant role on uptake and clearance of histamine.[245,246] Zhu et al. suggested that OCT3 is the responsible transporter for clearance of ischemia-induced histamine in the brain.[247] It was reported that inhibition of OCT3 was considered as an invalid approach for cancer treatment due to its absence in the glioblastoma cell membranes.[166]

All OCTN members (OCTN1–3) transport organic cations and carnitine.[248] However, they have different characteristics, such as affinity, specificity, and tissue distribution, and have a wide range of clinical importance. With regard to their expression in brain, rat Octn1 mRNA is expressed in the spinal cord CP, hippocampus, cortex, and cerebellum.[249,250] OCTN2 has been detected in primary brain capillary endothelial cells cultures of various species (human, rat, mouse, porcine, and bovine).[251] OCTN2 mRNA was detected in neurons from several regions, including hippocampus, cerebellum, spinal cord, and superior cervical ganglion.[250,252,253] Octn2 knockout mice are associated with decreased systemic and brain concentration of acetyl-L-carnitine, indicating that these transporters have luminal localization at the BBB and assist the entry of carnitine into the brain.[251]

2.2.4 SLC16A: Monocarboxylate Transporters

The solute carrier 16 (SLC16A) gene family, composed of 14 sequence-related isoforms, is known as the MCTs.[17] MCTs have 12 TMHs, intracellular N- and C-termini, and a large cytosolic loop between TMDs 6 and 7. MCT isoforms 1–4 are known to transport monocarboxylates, such as lactate, pyruvate, butyrate, and ketone bodies (e.g., hydroxybutyrate and acetoacetate).[254] MCT8 is known to facilitate proton-independent transport of thyroid hormones (T3 and T4). MCT10 (also known as TAT1) is an aromatic amino acid transporter.[255,256]

MCTs are expressed in various tissues, including liver, kidney, intestine, and brain. MCT1 was discovered as a wildtype protein that increased the uptake of mevalonate into Chinese hamster ovary cells.[257] In 1997, expression of MCT1 was detected in the luminal membrane of BMECs by Gerhart et al.[258] The role of MCT1 on benzoate uptake was also characterized in immortalized and primary cultures of BMECs.[174] It was reported that axonal and/or neuronal loss occurred with the downregulation of MCT1, suggesting that MCT1-expressing oligodendrocytes may have an important role in supplying energy to axons.[171,259] The latter has been supported by high expression levels of MCT1 in oligodendrocytes.[171] Also, MCT1 is coexpressed with another MCT in polarized cells, such as the CP.[172] MCT2 expression was reported in a variety of rodent tissues, including brain, liver, kidney, and testes. Although the major MCT present in rodent neurons is Mct2, its expression in the human brain appears to be minimal. MCT2 protein expression has been found mainly in neurons contributing to lactate uptake to maintain energy.[178] Lauritzen et al. reported localization of MCT2 in astrocytes of human epileptogenic hippocampus. In fact, loss of MCT2 on astrocyte end-feet has triggered MCT2 protein upregulation on astrocyte membranes facing the synapses, which was considered as a compensatory mechanism, highlighting the importance of MCT2 in transporting monocarboxylates within the parenchyma and the flux across the BBB.[179] Expression of MCT4 has been detected in astrocytes and recently in human glioblastoma (with MCT1). Especially in tumors showing rapid differentiation and proliferation, MCTs have been considered as important lactate suppliers.[181] Miranda-Goncalves et al. reported a decrease in tumor size with the use of MCT inhibitor α-cyano-4-hydroxycinnamic acid in an animal glioblastoma model, suggesting that MCT1 inhibition can be a potential approach for glioblastoma treatment.[176] MCT8 is known as a thyroid hormone transporter, which plays a role in neurodevelopment. Deficiency of MCT8 is considered to be related with delayed brain myelination, causing various mental retardation syndromes accompanied with psychomotor and cognitive impairments.[260–262] A decrease in the level of T3 hormone was also reported in Mct8-deficient mice.[263]

2.2.5 SLC15A: Oligopeptide Transporters

PEPTs belong to family 15 of the solute carrier superfamily (SLC15A, proton-dependent oligopeptide transporters, POT) and consist of four members: PEPT1 (SLC15A1), PEPT2 (SLC15A2), peptide–histidine transporter (PHT1 or SLC15A4), and PHT2 (SLC15A3). Except PEPT1/Pept1, the other three members are expressed in the brain. However Pept2 was found in rat astrocytes, subependymal cells, ependymal cells, and epithelial cells of the CP.[264] Also transcripts of PhT1 and PhT2 have been detected in the brain. However, their functional role has been understudied.[17,265]

Role of PEPT2 on the disposition of cefadroxil within the brain was demonstrated for the first time[184] by Shen et al. using Pept2 knockout mice. Smith et al. reported decreased glycylsarcosine and cefadroxil (a peptide drug) levels in CP epithelial cells due to the lack of Pept2-mediated clearance.[266] PEPT2 is considered as an uptake transporter of peptides/mimetics located at the BBB and in brain parenchymal cells. It can limit the exposition of such molecules in the brain due to its CSF-to-blood transport at endothelial cells of CP. Pept2-mediated uptake of 5-aminolevulinic acid and carnosine in astrocytes was demonstrated by Xiang el al.[267] When 5-aminolevulinic acid was administered to Pept2-deficient mice, 8–30 times higher CSF concentration was detected in comparison to the wildtype control group.[268]

Localization of Pht1 in several regions of the rat brain (hippocampus, cerebellum, and pontine nucleus, and in lower levels in the cerebral cortex, brain stem, thalamus, and hypothalamus) has been demonstrated by in situ hybridization studies.[269]

2.2.6 Nucleoside Transporters

Mammalian nucleoside transporters are classified into families based on their sodium dependency: CNTs (SLC28A) and ENTs (SLC29A).[270] The primary physiological function of CNTs and ENTs in mammalian cells is to facilitate cellular uptake of natural nucleosides derived from diet or produced by tissues, such as the liver, for nucleotide synthesis in the salvage pathways.

CNTs are found primarily in specialized cell types, including microglia and CP.[193,194] Based on substrate specificity, six functionally different transport activities have been described (*cit, cif, cib, cit-like, cs,* and *csg*).[271–273] Three different proteins (CNT1, CNT2, and CNT3) are responsible for the *cit, cif,* and *cib* activities. The Cnt1 gene transcript is expressed in many regions of the rat brain (cerebral cortex, cerebellum, hippocampus, striatum, brain stem, superior colliculus, posterior hypothalamus, and CP).[274] Although there is no evidence for the expression of the protein at the BBB,[275] expression of the *cit* nucleoside transport system at the BBB and BCSFB has been reported.[276] mRNA for human CNT2 (SLC28A2) has been detected in various tissues, including the brain.[201] CNT2 is considered as the BBB adenosine transporter. At the BBB and BCSFB, CNT2 localization is limited to the membrane facing interstitial fluid and CSF, respectively.[188] Human CNT3 (SLC28A3) mRNA has been identified in many tissues, including the brain.[277] In vitro studies using rabbit choroidal cells have shown *cib*-type transport of nucleosides. In addition, in vivo experiments with isolated rabbit choroidal cells have shown transport of nucleosides from the blood to CSF.[278]

Four isoforms of ENTs (ENT1–4/Ent1–4), have been identified and classified into two subtypes based on their sensitivity to NBMPR. Transport via ENTs is bidirectional and they can mediate uptake or efflux of substrate compounds, depending on the nucleoside concentration gradient across the plasma membrane. Furthermore, all ENTs transport adenosine, but have different capacities to transport other nucleosides and nucleobases.[200] Both Ent1 and ENT1 are located at pyramidal neurons of the hippocampus, granule neurons of the dentate gyrus, granule neurons of the cerebellum, cortical and striatal neurons, and astrocytes.[201,279] In humans, ENT1 protein has been localized at frontal and parietal lobes of the cerebral cortex, thalamus, midbrain, and basal ganglia.[280] Human ENT2 protein is highly expressed in the cerebellum, thalamus, medulla, midbrain, and brainstem regions, particularly the pons, and is expressed weakly in cerebral cortex and basal ganglia.[280] Also Ent2 has been identified in rat C6 glioma cells.[202] ENT3 (SLC29A3) gene transcript has been detected in human and rat brain.[281] ENT4 (SLC29A4) protein is expressed ubiquitously in human tissue, particularly in the brain, and is a low-affinity nucleoside transporter at the cell surface.[282]

3 CONCLUSIONS

The brain is a dynamic and vital organ, which is surrounded by the BBB and the BCSFB. Brain parenchymal cells (i.e., astrocytes, microglia, oligodendrocytes, and neurons) interact with each other in a very organized manner to ensure effective brain function. Each brain compartment contains enzymes, proteins, and secretory factors that maintain brain function. In the treatment of CNS diseases, several barriers need to be crossed for drug targeting to the brain. In some cases, they also have to permeate through cellular compartments of the brain parenchyma. Expression levels and localizations of these influx (i.e., OATs, OCTs, and OATPs) and efflux transporters (i.e., P-gp, MRPs, and BCRP) at the brain barriers and within brain parenchymal cells may influence the brain concentration of drugs used for the treatment of CNS diseases. Further studies are required to understand the role and importance of brain transporters in clinical practice.

Abbreviations

ABC	ATP-binding cassette
ATP	Adenosine triphosphate
BBB	Blood–brain barrier
BCRP	Breast cancer resistance protein
BCSFB	Blood–cerebrospinal fluid barrier
BDDCS	Biopharmaceutics drug disposition classification system
BMECs	Brain microvessel endothelial cells
CNS	Central nervous system
CP	Choroid plexus
CSF	Cerebrospinal fluid
ENTs	Equilibrative nucleoside transporters
MCTs	Monocarboxylate transporters
MDR	Multidrug resistance
MRP	Multidrug resistance–associated protein
NMD	Nucleotide-binding domain
OAT	Organic anion transporter
OATP	Organic anion–transporting polypeptide
OCT	Organic cation transporter
OCTN	Organic carnitine transporter
PEPT	Peptide transporter
P-gp	P-glycoprotein
PHT	Peptide–histidine transporter
POT	Proton-dependent oligopeptide transporter
SLC	Solute lipid carrier
TMD	Transmembrane domain
TMH	Transmembrane helix

References

1. Habgood MD, Begley DJ, Abbott NJ. Determinants of passive drug entry into the central nervous system. *Cell Mol Neurobiol.* 2000;20(2):231–253.
2. Abbott NJ. Astrocyte-endothelial interactions and blood–brain barrier permeability. *J Anat.* 2002;200(6):629–638.
3. Reese TS, Karnovsky MJ. Fine structural localization of a blood–brain barrier to exogenous peroxidase. *J Cell Biol.* 1967;34(1):207–217.

4. Lee G, Dallas S, Hong M, Bendayan R. Drug transporters in the central nervous system: brain barriers and brain parenchyma considerations. *Pharmacol Rev*. 2001;53(4):569–596.
5. Ghersi-Egea JF, Strazielle N. Brain drug delivery, drug metabolism, and multidrug resistance at the choroid plexus. *Microsc Res Tech*. 2001;52(1):83–88.
6. Spector R, Johanson CE. The mammalian choroid plexus. *Sci Am*. 1989;261(5):68–74.
7. Peters A, Josephson K, Vincent SL. Effects of aging on the neuroglial cells and pericytes within area 17 of the rhesus monkey cerebral cortex. *Anat Rec*. 1991;229(3):384–398.
8. Walz W. Controversy surrounding the existence of discrete functional classes of astrocytes in adult gray matter. *Glia*. 2000;31(2):95–103.
9. Wilson R, Brophy PJ. Role for the oligodendrocyte cytoskeleton in myelination. *J Neurosci Res*. 1989;22(4):439–448.
10. Rock RB, Gekker G, Hu S, Sheng WS, Cheeran M, Lokensgard JR, et al. Role of microglia in central nervous system infections. *Clin Microbiol Rev*. 2004;17(4):942–964.
11. Dickson DW, Mattiace LA, Kure K, Hutchins K, Lyman WD, Brosnan CF. Microglia in human disease, with an emphasis on acquired immune deficiency syndrome. *Lab Invest*. 1991;64(2):135–156.
12. Lee G, Bendayan R. Functional expression and localization of P-glycoprotein in the central nervous system: relevance to the pathogenesis and treatment of neurological disorders. *Pharm Res*. 2004;21(8):1313–1330.
13. You G, Morris ME. Overview of drug transporter families. Drug Transporters. USA: John Wiley & Sons, Inc.; 2006:[pp. 1–10].
14. Syvänen S, Xie R, Sahin S, Hammarlund-Udenaes M. Pharmacokinetic consequences of active drug efflux at the blood-brain barrier. *Pharm Res*. 2006;23(4):705–717.
15. Vasiliou V, Vasiliou K, Nebert DW. Human ATP-binding cassette (ABC) transporter family. *Hum Genomics*. 2009;3(3):281–290.
16. Daood M, Tsai C, Ahdab-Barmada M, Watchko JF. ABC transporter (P-gp/ABCB1, MRP1/ABCC1, BCRP/ABCG2) expression in the developing human CNS. *Neuropediatrics*. 2008;39(4):211–218.
17. Ashraf T, Kao A, Bendayan R. Functional expression of drug transporters in glial cells: potential role on drug delivery to the CNS. *Adv Pharmacol*. 2014;71:45–111.
18. Rao VV, Dahlheimer JL, Bardgett ME, Snyder AZ, Finch RA, Sartorelli AC, et al. Choroid plexus epithelial expression of MDR1 P glycoprotein and multidrug resistance-associated protein contribute to the blood-cerebrospinal-fluid drug-permeability barrier. *Proc Natl Acad Sci USA*. 1999;96(7):3900–3905.
19. Tsukimoto M, Ohashi R, Torimoto N, Togo Y, Suzuki T, Maeda T, et al. Effects of the inhibition of intestinal P-glycoprotein on aliskiren pharmacokinetics in cynomolgus monkeys. *Biopharm Drug Dispos*. 2015;36(1):15–33.
20. Spence R, Mandagere A, Richards DB, Magee MH, Dufton C, Boinpally R. Potential for pharmacokinetic interactions between ambrisentan and cyclosporine. *Clin Pharmacol Ther*. 2010;88(4):513–520.
21. Zhao R, Pollack GM. Regional differences in capillary density, perfusion rate, and P-glycoprotein activity: a quantitative analysis of regional drug exposure in the brain. *Biochem Pharmacol*. 2009;78(8):1052–1059.
22. Yates CR, Chang C, Kearbey JD, Yasuda K, Schuetz EG, Miller DD, et al. Structural determinants of P-glycoprotein-mediated transport of glucocorticoids. *Pharm Res*. 2003;20(11):1794–1803.
23. Ishiguro N, Kishimoto W, Volz A, Ludwig-Schwellinger E, Ebner T, Schaefer O. Impact of endogenous esterase activity on in vitro p-glycoprotein profiling of dabigatran etexilate in Caco-2 monolayers. *Drug Metab Dispos*. 2014;42(2):250–256.

24. Petropoulos S, Gibb W, Matthews SG. Developmental expression of multidrug resistance phosphoglycoprotein (P-gp) in the mouse fetal brain and glucocorticoid regulation. *Brain Res*. 2010;1357:9–18.

25. Ravaud A, Urva SR, Grosch K, Cheung WK, Anak O, Sellami DB. Relationship between everolimus exposure and safety and efficacy: meta-analysis of clinical trials in oncology. *Eur J Cancer*. 2014;50(3):486–495.

26. Tahara H, Kusuhara H, Fuse E, Sugiyama Y. P-glycoprotein plays a major role in the efflux of fexofenadine in the small intestine and blood-brain barrier, but only a limited role in its biliary excretion. *Drug Metab Dispos*. 2005;33(7):963–968.

27. van der Sandt IC, Smolders R, Nabulsi L, Zuideveld KP, de Boer AG, Breimer DD. Active efflux of the 5-HT(1A) receptor agonist flesinoxan via P-glycoprotein at the blood-brain barrier. *Eur J Pharm Sci*. 2001;14(1):81–86.

28. Kawamura K, Yamasaki T, Yui J, Hatori A, Konno F, Kumata K, et al. In vivo evaluation of P-glycoprotein and breast cancer resistance protein modulation in the brain using [(11)C] gefitinib. *Nucl Med Biol*. 2009;36(3):239–246.

29. Yu XY, Lin SG, Zhou ZW, Chen X, Liang J, Liu PQ, et al. Role of P-glycoprotein in the intestinal absorption of tanshinone IIA, a major active ingredient in the root of *Salvia miltiorrhiza* Bunge. *Curr Drug Metab*. 2007;8(4):325–340.

30. Breedveld P, Pluim D, Cipriani G, Wielinga P, van Tellingen O, Schinkel AH, et al. The effect of Bcrp1 (Abcg2) on the in vivo pharmacokinetics and brain penetration of imatinib mesylate (Gleevec): implications for the use of breast cancer resistance protein and P-glycoprotein inhibitors to enable the brain penetration of imatinib in patients. *Cancer Res*. 2005;65(7):2577–2582.

31. Kim RB, Fromm MF, Wandel C, Leake B, Wood AJ, Roden DM, et al. The drug transporter P-glycoprotein limits oral absorption and brain entry of HIV-1 protease inhibitors. *J Clin Invest*. 1998;101(2):289–294.

32. Schinkel AH, Wagenaar E, van Deemter L, Mol CA, Borst P. Absence of the mdr1a P-glycoprotein in mice affects tissue distribution and pharmacokinetics of dexamethasone, digoxin, and cyclosporin A. *J Clin Invest*. 1995;96(4):1698–1705.

33. Chu C, Noel-Hudson MS, Boige V, Goere D, Marion S, Polrot M, et al. Therapeutic efficiency of everolimus and lapatinib in xenograft model of human colorectal carcinoma with KRAS mutation. *Fundam Clin Pharmacol*. 2013;27(4):434–442.

34. Chen C, Hanson E, Watson JW, Lee JS. P-glycoprotein limits the brain penetration of nonsedating but not sedating H1-antagonists. *Drug Metab Dispos*. 2003;31(3):312–318.

35. Zembruski NC, Buchel G, Jodicke L, Herzog M, Haefeli WE, Weiss J. Potential of novel antiretrovirals to modulate expression and function of drug transporters in vitro. *J Antimicrob Chemother*. 2011;66(4):802–812.

36. Mahon FX, Hayette S, Lagarde V, Belloc F, Turcq B, Nicolini F, et al. Evidence that resistance to nilotinib may be due to BCR-ABL, Pgp, or Src kinase overexpression. *Cancer Res*. 2008;68(23):9809–9816.

37. Schinkel AH, Wagenaar E, Mol CA, van Deemter L. P-glycoprotein in the blood-brain barrier of mice influences the brain penetration and pharmacological activity of many drugs. *J Clin Invest*. 1996;97(11):2517–2524.

38. Fellner S, Bauer B, Miller DS, Schaffrik M, Fankhanel M, Spruss T, et al. Transport of paclitaxel (Taxol) across the blood-brain barrier in vitro and in vivo. *J Clin Invest*. 2002;110(9):1309–1318.

39. Trujillo TC. Advances in the management of stable angina. *J Manag Care Pharm*. 2006;12(8 suppl):S10–S16.

40. Lagas JS, Sparidans RW, van Waterschoot RA, Wagenaar E, Beijnen JH, Schinkel AH. P-glycoprotein limits oral availability, brain penetration, and toxicity of an anionic drug, the antibiotic salinomycin. *Antimicrob Agents Chemother*. 2008;52(3):1034–1039.

41. Ding W, Hou X, Cong S, Zhang Y, Chen M, Lei J, et al. Co-delivery of honokiol, a constituent of *Magnolia* species, in a self-microemulsifying drug delivery system for improved oral transport of lipophilic sirolimus. *Drug Deliv.* 2016;23(7):2513–2523.
42. Krishna R, Bergman A, Larson P, Cote J, Lasseter K, Dilzer S, et al. Effect of a single cyclosporine dose on the single-dose pharmacokinetics of sitagliptin (MK-0431), a dipeptidyl peptidase-4 inhibitor, in healthy male subjects. *J Clin Pharmacol.* 2007;47(2):165–174.
43. de Lange EC, Marchand S, van den Berg D, van der Sandt IC, de Boer AG, Delon A, et al. In vitro and in vivo investigations on fluoroquinolones; effects of the P-glycoprotein efflux transporter on brain distribution of sparfloxacin. *Eur J Pharm Sci.* 2000;12(2):85–93.
44. El Ela AA, Hartter S, Schmitt U, Hiemke C, Spahn-Langguth H, Langguth P. Identification of P-glycoprotein substrates and inhibitors among psychoactive compounds—implications for pharmacokinetics of selected substrates. *J Pharm Pharmacol.* 2004;56(8):967–975.
45. Shoaf SE, Ohzone Y, Ninomiya S, Furukawa M, Bricmont P, Kashiyama E, et al. In vitro P-glycoprotein interactions and steady-state pharmacokinetic interactions between tolvaptan and digoxin in healthy subjects. *J Clin Pharmacol.* 2011;51(5):761–769.
46. Shen J, Carcaboso AM, Hubbard KE, Tagen M, Wynn HG, Panetta JC, et al. Compartment-specific roles of ATP-binding cassette transporters define differential topotecan distribution in brain parenchyma and cerebrospinal fluid. *Cancer Res.* 2009;69(14):5885–5892.
47. Huang Y, Okochi H, May BC, Legname G, Prusiner SB, Benet LZ, et al. Quinacrine is mainly metabolized to mono-desethyl quinacrine by CYP3A4/5 and its brain accumulation is limited by P-glycoprotein. *Drug Metab Dispos.* 2006;34(7):1136–1144.
48. Schinkel AH, Smit JJ, van Tellingen O, Beijnen JH, Wagenaar E, van Deemter L, et al. Disruption of the mouse mdr1a P-glycoprotein gene leads to a deficiency in the blood-brain barrier and to increased sensitivity to drugs. *Cell.* 1994;77(4):491–502.
49. Mendell J, Zahir H, Matsushima N, Noveck R, Lee F, Chen S, et al. Drug-drug interaction studies of cardiovascular drugs involving P-glycoprotein, an efflux transporter, on the pharmacokinetics of edoxaban, an oral factor Xa inhibitor. *Am J Cardiovasc Drugs.* 2013;13(5):331–342.
50. Bachmakov I, Werner U, Endress B, Auge D, Fromm MF. Characterization of beta-adrenoceptor antagonists as substrates and inhibitors of the drug transporter P-glycoprotein. *Fundam Clin Pharmacol.* 2006;20(3):273–282.
51. Hughes J, Crowe A. Inhibition of P-glycoprotein-mediated efflux of digoxin and its metabolites by macrolide antibiotics. *J Pharm Sci.* 2010;113(4):315–324.
52. Tun NM, Oo TH. Prevention and treatment of venous thromboembolism with new oral anticoagulants: a practical update for clinicians. *Thrombosis.* 2013;2013:183616.
53. Hsiao P, Bui T, Ho RJ, Unadkat JD. In vitro-to-in vivo prediction of P-glycoprotein-based drug interactions at the human and rodent blood-brain barrier. *Drug Metab Dispos.* 2008;36(3):481–484.
54. Frost CE, Byon W, Song Y, Wang J, Schuster AE, Boyd RA, et al. Effect of ketoconazole and diltiazem on the pharmacokinetics of apixaban, an oral direct factor Xa inhibitor. *Br J Clin Pharm.* 2015;79(5):838–846.
55. Choi JS, Piao YJ, Kang KW. Effects of quercetin on the bioavailability of doxorubicin in rats: role of CYP3A4 and P-gp inhibition by quercetin. *Arch Pharm Res.* 2011;34(4):607–613.
56. Borst P, Evers R, Kool M, Wijnholds J. A family of drug transporters: the multidrug resistance-associated proteins. *J Natl Cancer Inst.* 2000;92(16):1295–1302.
57. Deeley RG, Cole SP. Substrate recognition and transport by multidrug resistance protein 1 (ABCC1). *FEBS Lett.* 2006;580(4):1103–1111.
58. Bakos É, Homolya L. Portrait of multifaceted transporter, the multidrug resistance-associated protein 1 (MRP1/ABCC1). *Pflugers Arch Eur J Physiol.* 2007;453(5):621–641.

59. Tivnan A, Zakaria Z, O'Leary C, Kogel D, Pokorny JL, Sarkaria JN, et al. Inhibition of multidrug resistance protein 1 (MRP1) improves chemotherapy drug response in primary and recurrent glioblastoma multiforme. *Front Neurosci*. 2015;9:218.
60. Dombrowski SM, Desai SY, Marroni M, Cucullo L, Goodrich K, Bingaman W, et al. Overexpression of multiple drug resistance genes in endothelial cells from patients with refractory epilepsy. *Epilepsia*. 2001;42(12):1501–1506.
61. Miller DS, Nobmann SN, Gutmann H, Toeroek M, Drewe J, Fricker G. Xenobiotic transport across isolated brain microvessels studied by confocal microscopy. *Mol Pharmacol*. 2000;58(6):1357–1367.
62. Miller DS, Graeff C, Droulle L, Fricker S, Fricker G. Xenobiotic efflux pumps in isolated fish brain capillaries. *Am J Physiol Regul Integr Comp Physiol*. 2002;282(1):R191–R198.
63. Miller DS. Regulation of ABC transporters blood-brain barrier: the good, the bad, and the ugly. *Adv Cancer Res*. 2015;125:43–70.
64. Hesselink DA, Van Hest RM, Mathot RAA, Bonthuis F, Weimar W, De Bruin RWF, et al. Cyclosporine interacts with mycophenolic acid by inhibiting the multidrug resistance-associated protein 2. *Am J Transplant*. 2005;5(5):987–994.
65. Leggas M, Adachi M, Scheffer GL, Sun D, Wielinga P, Du G, et al. Mrp4 confers resistance to topotecan and protects the brain from chemotherapy. *Mol Cell Biol*. 2004;24(17):7612–7621.
66. Dallas S, Schlichter L, Bendayan R. Multidrug resistance protein (MRP) 4- and MRP 5-mediated efflux of 9-(2-phosphonylmethoxyethyl)adenine by microglia. *J Pharmacol Exp Ther*. 2004;309(3):1221–1229.
67. Ose A, Ito M, Kusuhara H, Yamatsugu K, Kanai M, Shibasaki M, et al. Limited brain distribution of [3R,4R,5S]-4-acetamido-5-amino-3-(1-ethylpropoxy)-1-cyclohexene-1-carboxylate phosphate (Ro 64-0802), a pharmacologically active form of oseltamivir, by active efflux across the blood-brain barrier mediated by organic anion transporter 3 (Oat3/Slc22a8) and multidrug resistance-associated protein 4 (Mrp4/Abcc4). *Drug Metab Dispos*. 2009;37(2):315–321.
68. Cooray HC, Blackmore CG, Maskell L, Barrand MA. Localisation of breast cancer resistance protein in microvessel endothelium of human brain. *Neuroreport*. 2002;13(16):2059–2063.
69. Enokizono J, Kusuhara H, Sugiyama Y. Effect of breast cancer resistance protein (Bcrp/Abcg2) on the disposition of phytoestrogens. *Mol Pharmacol*. 2007;72(4):967–975.
70. Lee YJ, Kusuhara H, Jonker JW, Schinkel AH, Sugiyama Y. Investigation of efflux transport of dehydroepiandrosterone sulfate and mitoxantrone at the mouse blood-brain barrier: a minor role of breast cancer resistance protein. *J Pharmacol Exp Ther*. 2005;312(1):44–52.
71. Mori S, Ohtsuki S, Takanaga H, Kikkawa T, Kang YS, Terasaki T. Organic anion transporter 3 is involved in the brain-to-blood efflux transport of thiopurine nucleobase analogs. *J Neurochem*. 2004;90(4):931–941.
72. Imai Y, Yoshimori M, Fukuda K, Yamagishi H, Ueda Y. The PI3K/Akt inhibitor LY294002 reverses BCRP-mediated drug resistance without affecting BCRP translocation. *Oncol Rep*. 2012;27(6):1703–1709.
73. Polli JW, Humphreys JE, Harmon KA, Castellino S, O'Mara MJ, Olson KL, et al. The role of efflux and uptake transporters in [N-{3-chloro-4-[(3-fluorobenzyl)oxy]phenyl}-6-[5-({[2-(methylsulfonyl)ethyl]amino }methyl)-2-furyl]-4-quinazolinamine (GW572016, lapatinib) disposition and drug interactions. *Drug Metab Dispos*. 2008;36(4):695–701.
74. Zhao R, Raub TJ, Sawada GA, Kasper SC, Bacon JA, Bridges AS, et al. Breast cancer resistance protein interacts with various compounds in vitro, but plays a minor role in substrate efflux at the blood-brain barrier. *Drug Metab Dispos*. 2009;37(6):1251–1258.
75. Lee CA, O'Connor MA, Ritchie TK, Galetin A, Cook JA, Ragueneau-Majlessi I, et al. Breast cancer resistance protein (ABCG2) in clinical pharmacokinetics and drug interactions: practical recommendations for clinical victim and perpetrator drug-drug interaction study design. *Drug Metab Dispos*. 2015;43(4):490–509.

76. de Vries NA, Zhao J, Kroon E, Buckle T, Beijnen JH, van Tellingen O. P-glycoprotein and breast cancer resistance protein: two dominant transporters working together in limiting the brain penetration of topotecan. *Clin Cancer Res*. 2007;13(21):6440–6449.

77. Gupta A, Dai Y, Vethanayagam RR, Hebert MF, Thummel KE, Unadkat JD, et al. Cyclosporin A, tacrolimus and sirolimus are potent inhibitors of the human breast cancer resistance protein (ABCG2) and reverse resistance to mitoxantrone and topotecan. *Cancer Chemother Pharmacol*. 2006;58(3):374–383.

78. Allred AJ, Bowen CJ, Park JW, Peng B, Williams DD, Wire MB, et al. Eltrombopag increases plasma rosuvastatin exposure in healthy volunteers. *Br J Clin Pharmacol*. 2011;72(2): 321–329.

79. Yanase K, Tsukahara S, Asada S, Ishikawa E, Imai Y, Sugimoto Y. Gefitinib reverses breast cancer resistance protein-mediated drug resistance. *Mol Cancer Ther*. 2004;3(9):1119–1125.

80. Perry J, Ghazaly E, Kitromilidou C, McGrowder EH, Joel S, Powles T. A synergistic interaction between lapatinib and chemotherapy agents in a panel of cell lines is due to the inhibition of the efflux pump BCRP. *Mol Cancer Ther*. 2010;9(12):3322–3329.

81. Mairinger S, Erker T, Muller M, Langer O. PET and SPECT radiotracers to assess function and expression of ABC transporters in vivo. *Curr Drug Metab*. 2011;12(8):774–792.

82. Leslie EM, Deeley RG, Cole SP. Multidrug resistance proteins: role of P-glycoprotein, MRP1, MRP2, and BCRP (ABCG2) in tissue defense. *Toxicol Appl Pharmacol*. 2005;204 (3):216–237.

83. Shitara Y, Horie T, Sugiyama Y. Transporters as a determinant of drug clearance and tissue distribution. *Eur J Pharm Sci*. 2006;27(5):425–446.

84. Glavinas H, Krajcsi P, Cserepes J, Sarkadi B. The role of ABC transporters in drug resistance, metabolism and toxicity. *Curr Drug Deliv*. 2004;1(1):27–42.

85. Choudhuri S, Klaassen CD. Structure, function, expression, genomic organization, and single nucleotide polymorphisms of human ABCB1 (MDR1), ABCC (MRP), and ABCG2 (BCRP) efflux transporters. *Int J Toxicol*. 2006;25(4):231–259.

86. Strazielle N, Ghersi-Egea JF. Factors affecting delivery of antiviral drugs to the brain. *Rev Med Virol*. 2005;15(2):105–133.

87. Hartz AM, Bauer B. ABC transporters in the CNS—an inventory. *Curr Pharm Biotechnol*. 2011;12(4):656–673.

88. Neuwelt EA, Bauer B, Fahlke C, Fricker G, Iadecola C, Janigro D, et al. Engaging neuroscience to advance translational research in brain barrier biology. *Nat Rev Neurosci*. 2011;12(3):169–182.

89. Zhang Y, Schuetz JD, Elmquist WF, Miller DW. Plasma membrane localization of multidrug resistance-associated protein homologs in brain capillary endothelial cells. *J Pharmacol Exp Ther*. 2004;311(2):449–455.

90. Pardridge WM, Golden PL, Kang YS, Bickel U. Brain microvascular and astrocyte localization of P-glycoprotein. *J Neurochem*. 1997;68(3):1278–1285.

91. Soontornmalai A, Vlaming ML, Fritschy JM. Differential, strain-specific cellular and subcellular distribution of multidrug transporters in murine choroid plexus and blood-brain barrier. *Neuroscience*. 2006;138(1):159–169.

92. Kilic E, Spudich A, Kilic U, Rentsch KM, Vig R, Matter CM, et al. ABCC1: a gateway for pharmacological compounds to the ischaemic brain. *Brain*. 2008;131(pt 10):2679–2689.

93. Bendayan R, Ronaldson PT, Gingras D, Bendayan M. In situ localization of P-glycoprotein (ABCB1) in human and rat brain. *J Histochem Cytochem*. 2006;54(10):1159–1167.

94. Szakacs G, Paterson JK, Ludwig JA, Booth-Genthe C, Gottesman MM. Targeting multidrug resistance in cancer. *Nat Rev Drug Discov*. 2006;5(3):219–234.

95. Loscher W, Potschka H. Role of drug efflux transporters in the brain for drug disposition and treatment of brain diseases. *Prog Neurobiol*. 2005;76(1):22–76.

96. Loscher W, Potschka H. Drug resistance in brain diseases and the role of drug efflux transporters. *Nat Rev Neurosci*. 2005;6(8):591–602.

97. Gottesman MM, Fojo T, Bates SE. Multidrug resistance in cancer: role of ATP-dependent transporters. *Nat Rev Cancer*. 2002;2(1):48–58.

98. Zhang W, Mojsilovic-Petrovic J, Andrade MF, Zhang H, Ball M, Stanimirovic DB. The expression and functional characterization of ABCG2 in brain endothelial cells and vessels. *FASEB J*. 2003;17(14):2085–2087.

99. Lam FC, Liu R, Lu P, Shapiro AB, Renoir JM, Sharom FJ, et al. beta-Amyloid efflux mediated by p-glycoprotein. *J Neurochem*. 2001;76(4):1121–1128.

100. Vogelgesang S, Warzok RW, Cascorbi I, Kunert-Keil C, Schroeder E, Kroemer HK, et al. The role of P-glycoprotein in cerebral amyloid angiopathy; implications for the early pathogenesis of Alzheimer's disease. *Curr Alzheimer Res*. 2004;1(2):121–125.

101. Ohtsuki S, Ito S, Terasaki T. Is P-glycoprotein involved in amyloid-beta elimination across the blood-brain barrier in Alzheimer's disease?. *Clin Pharmacol Ther*. 2010;88(4):443–445.

102. Jedlitschky G, Vogelgesang S, Kroemer HK. MDR1-P-glycoprotein (ABCB1)-mediated disposition of amyloid-beta peptides: implications for the pathogenesis and therapy of Alzheimer's disease. *Clin Pharmacol Ther*. 2010;88(4):441–443.

103. Vautier S, Fernandez C. ABCB1: the role in Parkinson's disease and pharmacokinetics of antiparkinsonian drugs. *Expert Opin Drug Metab Toxicol*. 2009;5(11):1349–1358.

104. Langford D, Grigorian A, Hurford R, Adame A, Ellis RJ, Hansen L, et al. Altered P-glycoprotein expression in AIDS patients with HIV encephalitis. *J Neuropathol Exp Neurol*. 2004;63(10):1038–1047.

105. Jablonski M, Miller DS, Pasinelli P, Trotti D. ABC transporter-driven pharmacoresistance in amyotrophic lateral sclerosis. *Brain Res*. 2015;1607:1–14.

106. Juliano RL, Ling V. A surface glycoprotein modulating drug permeability in Chinese hamster ovary cell mutants. *Biochim Biophys Acta*. 1976;455(1):152–162.

107. Choudhuri S, Cherrington NJ, Li N, Klaassen CD. Constitutive expression of various xenobiotic and endobiotic transporter mRNAs in the choroid plexus of rats. *Drug Metab Dispos*. 2003;31(11):1337–1345.

108. Gribar JJ, Ramachandra M, Hrycyna CA, Dey S, Ambudkar SV. Functional characterization of glycosylation-deficient human P-glycoprotein using a vaccinia virus expression system. *J Membr Biol*. 2000;173(3):203–214.

109. Siarheyeva A, Liu R, Sharom FJ. Characterization of an asymmetric occluded state of P-glycoprotein with two bound nucleotides: implications for catalysis. *J Biol Chem*. 2010;285(10):7575–7586.

110. Sharom FJ. The P-glycoprotein multidrug transporter. *Essays Biochem*. 2011;50(1):161–178.

111. Higgins CF, Gottesman MM. Is the multidrug transporter a flippase?. *Trends Biochem Sci*. 1992;17(1):18–21.

112. Romsicki Y, Sharom FJ. Phospholipid flippase activity of the reconstituted P-glycoprotein multidrug transporter. *Biochemistry*. 2001;40(23):6937–6947.

113. Garrigues A, Escargueil AE, Orlowski S. The multidrug transporter, P-glycoprotein, actively mediates cholesterol redistribution in the cell membrane. *Proc Natl Acad Sci USA*. 2002;99(16):10347–10352.

114. Berezowski V, Landry C, Dehouck M-P, Cecchelli R, Fenart L. Contribution of glial cells and pericytes to the mRNA profiles of P-glycoprotein and multidrug resistance-associated proteins in an in vitro model of the blood–brain barrier. *Brain Res*. 2004;1018(1):1–9.

115. Smyth MJ, Krasovskis E, Sutton VR, Johnstone RW. The drug efflux protein, P-glycoprotein, additionally protects drug-resistant tumor cells from multiple forms of caspase-dependent apoptosis. *Proc Natl Acad Sci USA*. 1998;95(12):7024–7029.

116. Broccatelli F, Larregieu CA, Cruciani G, Oprea TI, Benet LZ. Improving the prediction of the brain disposition for orally administered drugs using BDDCS. *Adv Drug Deliv Rev*. 2012;64(1):95–109.

117. Nies AT, Jedlitschky G, Konig J, Herold-Mende C, Steiner HH, Schmitt HP, et al. Expression and immunolocalization of the multidrug resistance proteins, MRP1-MRP6 (ABCC1-ABCC6), in human brain. *Neuroscience*. 2004;129(2):349–360.

118. Dallas S, Miller DS, Bendayan R. Multidrug resistance-associated proteins: expression and function in the central nervous system. *Pharmacol Rev*. 2006;58(2):140–161.

119. Hirrlinger J, Konig J, Dringen R. Expression of mRNAs of multidrug resistance proteins (Mrps) in cultured rat astrocytes, oligodendrocytes, microglial cells and neurones. *J Neurochem*. 2002;82(3):716–719.

120. Beck K, Hayashi K, Dang K, Hayashi M, Boyd CD. Analysis of ABCC6 (MRP6) in normal human tissues. *Histochem Cell Biol*. 2005;123(4–5):517–528.

121. Doyle LA, Yang W, Abruzzo LV, Krogmann T, Gao Y, Rishi AK, et al. A multidrug resistance transporter from human MCF-7 breast cancer cells. *Proc Natl Acad Sci USA*. 1998;95(26):15665–15670.

122. Miyake K, Mickley L, Litman T, Zhan Z, Robey R, Cristensen B, et al. Molecular cloning of cDNAs which are highly overexpressed in mitoxantrone-resistant cells: demonstration of homology to ABC transport genes. *Cancer Res*. 1999;59(1):8–13.

123. Paturi DK, Kwatra D, Ananthula HK, Pal D, Mitra AK. Identification and functional characterization of breast cancer resistance protein in human bronchial epithelial cells (Calu-3). *Int J Pharm*. 2010;384(1–2):32–38.

124. McDevitt CA, Collins RF, Conway M, Modok S, Storm J, Kerr ID, et al. Purification and 3D structural analysis of oligomeric human multidrug transporter ABCG2. *Structure*. 2006;14(11):1623–1632.

125. Mao Q, Unadkat JD. Role of the breast cancer resistance protein (ABCG2) in drug transport. *AAPS J*. 2005;7(1):E118–E133.

126. Dauchy S, Dutheil F, Weaver RJ, Chassoux F, Daumas-Duport C, Couraud PO, et al. ABC transporters, cytochromes P450 and their main transcription factors: expression at the human blood-brain barrier. *J Neurochem*. 2008;107(6):1518–1528.

127. Lee G, Babakhanian K, Ramaswamy M, Prat A, Wosik K, Bendayan R. Expression of the ATP-binding cassette membrane transporter, ABCG2, in human and rodent brain microvessel endothelial and glial cell culture systems. *Pharm Res*. 2007;24(7):1262–1274.

128. Schlessinger A, Khuri N, Giacomini KM, Sali A. Molecular modeling and ligand docking for solute carrier (SLC) transporters. *Curr Top Med Chem*. 2013;13(7):843–856.

129. Gao B, Hagenbuch B, Kullak-Ublick GA, Benke D, Aguzzi A, Meier PJ. Organic anion-transporting polypeptides mediate transport of opioid peptides across blood-brain barrier. *J Pharmacol Exp Ther*. 2000;294(1):73–79.

130. Cvetkovic M, Leake B, Fromm MF, Wilkinson GR, Kim RB. OATP P-glycoprotein transporters mediate the cellular uptake and excretion of fexofenadine. *Drug Metab Dispos*. 1999;27(8):866–871.

131. Maeda T, Takahashi K, Ohtsu N, Oguma T, Ohnishi T, Atsumi R, et al. Identification of influx transporter for the quinolone antibacterial agent levofloxacin. *Mol Pharm*. 2007;4(1):85–94.

132. Badagnani I, Castro RA, Taylor TR, Brett CM, Huang CC, Stryke D, et al. Interaction of methotrexate with organic-anion transporting polypeptide 1A2 and its genetic variants. *J Pharm Exp Ther*. 2006;318(2):521–529.

133. Fischer WJ, Altheimer S, Cattori V, Meier PJ, Dietrich DR, Hagenbuch B. Organic anion transporting polypeptides expressed in liver and brain mediate uptake of microcystin. *Toxicol Appl Pharmacol*. 2005;203(3):257–263.

134. Alebouyeh M, Takeda M, Onozato ML, Tojo A, Noshiro R, Hasannejad H, et al. Expression of human organic anion transporters in the choroid plexus and their interactions with neurotransmitter metabolites. *J Pharmacol Sci.* 2003;93(4):430–436.

135. Ye J, Liu Q, Wang C, Meng Q, Sun H, Peng J, et al. Benzylpenicillin inhibits the renal excretion of acyclovir by OAT1 and OAT3. *Pharmacol Rep.* 2013;65:505–512.

136. Maeda K, Tian Y, Fujita T, Ikeda Y, Kumagai Y, Kondo T, et al. Inhibitory effects of *p*-aminohippurate and probenecid on the renal clearance of adefovir and benzylpenicillin as probe drugs for organic anion transporter (OAT) 1 and OAT3 in humans. *Eur J Pharm Sci.* 2014;59:94–103.

137. Ebner T, Ishiguro N, Taub ME. The use of transporter probe drug cocktails for the assessment of transporter-based drug-drug interactions in a clinical setting-proposal of a four component transporter cocktail. *J Pharm Sci.* 2015;104(9):3220–3228.

138. Wada S, Tsuda M, Sekine T, Cha SH, Kimura M, Kanai Y, et al. Rat multispecific organic anion transporter 1 (rOAT1) transports zidovudine, acyclovir, and other antiviral nucleoside analogs. *J Pharmacol Exp Ther.* 2000;294(3):844–849.

139. Ahn SY, Jamshidi N, Mo ML, Wu W, Eraly SA, Dnyanmote A, et al. Linkage of organic anion transporter-1 to metabolic pathways through integrated "omics"-driven network and functional analysis. *J Biol Chem.* 2011;286(36):31522–31531.

140. Tahara H, Kusuhara H, Endou H, Koepsell H, Imaoka T, Fuse E, et al. A species difference in the transport activities of H2 receptor antagonists by rat and human renal organic anion and cation transporters. *J Pharmacol Exp Ther.* 2005;315(1):337–345.

141. Zhang X, Wang R, Piotrowski M, Zhang H, Leach KL. Intracellular concentrations determine the cytotoxicity of adefovir, cidofovir and tenofovir. *Toxicol In Vitro.* 2015;29(1):251–258.

142. Lin DW. The Role of Transporters in Nephrotoxicity: An Investigation From the Bench to Populations-Based Studies. Doctoral Dissertation. San Francisco: University of California; 2008

143. Takeda M, Babu E, Narikawa S, Endou H. Interaction of human organic anion transporters with various cephalosporin antibiotics. *Eur J Pharmacol.* 2002;438(3):137–142.

144. Sugiyama D, Kusuhara H, Shitara Y, Abe T, Meier PJ, Sekine T, et al. Characterization of the efflux transport of 17 beta-estradiol-D-17 beta-glucuronide from the brain across the blood-brain barrier. *J Pharmacol Exp Ther.* 2001;298(1):316–322.

145. Mori N, Yokooji T, Kamio Y, Murakami T. Characterization of intestinal absorption of mizoribine mediated by concentrative nucleoside transporters in rats. *Eur J Pharmacol.* 2008;586(1–3):52–58.

146. Ye J, Liu Q, Wang C, Meng Q, Sun H, Peng J, et al. Benzylpenicillin inhibits the renal excretion of acyclovir by OAT1 and OAT3. *Pharmacol Rep.* 2013;65(2):505–512.

147. Ohtsuki S, Asaba H, Takanaga H, Deguchi T, Hosoya K, Otagiri M, et al. Role of blood-brain barrier organic anion transporter 3 (OAT3) in the efflux of indoxyl sulfate, a uremic toxin: its involvement in neurotransmitter metabolite clearance from the brain. *J Neurochem.* 2002;83(1):57–66.

148. Watanabe T, Kusuhara H, Watanabe T, Debori Y, Maeda K, Kondo T, et al. Prediction of the overall renal tubular secretion and hepatic clearance of anionic drugs and a renal drug-drug interaction involving organic anion transporter 3 in humans by in vitro uptake experiments. *Drug Metab Dispos.* 2011;39(6):1031–1038.

149. Maeda A, Tsuruoka S, Ushijima K, Kanai Y, Endou H, Saito K, et al. Drug interaction between celecoxib and methotrexate in organic anion transporter 3-transfected renal cells and in rats in vivo. *Eur J Pharmacol.* 2010;640(1–3):168–171.

150. Sykes D, Sweet DH, Lowes S, Nigam SK, Pritchard JB, Miller DS. Organic anion transport in choroid plexus from wild-type and organic anion transporter 3 (Slc22a8)-null mice. *Am J Physiol Renal.* 2004;286(5):F972–F978.

151. Nagle MA, Wu W, Eraly SA, Nigam SK. Organic anion transport pathways in antiviral handling in choroid plexus in Oat1 (Slc22a6) and Oat3 (Slc22a8) deficient tissue. *Neurosci Lett.* 2013;534:133–138.

152. Morita N, Kusuhara H, Sekine T, Endou H, Sugiyama Y. Functional characterization of rat organic anion transporter 2 in LLC-PK1 cells. *J Pharmacol Exp Ther.* 2001;298(3): 1179–1184.

153. Lin CJ, Tai Y, Huang MT, Tsai YF, Hsu HJ, Tzen KY, et al. Cellular localization of the organic cation transporters, OCT1 and OCT2, in brain microvessel endothelial cells and its implication for MPTP transport across the blood-brain barrier and MPTP-induced dopaminergic toxicity in rodents. *J Neurochem.* 2010;114(3):717–727.

154. Ito N, Ito K, Ikebuchi Y, Kito T, Miyata H, Toyoda Y, et al. Organic cation transporter/solute carrier family 22a is involved in drug transfer into milk in mice. *J Pharm Sci.* 2014;103(10): 3342–3348.

155. Zhang S, Lovejoy KS, Shima JE, Lagpacan LL, Shu Y, Lapuk A, et al. Organic cation transporters are determinants of oxaliplatin cytotoxicity. *Cancer Res.* 2006;66(17):8847–8857.

156. Koepsell H. The SLC22 family with transporters of organic cations, anions and zwitterions. *Mol Aspects Med.* 2013;34(2–3):413–435.

157. Bacq A, Balasse L, Biala G, Guiard B, Gardier AM, Schinkel A, et al. Organic cation transporter 2 controls brain norepinephrine and serotonin clearance and antidepressant response. *Mol Psych.* 2012;17(9):926–939.

158. Busch AE, Karbach U, Miska D, Gorboulev V, Akhoundova A, Volk C, et al. Human neurons express the polyspecific cation transporter hOCT2, which translocates monoamine neurotransmitters, amantadine, and memantine. *Mol Pharmacol.* 1998;54:342–352.

159. Wessler I, Roth E, Deutsch C, Brockerhoff P, Bittinger F, Kirkpatrick C, et al. Release of non-neuronal acetylcholine from the isolated human placenta is mediated by organic cation transporters. *Br J Pharmacol.* 2001;134(5):951–956.

160. Nakata T, Matsui T, Kobayashi K, Kobayashi Y, Anzai N. Organic cation transporter 2 (SLC22A2), a low-affinity and high-capacity choline transporter, is preferentially enriched on synaptic vesicles in cholinergic neurons. *Neuroscience.* 2013;252:212–221.

161. Fujita T, Urban TJ, Leabman MK, Fujita K, Giacomini KM. Transport of drugs in the kidney by the human organic cation transporter, OCT2 and its genetic variants. *J Pharm Sci.* 2006;95(1):25–36.

162. Kimura N, Masuda S, Tanihara Y, Harumasa Ueo, Okuda M, Katsura T, et al. Metformin is a superior substrate for renal organic cation transporter OCT2 rather than hepatic OCT1. *Drug Metab Pharmacokinet.* 2005;50(5):379–386.

163. Bauer LA, Black DJ, Lill JS, Garrison J, Raisys VA, Hooton TM. Levofloxacin and ciprofloxacin decrease procainamide and *N*-acetylprocainamide renal clearances. *Antimicrob Agents Chemother.* 2005;49(4):1649–1651.

164. Bergen AW, Javitz HS, Krasnow R, Michel M, Nishita D, Conti DV, et al. Organic cation transporter variation and response to smoking cessation therapies. *Nicotine Tob Res.* 2014;16(12):1638–1646.

165. Wang K, Sun S, Li L, Tu M, Jiang H. Involvement of organic cation transporter 2 inhibition in potential mechanisms of antidepressant action. *Progr Neuropsychopharmacol Biol Psych.* 2014;53:90–98.

166. Kucheryavykh LY, Rolon-Reyes K, Kucheryavykh YV, Skatchkov S, Eaton MJ, Sanabria P, et al. Glioblastoma development in mouse brain: general reduction of OCTs and mislocalization of OCT3 transporter and subsequent uptake of ASP substrate to the nuclei. *J Neurosci Neuroeng.* 2014;3(1):3–9.

167. Duan H, Wang J. Selective transport of monoamine neurotransmitters by human plasma membrane monoamine transporter and organic cation transporter 3. *J Pharmacol Exp Ther.* 2010;335(3):743–753.

168. Chen L, Pawlikowski B, Schlessinger A, More SS, Stryke D, Johns SJ, et al. Role of organic cation transporter 3 (SLC22A3) and its missense variants in the pharmacologic action of metformin. *Pharmacogenet Genomics.* 2010;20(11):687–699.

169. Zhu HJ, Appel DI, Grundemann D, Richelson E, Markowitz JS. Evaluation of organic cation transporter 3 (SLC22A3) inhibition as a potential mechanism of antidepressant action. *Pharmacol Res.* 2012;65(4):491–496.

170. Urquhart BL, Kim RB. Blood-brain barrier transporters and response to CNS-active drugs. *Eur J Clin Pharmacol.* 2009;65(11):1063–1070.

171. Morrison BM, Lee Y, Rothstein JD. Oligodendroglia: metabolic supporters of axons. *Trends Cell Biol.* 2013;23(12):644–651.

172. Spanier JA, Drewes LR. Monocarboxylate transporters. Drug Transporters. USA: John Wiley & Sons, Inc.; 2006:[pp. 147–170].

173. Bhattacharya I, Boje KM. GHB (gamma-hydroxybutyrate) carrier-mediated transport across the blood-brain barrier. *J Pharmacol Exp Ther.* 2004;311(1):92–98.

174. Kido Y, Tamai I, Okamoto M, Suzuki F, Tsuji A. Functional clarification of MCT1-mediated transport of monocarboxylic acids at the blood-brain barrier using in vitro cultured cells and in vivo BUI studies. *Pharm Res.* 2000;17(1):55–62.

175. Vijay N, Morris ME. Role of monocarboxylate transporters in drug delivery to the brain. *Curr Pharm Design.* 2014;20(10):1487–1498.

176. Miranda-Goncalves V, Honavar M, Pinheiro C, Martinho O, Pires MM, Pinheiro C, et al. Monocarboxylate transporters (MCTs) in gliomas: expression and exploitation as therapeutic targets. *Neuro Oncol.* 2013;15(2):172–188.

177. Moschen I, Broer A, Galic S, Lang F, Broer S. Significance of short chain fatty acid transport by members of the monocarboxylate transporter family (MCT). *Neurochem Res.* 2012;37(11):2562–2568.

178. Wilson MC, Kraus M, Marzban H, Sarna JR, Wang Y, Hawkes R, et al. The neuroplastin adhesion molecules are accessory proteins that chaperone the monocarboxylate transporter MCT2 to the neuronal cell surface. *PloS One.* 2013;8(11):e78654.

179. Lauritzen F, Heuser K, de Lanerolle NC, Lee T-SW, Spencer DD, Kim JH, et al. Redistribution of monocarboxylate transporter 2 on the surface of astrocytes in the human epileptogenic hippocampus. *Glia.* 2012;60(7):1172–1181.

180. Jackson VN, Halestrap AP. The kinetics, substrate, and inhibitor specificity of the monocarboxylate (lactate) transporter of rat liver cells determined using the fluorescent intracellular pH indicator, 2′,7′-bis(carboxyethyl)-5(6)-carboxyfluorescein. *J Biol Chem.* 1996;271(2):861–868.

181. Pinheiro C, Longatto-Filho A, Azevedo-Silva J, Casal M, Schmitt FC, Baltazar F. Role of monocarboxylate transporters in human cancers: state of the art. *J Bioenerg Biomembr.* 2012;44(1):127–139.

182. Sala-Rabanal M, Loo DD, Hirayama BA, Wright EM. Molecular mechanism of dipeptide and drug transport by the human renal H^+ oligopeptide cotransporter hPEPT2. *Am J Physiol Renal Physiol.* 2008;294(6):F1422–F1432.

183. Li M. Interactions of amoxicillin and cefaclor with human renal organic anion and peptide transporters. *Drug Metab Dispos.* 2006;34(4):547–555.

184. Shen H, Ocheltree SM, Hu Y, Keep RF, Smith DE. Impact of genetic knockout of PEPT2 on cefadroxil pharmacokinetics, renal tubular reabsorption, and brain penetration in mice. *Drug Metabolism Dispos.* 2007;35(7):1209–1216.

185. Zhu T, Chen XZ, Steel A, Hediger MA, Smith DE. Differential recognition of ACE inhibitors in *Xenopus laevis* oocytes expressing rat PEPT1 and PEPT2. *Pharm Res*. 2000;17(5):526–532.

186. Ganapathy ME, Leibach FH, Huang W, Wang H, Ganapathy V. Valacyclovir: a substrate for the intestinal and renal peptide transporters PEPT1 and PEPT2. *Biochem Biophys Res Commun*. 1998;246:470–475.

187. Chen X, Loryan I, Payan M, Keep RF, Smith DE, Hammarlund-Udenaes M. Effect of transporter inhibition on the distribution of cefadroxil in rat brain. *Fluids Barriers CNS*. 2014;11(1):25.

188. Redzic ZB, Biringer J, Barnes K, Baldwin SA, Al-Sarraf H, Nicola PA, et al. Polarized distribution of nucleoside transporters in rat brain endothelial and choroid plexus epithelial cells. *J Neurochem*. 2005;94(5):1420–1426.

189. Li B, Gu L, Hertz L, Peng L. Expression of nucleoside transporter in freshly isolated neurons and astrocytes from mouse brain. *Neurochem Res*. 2013;38(11):2351–2358.

190. Peng L, Huang R, Yu ACH, Fung KY, Rathbone MP, Hertz L. Nucleoside transporter expression and function in cultured mouse astrocytes. *Glia*. 2005;52(1):25–35.

191. Dragan Y, Valdes R, Gomez-Angelats M, Felipe A, Javier Casado F, Pitot H, et al. Selective loss of nucleoside carrier expression in rat hepatocarcinomas. *Hepatology*. 2000;32(2):239–246.

192. Mori N, Shimomukai Y, Yokooji T, Ishiguro M, Kamio Y, Murakami T. Modulation in concentrative nucleoside transporters-mediated intestinal absorption of mizoribine, an immunosuppressive agent, in lipopolysaccharide-treated rats. *Die Pharm*. 2011;66(3):207–211.

193. Hong M, Schlichter L, Bendayan R. A Na(+)-dependent nucleoside transporter in microglia. *J Pharmacol Exp Ther*. 2000;292(1):366–374.

194. Wu X, Yuan G, Brett CM, Hui AC, Giacomini KM. Sodium-dependent nucleoside transport in choroid plexus from rabbit. Evidence for a single transporter for purine and pyrimidine nucleosides. *J Biol Chem*. 1992;267(13):8813–8818.

195. Furihata T, Fukuchi Y, Iikura M, Hashizume M, Miyajima A, Nagai M, et al. Striking species difference in the contribution of concentrative nucleoside transporter 2 to nucleoside uptake between mouse and rat hepatocytes. *Antimicrob Agents Chemother*. 2010;54(7):3035–3038.

196. Klaassen CD, Aleksunes LM. Xenobiotic, bile acid, and cholesterol transporters: function and regulation. *Pharmacol Revi*. 2010;62(1):1–96.

197. Schlessinger A, Yee SW, Sali A, Giacomini KM. SLC classification: an update. *Clin Pharmacol Ther*. 2013;94(1):19–23.

198. Medina-Pulido L, Molina-Arcas M, Justicia C, Soriano E, Burgaya F, Planas AM, et al. Hypoxia and P1 receptor activation regulate the high-affinity concentrative adenosine transporter CNT2 in differentiated neuronal PC12 cells. *Biochem J*. 2013;454(3):437–445.

199. Toan SV, To KK, Leung GP, de Souza MO, Ward JL, Tse CM. Genomic organization and functional characterization of the human concentrative nucleoside transporter-3 isoform (hCNT3) expressed in mammalian cells. *Pflugers Arch Eur J Physiol*. 2003;447(2):195–204.

200. Baldwin SA, Beal PR, Yao SY, King AE, Cass CE, Young JD. The equilibrative nucleoside transporter family SLC29. *Pflugers Arch Eur J Physiol*. 2004;447(5):735–743.

201. Nagai K, Nagasawa K, Fujimoto S. Transport mechanisms for adenosine and uridine in primary-cultured rat cortical neurons and astrocytes. *Biochem Biophys Res Commun*. 2005;334(4):1343–1350.

202. Sinclair CJ, LaRiviere CG, Young JD, Cass CE, Baldwin SA, Parkinson FE. Purine uptake and release in rat C6 glioma cells: nucleoside transport and purine metabolism under ATP-depleting conditions. *J Neurochem*. 2000;75(4):1528–1538.

203. Bronger H, Konig J, Kopplow K, Steiner HH, Ahmadi R, Herold-Mende C, et al. ABCC drug efflux pumps and organic anion uptake transporters in human gliomas and the blood-tumor barrier. *Cancer Res*. 2005;65(24):11419–11428.

204. Hong M, Schlichter L, Bendayan R. A novel zidovudine uptake system in microglia. *J Pharm Exp Ther*. 2001;296(1):141–149.

205. Cano-Soldado P, Lorrayoz IM, Molina-Arcas M, Casado FJ, Martinez-Picado J, Lostao MP, et al. Interaction of nucleoside inhibitors of HIV-1 reverse transcriptase with the concentrative nucleoside transporter-1 (SLC28A1). *Antivir Ther*. 2004;9(6):993–1002.

206. Hagenbuch B, Meier PJ. The superfamily of organic anion transporting polypeptides. *Biochim Biophys Acta*. 2003;1609(1):1–18.

207. Butt AM, Jones HC, Abbott NJ. Electrical resistance across the blood-brain barrier in anaesthetized rats: a developmental study. *J Physiol*. 1990;429:47–62.

208. Kullak-Ublick GA, Hagenbuch B, Stieger B, Wolkoff AW, Meier PJ. Functional characterization of the basolateral rat liver organic anion transporting polypeptide. *Hepatology*. 1994;20(2):411–416.

209. Adachi H, Suzuki T, Abe M, Asano N, Mizutamari H, Tanemoto M, et al. Molecular characterization of human and rat organic anion transporter OATP-D. *Am J Physiol Renal Physiol*. 2003;285(6):F1188–F1197.

210. Fujiwara K, Adachi H, Nishio T, Unno M, Tokui T, Okabe M, et al. Identification of thyroid hormone transporters in humans: different molecules are involved in a tissue-specific manner. *Endocrinology*. 2001;142(5):2005–2012.

211. Pizzagalli F, Hagenbuch B, Stieger B, Klenk U, Folkers G, Meier PJ. Identification of a novel human organic anion transporting polypeptide as a high affinity thyroxine transporter. *Mol Endocrinol*. 2002;16(10):2283–2296.

212. Huber RD, Gao B, Sidler Pfandler MA, Zhang-Fu W, Leuthold S, Hagenbuch B, et al. Characterization of two splice variants of human organic anion transporting polypeptide 3A1 isolated from human brain. *Am J Physiol Cell Physiol*. 2007;292(2):C795–C806.

213. Angeletti RH, Bergwerk AJ, Novikoff PM, Wolkoff AW. Dichotomous development of the organic anion transport protein in liver and choroid plexus. *Am J Physiol*. 1998;275(3 pt 1):C882–C887.

214. Ohtsuki S, Takizawa T, Takanaga H, Terasaki N, Kitazawa T, Sasaki M, et al. In vitro study of the functional expression of organic anion transporting polypeptide 3 at rat choroid plexus epithelial cells and its involvement in the cerebrospinal fluid-to-blood transport of estrone-3-sulfate. *Mol Pharmacol*. 2003;63(3):532–537.

215. Sugiyama D, Kusuhara H, Taniguchi H, Ishikawa S, Nozaki Y, Aburatani H, et al. Functional characterization of rat brain-specific organic anion transporter (Oatp14) at the blood-brain barrier: high affinity transporter for thyroxine. *J Biol Chem*. 2003;278(44):43489–43495.

216. Nishio T, Adachi H, Nakagomi R, Tokui T, Sato E, Tanemoto M, et al. Molecular identification of a rat novel organic anion transporter moat1, which transports prostaglandin D(2), leukotriene C(4), and taurocholate. *Biochem Biophys Res Commun*. 2000;275(3):831–888.

217. Abe T, Kakyo M, Sakagami H, Tokui T, Nishio T, Tanemoto M, et al. Molecular characterization and tissue distribution of a new organic anion transporter subtype (oatp3) that transports thyroid hormones and taurocholate and comparison with oatp2. *J Biol Chem*. 1998;273(35):22395–22401.

218. Sioka C, Kyritsis AP. Central and peripheral nervous system toxicity of common chemotherapeutic agents. *Cancer Chemother Pharmacol*. 2009;63(5):761–767.

219. Sekine T, Watanabe N, Hosoyamada M, Kanai Y, Endou H. Expression cloning and characterization of a novel multispecific organic anion transporter. *J Biol Chem*. 1997;272(30):18526–18529.

220. Sekine T, Cha SH, Tsuda M, Apiwattanakul N, Nakajima N, Kanai Y, et al. Identification of multispecific organic anion transporter 2 expressed predominantly in the liver. *FEBS Lett*. 1998;429(2):179–182.

221. Kusuhara H, Sekine T, Utsunomiya-Tate N, Tsuda M, Kojima R, Cha SH, et al. Molecular cloning and characterization of a new multispecific organic anion transporter from rat brain. *J Biol Chem*. 1999;274(19):13675–13680.

222. Bahn A, Hagos Y, Reuter S, Balen D, Brzica H, Krick W, et al. Identification of a new urate and high affinity nicotinate transporter, hOAT10 (SLC22A13). *J Biol Chem*. 2008;283(24):16332–16341.

223. Kikuchi R, Kusuhara H, Sugiyama D, Sugiyama Y. Contribution of organic anion transporter 3 (Slc22a8) to the elimination of *p*-aminohippuric acid and benzylpenicillin across the blood-brain barrier. *J Pharmacol Exp Ther*. 2003;306(1):51–58.

224. Bahn A, Ljubojevic M, Lorenz H, Schultz C, Ghebremedhin E, Ugele B, et al. Murine renal organic anion transporters mOAT1 and mOAT3 facilitate the transport of neuroactive tryptophan metabolites. *Am J Physiol Cell Physiol*. 2005;289(5):C1075–C1084.

225. Nagata Y, Kusuhara H, Endou H, Sugiyama Y. Expression and functional characterization of rat organic anion transporter 3 (rOat3) in the choroid plexus. *Mol Pharmacol*. 2002;61(5):982–988.

226. Kusch-Poddar M, Drewe J, Fux I, Gutmann H. Evaluation of the immortalized human brain capillary endothelial cell line BB19 as a human cell culture model for the blood–brain barrier. *Brain Res*. 2005;1064(1–2):21–31.

227. Xu G, Bhatnagar V, Wen G, Hamilton BA, Eraly SA, Nigam SK. Analyses of coding region polymorphisms in apical and basolateral human organic anion transporter (OAT) genes [OAT1 (NKT), OAT2, OAT3, OAT4, URAT (RST)]. *Kidney Int*. 2005;68(4):1491–1499.

228. Nishiwaki T, Daigo Y, Tamari M, Fujii Y, Nakamura Y. Molecular cloning, mapping, and characterization of two novel human genes, ORCTL3 and ORCTL4, bearing homology to organic-cation transporters. *Cytogenet Cell Genet*. 1998;83(3–4):251–255.

229. Imaoka T, Kusuhara H, Adachi-Akahane S, Hasegawa M, Morita N, Endou H, et al. The renal-specific transporter mediates facilitative transport of organic anions at the brush border membrane of mouse renal tubules. *J Am Soc Nephrol*. 2004;15(8):2012–2022.

230. Breen CM, Sykes DB, Fricker G, Miller DS. Confocal imaging of organic anion transport in intact rat choroid plexus. *Am J Physiol Renal Physiol*. 2002;282(5):F877–F885.

231. Ullrich KJ. Specificity of transporters for 'organic anions' and 'organic cations' in the kidney. *Biochim Biophys Acta*. 1994;1197(1):45–62.

232. Sekine T, Kusuhara H, Utsunomiya-Tate N, Tsuda M, Sugiyama Y, Kanai Y, et al. Molecular cloning and characterization of high-affinity carnitine transporter from rat intestine. *Biochem Biophys Res Commun*. 1998;251(2):586–591.

233. Tamai I, Yabuuchi H, Nezu J, Sai Y, Oku A, Shimane M, et al. Cloning and characterization of a novel human pH-dependent organic cation transporter OCTN1. *FEBS Lett*. 1997;419(1):107–111.

234. Tamai I, Ohashi R, Nezu J, Yabuuchi H, Oku A, Shimane M, et al. Molecular and functional identification of sodium ion-dependent, high affinity human carnitine transporter OCTN2. *J Biol Chem*. 1998;273(32):20378–20382.

235. Wu X, Prasad PD, Leibach FH, Ganapathy V. cDNA sequence, transport function, and genomic organization of human OCTN2, a new member of the organic cation transporter family. *Biochem Biophys Res Commun*. 1998;246(3):589–595.

236. Wu X, Huang W, Ganapathy ME, Wang H, Kekuda R, Conway SJ, et al. Structure, function, and regional distribution of the organic cation transporter OCT3 in the kidney. *Am J Physiol Renal Physiol*. 2000;279(3):F449–F458.

237. Hayer M, Bonisch H, Bruss M. Molecular cloning, functional characterization and genomic organization of four alternatively spliced isoforms of the human organic cation transporter 1 (hOCT1/SLC22A1). *Ann Human Genet*. 1999;63(Pt 6):473–482.

238. Hayer-Zillgen M, Bruss M, Bonisch H. Expression and pharmacological profile of the human organic cation transporters hOCT1, hOCT2 and hOCT3. *Br J Pharmacol.* 2002;136(6):829–836.

239. Inazu M, Takeda H, Matsumiya T. Molecular and functional characterization of an Na^+-independent choline transporter in rat astrocytes. *J Neurochem.* 2005;94(5):1427–1437.

240. Grundemann D, Koster S, Kiefer N, Breidert T, Engelhardt M, Spitzenberger F, et al. Transport of monoamine transmitters by the organic cation transporter type 2 OCT2. *J Biol Chem.* 1998;273(47):30915–30920.

241. Kekuda R, Prasad PD, Wu X, Wang H, Fei YJ, Leibach FH, et al. Cloning and functional characterization of a potential-sensitive, polyspecific organic cation transporter (OCT3) most abundantly expressed in placenta. *J Biol Chem.* 1998;273(26):15971–15979.

242. Wu X, Kekuda R, Huang W, Fei YJ, Leibach FH, Chen J, et al. Identity of the organic cation transporter OCT3 as the extraneuronal monoamine transporter (uptake2) and evidence for the expression of the transporter in the brain. *J Biol Chem.* 1998;273(49):32776–32786.

243. Vialou V, Balasse L, Callebert J, Launay JM, Giros B, Gautron S. Altered aminergic neurotransmission in the brain of organic cation transporter 3-deficient mice. *J Neurochem.* 2008;106(3):1471–1482.

244. Vialou V, Amphoux A, Zwart R, Giros B, Gautron S. Organic cation transporter 3 (Slc22a3) is implicated in salt-intake regulation. *J Neurosci.* 2004;24(11):2846–2851.

245. Ogasawara M, Yamauchi K, Satoh Y, Yamaji R, Inui K, Jonker JW, et al. Recent advances in molecular pharmacology of the histamine systems: organic cation transporters as a histamine transporter and histamine metabolism. *J Pharmacol Sci.* 2006;101(1):24–30.

246. Schneider E, Machavoine F, Pleau JM, Bertron AF, Thurmond RL, Ohtsu H, et al. Organic cation transporter 3 modulates murine basophil functions by controlling intracellular histamine levels. *J Exp Med.* 2005;202(3):387–393.

247. Zhu P, Hata R, Ogasawara M, Cao F, Kameda K, Yamauchi K, et al. Targeted disruption of organic cation transporter 3 (Oct3) ameliorates ischemic brain damage through modulating histamine and regulatory T cells. *J Cereb Blood Flow Metab.* 2012;32(10):1897–1908.

248. Tamai I, Ohashi R, Nezu JI, Sai Y, Kobayashi D, Oku A, et al. Molecular and functional characterization of organic cation/carnitine transporter family in mice. *J Biol Chem.* 2000;275(51):40064–40072.

249. Wu X, George RL, Huang W, Wang H, Conway SJ, Leibach FH, et al. Structural and functional characteristics and tissue distribution pattern of rat OCTN1, an organic cation transporter, cloned from placenta. *Biochim Biophys Acta.* 2000;1466(1–2):315–327.

250. Slitt AL, Cherrington NJ, Hartley DP, Leazer TM, Klaassen CD. Tissue distribution and renal developmental changes in rat organic cation transporter mRNA levels. *Drug Metab Dispos.* 2002;30(2):212–219.

251. Kido Y, Tamai I, Ohnari A, Sai Y, Kagami T, Nezu J, et al. Functional relevance of carnitine transporter OCTN2 to brain distribution of L-carnitine and acetyl-L-carnitine across the blood-brain barrier. *J Neurochem.* 2001;79(5):959–969.

252. Kristufek D, Rudorfer W, Pifl C, Huck S. Organic cation transporter mRNA and function in the rat superior cervical ganglion. *J Physiol.* 2002;543(pt 1):117–134.

253. Wu X, Huang W, Prasad PD, Seth P, Rajan DP, Leibach FH, et al. Functional characteristics and tissue distribution pattern of organic cation transporter 2 (OCTN2), an organic cation/carnitine transporter. *J Pharmacol Exp Ther.* 1999;290(3):1482–1492.

254. Enerson BE, Drewes LR. Molecular features, regulation, and function of monocarboxylate transporters: implications for drug delivery. *J Pharm Sci.* 2003;92(8):1531–1544.

255. Bröer S, Schneider HP, Bröer A, Rahman B, Hamprecht B, Deitmer JW. Characterization of the monocarboxylate transporter 1 expressed in *Xenopus laevis* oocytes by changes in cytosolic pH. *Biochem J.* 1998;333(pt 1):167–174.

256. Halestrap AP, Wilson MC. The monocarboxylate transporter family—role and regulation. *IUBMB Life*. 2012;64(2):109–119.

257. Kim CM, Goldstein JL, Brown MS. cDNA cloning of MEV, a mutant protein that facilitates cellular uptake of mevalonate, and identification of the point mutation responsible for its gain of function. *J Biol Chem*. 1992;267(32):23113–23121.

258. Gerhart DZ, Enerson BE, Zhdankina OY, Leino RL, Drewes LR. Expression of monocarboxylate transporter MCT1 by brain endothelium and glia in adult and suckling rats. *Am J Physiol*. 1997;273(1 pt 1):E207–E213.

259. Lee Y, Morrison BM, Li Y, Lengacher S, Farah MH, Hoffman PN, et al. Oligodendroglia metabolically support axons and contribute to neurodegeneration. *Nature*. 2012;487(7408):443–448.

260. Holden KR, Zuñiga OF, May MM, Su H, Molinero MR, Rogers RC, et al. X-linked MCT8 gene mutations: characterization of the pediatric neurologic phenotype. *J Child Neurol*. 2005;20(10):852–857.

261. Tonduti D, Vanderver A, Berardinelli A, Schmidt JL, Collins CD, Novara F, et al. MCT8 deficiency: extrapyramidal symptoms and delayed myelination as prominent features. *J Child Neurol*. 2013;28(6):795–800.

262. Schwartz CE, May MM, Carpenter NJ, Rogers RC, Martin J, Bialer MG, et al. Allan-Herndon-Dudley syndrome and the monocarboxylate transporter 8 (MCT8) gene. *Am J Human Genet*. 2005;77(1):41–53.

263. Ceballos A, Belinchon MM, Sanchez-Mendoza E, Grijota-Martinez C, Dumitrescu AM, Refetoff S, et al. Importance of monocarboxylate transporter 8 for the blood-brain barrier-dependent availability of 3,5,3′-triiodo-L-thyronine. *Endocrinology*. 2009;150(5):2491–2496.

264. Berger UV, Hediger MA. Distribution of peptide transporter PEPT2 mRNA in the rat nervous system. *Anat Embryol*. 1999;199(5):439–449.

265. Sakata K, Yamashita T, Maeda M, Moriyama Y, Shimada S, Tohyama M. Cloning of a lymphatic peptide/histidine transporter. *Biochem J*. 2001;356(1):53–60.

266. Smith DE, Hu Y, Shen H, Nagaraja TN, Fenstermacher JD, Keep RF. Distribution of glycylsarcosine and cefadroxil among cerebrospinal fluid, choroid plexus, and brain parenchyma after intracerebroventricular injection is markedly different between wild-type and Pept2 null mice. *J Cereb Blood Flow Metab*. 2011;31(1):250–261.

267. Xiang J, Hu Y, Smith DE, Keep RF. PEPT2-mediated transport of 5-aminolevulinic acid and carnosine in astrocytes. *Brain Res*. 2006;1122(1):18–23.

268. Hu Y, Shen H, Keep RF, Smith DE. Peptide transporter 2 (PEPT2) expression in brain protects against 5-aminolevulinic acid neurotoxicity. *J Neurochem*. 2007;103(5):2058–2065.

269. Yamashita T, Shimada S, Guo W, Sato K, Kohmura E, Hayakawa T, et al. Cloning and functional expression of a brain peptide/histidine transporter. *J Biol Chem*. 1997;272(15):10205–10211.

270. Baldwin SA, Mackey JR, Cass CE, Young JD. Nucleoside transporters: molecular biology and implications for therapeutic development. *Mol Med Today*. 1999;5(5):216–224.

271. Ritzel MW, Yao SY, Huang MY, Elliott JF, Cass CE, Young JD. Molecular cloning and functional expression of cDNAs encoding a human Na^+-nucleoside cotransporter (hCNT1). *Am J Physiol*. 1997;272(2 pt 1):C707–C714.

272. Ritzel MW, Yao SY, Ng AM, Mackey JR, Cass CE, Young JD. Molecular cloning, functional expression and chromosomal localization of a cDNA encoding a human Na^+ nucleoside cotransporter (hCNT2) selective for purine nucleosides and uridine. *Mol Membr Biol*. 1998;15(4):203–211.

273. Ritzel MW, Ng AM, Yao SY, Graham K, Loewen SK, Smith KM, et al. Molecular identification and characterization of novel human and mouse concentrative Na^+-nucleoside cotransporter proteins (hCNT3 and mCNT3) broadly selective for purine and pyrimidine nucleosides (system cib). *J Biol Chem*. 2001;276(4):2914–2927.

274. Anderson CM, Xiong W, Young JD, Cass CE, Parkinson FE. Demonstration of the existence of mRNAs encoding N1/cif and N2/cit sodium/nucleoside cotransporters in rat brain. *Brain Res Mol Brain Res.* 1996;42(2):358–361.

275. Pardridge WM. The blood-brain barrier: bottleneck in brain drug development. *NeuroRx.* 2005;2(1):3–14.

276. Thomas SA, Segal MB. Saturation kinetics, specificity and NBMPR sensitivity of thymidine entry into the central nervous system. *Brain Res.* 1997;760(1–2):59–67.

277. Ritzel MW, Ng AM, Yao SY, Graham K, Loewen SK, Smith KM, et al. Recent molecular advances in studies of the concentrative Na^+-dependent nucleoside transporter (CNT) family: identification and characterization of novel human and mouse proteins (hCNT3 and mCNT3) broadly selective for purine and pyrimidine nucleosides (system cib). *Mol Membr Biol.* 2001;18(1):65–72.

278. Spector R. Nucleoside transport in choroid plexus: mechanism and specificity. *Arch Biochem Biophys.* 1982;216(2):693–703.

279. Anderson CM, Xiong W, Geiger JD, Young JD, Cass CE, Baldwin SA, et al. Distribution of equilibrative, nitrobenzylthioinosine-sensitive nucleoside transporters (ENT1) in brain. *J Neurochem.* 1999;73(2):867–873.

280. Jennings LL, Hao C, Cabrita MA, Vickers MF, Baldwin SA, Young JD, et al. Distinct regional distribution of human equilibrative nucleoside transporter proteins 1 and 2 (hENT1 and hENT2) in the central nervous system. *Neuropharmacolog.* 2001;40(5):722–731.

281. Podgorska M, Kocbuch K, Pawelczyk T. Recent advances in studies on biochemical and structural properties of equilibrative and concentrative nucleoside transporters. *Acta Biochim Pol.* 2005;52(4):749–758.

282. Engel K, Zhou M, Wang J. Identification and characterization of a novel monoamine transporter in the human brain. *J Biol Chem.* 2004;279(48):50042–50049.

CHAPTER 4

Drug Delivery to the Brain: Pharmacokinetic Concepts

Tugba Copur, BSc, Levent Oner, PhD
Hacettepe University, Ankara, Turkey

CONTENTS

1 INTRODUCTION

Drug discovery and development are particularly challenging for the treatment of central nervous system (CNS) diseases when compared with other therapeutic categories. One of the most important stages of CNS drug research is to obtain an effective and reliable concentration profile in the brain; however, it is also one of the most difficult stages of CNS drug discovery because of the blood–brain barrier (BBB).[1–3] Therefore, the success rate of drug development in this field is low due to the complexity of the tissues and cells of the brain.

There has been a significant increase in the neurodegenerative disorders, such as stroke, epilepsy, Alzheimer's disease, attention deficit hyperactivity disorder, Parkinson's disease, schizophrenia and bipolar disorders, depression, pain, insomnia, and brain tumors and it is expected to increase further in the 21st century.[4–6] Of these diseases, schizophrenia and bipolar disorders, depression, pain, insomnia, epilepsy, attention deficit hyperactivity disorder, and other similar disorders can be treated with small molecules. One of the most important problems encountered in the treatment of these CNS-related diseases is the delivery of these drugs into the brain. The CNS microvascular structure that forms the BBB limits the transport of molecules and ions into the brain.[7] In previous years of CNS drug discovery, total drug levels in animal models were measured to determine the penetration degree of the drug into the brain.[8–11] Nowadays, in vitro, in vivo, and in silico methods are used for this purpose.[10,12–15] In these recent studies, penetration through BBB and pharmacokinetics (PK) of the drug distribution in the brain are critical factors that control drug concentrations at the target sites; the prerequisite for a drug hereby is to bind to molecular target proteins of the BBB.[16–18] Therefore, extensive research is performed to explain the penetration of CNS drugs into the brain, together with their distribution profile in a suitable model. The relationship between unbound drug in plasma and brain and the rate and extent of CNS drug penetration is investigated in PK studies.[3,19,20]

Nanotechnology Methods for Neurological Diseases and Brain Tumors. http://dx.doi.org/10.1016/B978-0-12-803796-6.00004-6

PK factors, such as absorption from the intestine, drug distribution (including the distribution in the brain), metabolism, and excretion (ADME) from the body, are involved in the process of reaching the target of orally administered CNS drugs. Accordingly, to ensure that the drug is effective and reliable at the target site of CNS, the ADME parameters need to be examined in detail in the early stages of drug discovery.[8,21,22] In silico approaches have been used to predict the ADME parameters for the identification of candidate molecules in CNS drug discovery, but these methods need to be validated in advance and should be combined with in vivo results.[23]

In CNS-targeting studies, the BBB, drug penetration, and drug distribution are critical factors for the drug concentration at the target site. Considering these critical factors, three main parameters control the drug concentration at the target site of CNS: (1) the rate of drug delivery to the brain, (2) the extent of delivery to the brain, and (3) the PK of intracerebral drug distribution.[24–26]

2 CNS DRUG DELIVERY

2.1 Barriers to CNS Drug Delivery

The barriers provide a protective shield for the brain, and prevent about 98% of small molecules and 100% of large molecules, such as recombinant peptides–proteins, monoclonal antibodies (mAbs), nucleic acids, and antisense oligonucleotides. These small- and large-molecule drugs need to cross the barriers to reach a sufficient drug concentration in the brain.[27–31] There are three barrier sites between the blood and the brain. First, the BBB is the major barrier that is composed of capillary endothelial cells (ECs), astrocytes, pericytes, neurons, and microglia. Second, the blood–cerebrospinal fluid (CSF) barrier (BCSFB) formed by the tightly joined epithelial cells, facing the CSF. Third, the arachnoid epithelium, underlying the dura and completely enclosing the CNS, which has an important role for CNS drug transport into brain.[32–38] As the BBB has a surface area that is 5000 times larger than the BCSFB, it is considered to be the most important barrier and an interface between the blood and the brain interstitial fluid (ISF) (which bathes the synaptic connections within the parenchyma of the brain and causes a buffering effect).[15,32,39]

BCSFB is localized between the epithelial cells of the choroid plexus, which is the main interface between the blood and CSF.[40–42] The CSF is produced in the choroid plexuses of the brain ventricles. About 100–140 μL of CSF is present in the human brain, and this entire volume is completely turned over every 4–5 h.[42–44] The capillaries in the human brain have a total length of 650 km and a volume of 1 mL.[32,35] An overview of the physiological properties of CNS in human and rats is shown in Table 4.1.

Table 4.1 An Overview of the Physiological Parameters for Human and Rat (Gathered From the Literature)

Parameters	Values	
	Human	**Rat**
Brain weight (g)	70,000[45–47]	250[24,45–47]
Brain weight (g)	1,400[45–47]	1.8[24,45–47]
Capillary surface area (cm^2/g brain)	100–200[48–50]	140[47,53]
	100–150[3,35,51,52]	100[24,51]
Brain capillary length (km)	650[3,35,54,55]	Nda
Brain endothelial cell thickness (BBB) (nm)	Nda	200–500[24,56,57]
Brain capillary volume	I mL[3,35]	11 μL/g brain[3,32]
Brain water content (mL/g brain)	Nda	0.788[24,58]
Brain ECF production rate (μL/min/g brain)	0.11–0.14[59–61]	0.11–0.28[33,61]
ECF turnover rate (h)	20–27[61,62]	10–24[61,62]
CSF production rate (μL/min/brain)	0.29[61,63]	1.22[61,64]
		2.1–5.4[24,54]
		3.7 ± 0.1[24,65]
		2.9[24,66]
		2.2[24,64]
CSF turnover rate (h)	6[61,62]	2[61,62]
Cerebral blood flow (mL/min/g brain)	0.5–0.6[61,67,68]	1.05–1.10[61,69,70]
		1.01 ± 0.24[24,71]
		1.44 ± 14[24,72]
		0.93[24,73]
Brain ECF flow	0.15–0.2 mL/min[60,74]	0.2–0.5 μL/min[33,75]
	0.2 mL/min[62]	0.2 μL/min[62]
Brain CSF flow	0.4 mL/min[62,76]	2.2 μL/min[62,64]
BrainISF bulk flow (μL/g brain)	Nda	0.18–0.29[24,77]
		0.11[24,75]
ISF Bulk flow rate (μL/min/g)	Nda	0.1–0.3[33]
		0.2[3,33]
Mean transit time (s)	Nda	1.41 ± 0.07[24,72]
		2.8 ± 0.8[24,78]
Brain ECF volume	240 mL[59,62,79]	290 μL[62,75,79,80]
CSF volume	140 mL[62,79,81]	300 μL[79]
	160 mL[3,33]	250 μL[3,33,62,82]
	100–140 mL[42,43]	
BrainISF volume (mL/g brain)	225[54,83]	0.21[24,84]
		13.5–14.5[24,85]

BBB, Blood–brain barrier; CSF, cerebrospinal fluid; ECF, extracellular fluid; ISF, interstitial fluid; Nda, no data available.

The ECs in the brain differ in two ways from ECs of other organs. First, there are continuous tight junctions (TJs) present between brain ECs, and these TJs prevent paracellular movement of molecules. Second, there are no detectable transendothelial pathways (e.g., intracellular vesicles) present between brain ECs.[86] On molecular level, two types of cellular junctions have been characterized in the brain ECs: the intercellular adherens junction and the paracellular TJs.[87,88] The transendothelial electrical resistance caused by the TJs between the ECs in the brain is 1500–2000 $\Omega \cdot cm^2$. Compared to an electrical resistance of 3–33 $\Omega \cdot cm^2$ in other tissues, these TJs reduce the hydrophilic paracellular diffusion that is seen in other organs.[89–92]

2.2 BBB Transport Mechanism

Similar to cell membranes in other tissues, permeation across the BBB may occur by passive diffusion, endocytosis, or active transport.[93]

2.2.1 Passive Diffusion

Essentially, brain microvessels are considered to be a passive anatomical lipid barrier, allowing the uptake of hydrophilic compounds into the brain (except for some nutrients). Meanwhile, hydrophobic compounds pass through the BBB by simple diffusion, which is the major entry mechanism for most CNS drugs. Passive diffusion is the movement of compounds in the direction of a concentration gradient without need of energy and it is not saturable. There are two mechanisms responsible for passive diffusion: passive paracellular diffusion and transcellular diffusion. Paracellular diffusion is limited by the TJs between ECs. For example, the brain permeabilities of both mannitol and sucrose are very low, to such a degree that they can be ignored.[94–97] On the other hand, alcohol passes through the BBB by means of passive transcellular diffusion due to its high permeability.[98–102]

The ability to pass through the BBB is closely related to the molecular properties of the drug. The most important characteristics are the role of molecular weight, lipophilicity, polarity, and hydrogen-bonding capacity during the transport process. These features can be compared with the Rules of Lipinski used for oral absorption. It should be noted that all molecular properties of the drug should be considered collectively, rather than just a single characteristic, during transport across the BBB.[31,93,100,103–107] In general, small molecules with molecular weight less than 400–500 Da are suitable and have a better potential to cross the BBB.[93,108,109] The permeability of the BBB can be indicated by using the logarithm of the octanol/water partition coefficient (Log P), which provides information about the lipophilicity of a molecule. In the 1970s, it was reported that lipophilic molecules can easily move from the blood to the EC membrane, as seen with diazepam, which has a rapid transport into the brain due to its lipophilic nature.[110,111] Besides, hydrophilicity and brain drug penetration are inversely proportional; the increase in polarity reduces the penetration of drugs into the brain.[103,106,112] Additionally hydrogen-bonding capacity and interactions are believed to play an active role in brain drug penetration and distribution.[103,113]

2.2.2 Endocytosis

The endocytic process can work in two ways; either through the fluid phase (pinocytosis) or via receptors (absorptive endocytosis). In the fluid-phase endocytosis process, the components are first enclosed in the membrane of the cell, transported across the cell membrane via vesicles through the cell, and released on the other side of the cell.[114]

In receptor-mediated endocytosis process, specific receptors on the luminal membrane are used to facilitate the uptake of hormones, enzymes, peptides, growth factors, and other macromolecular drugs to the brain.[115–119]

2.2.3 Active Transport

The BBB expresses several different transport pathways to control the permeation of molecules from the blood into the brain. Molecular transporters are integral membrane proteins, which generally transport the drug into and out of the cell against the concentration gradient of the drug. There are two groups of active transporters: adenosine triphosphate (ATP)–binding cassette (ABC) transporters and solute carrier (SLC) transporters.[120–122] The primary active ABC transporters are involved in different efflux events of P-glycoprotein (P-gp), breast cancer resistance protein (BCRP), and multidrug resistance proteins (MRPs). P-gp is responsible for the immediate pumping out of the majority of drugs and xenobiotics into the blood and thus reduces the total brain penetration.[123–125] SLC transporters or secondary active transporters are membrane proteins that transport solutes, such as drugs, metabolites, and ions, across the membranes using various energy coupling mechanisms. As most influx transporters belong to the SLC family, they are responsible for the uptake of a wide range of substrates, such as amino acids, nucleosides, glucose, fatty acids, minerals, and vitamins.[126–129]

Before investigating the drug distribution in the brain, PK parameters of classic (conventional) drugs and CNS drugs need to be compared with each other to understand the drug distribution and drug penetration rate and extent into the brain. To this end, Reichel's table[3] (Table 4.2) shows important information about CNS drug research. The analogy is related to the principal considerations of rate, extent, and distribution and is based on the concept that unbound concentrations drive the pharmacological drug action.

2.3 Rate of Drug Delivery to the Brain

Cerebral blood flow rate (F) and drug permeability can affect the CNS penetration rate.[31] However, the CNS penetration rate, independent of the drug amount delivered to the brain, does not guarantee a consistent drug concentration in the brain.[30,130,131] CSF changes that affect the perfusion can also affect the penetration rate.[132]

Table 4.2 Conceptual Analogy of "Classic" Pharmacokinetics (PK) and Central Nervous System (CNS) PK of the New Concept of CNS Penetration and Distribution

	"Classic" PK	CNS PK
1. Rate	"Elimination" clearance $CL = \frac{Dose}{AUC_{plasma}}$	"Uptake" clearance $K_{in} = \frac{A_{brain}}{AUC_{plasma}}$
2. Extent	Extent of oral bioavailability $F = \frac{AUC_{po}}{AUC_{iv}}$	Extent of brain uptake $K_{p,brain} = \frac{AUC_{brain}}{AUC_{plasma}}$
3. Distribution	Concept of total versus unbound concentrations V_{ss}, $f_{u,plasma}$, and $C_{u,plasma}$	$V_{u,brain}$, $f_{u,brain}$, and $C_{u,brain}$
4. Half-life	Half-life of elimination $t_{1/2,el} = \frac{\ln 2 \times V_{ss}}{CL}$	Half-life to equilibrium $t_{1/2,eq,in} = \frac{\ln 2 \times V_{u,brain}}{PS \times f_{u,brain}}$

From Reichel A. Addressing central nervous system (CNS) penetration in drug discovery: basics and implications of the evolving new concept. Chem Biodivers. *2009;6(11):2030–2049.*[3]

The brain uptake index (BUI) or the carotid artery injection method provides data for the total concentration of the drug in the brain tissues at initial time points (5–15 s)[36,133] and is calculated by Eq. (4.1). The relative uptake of a compound compared to the reference is represented by the BUI value.[134–136] In this equation BUI value is 100 for water, ^{3}H-labeled water (i.e., a sodium chloride solution in ^{3}H-labeled water), and ^{14}C-labeled compound.

$$BUI = 100 \times \frac{\left(^{14}C/^{3}H\right)\ \text{tissue}}{\left(^{14}C/^{3}H\right)\ \text{saline}} \qquad (4.1)$$

Permeability is related to the transport rate of drugs across the BBB and its measurement techniques are similar to those used in gastrointestinal tract absorption studies. The permeability of small molecules may vary in gastrointestinal tract permeability measurement methods due to the membrane properties and the physicochemical properties of the drug. Membrane characteristics of the BBB are therefore very important in CNS permeability as well. Although paracellular diffusion is important during drug penetration in peripheral tissues, this diffusion is limited due to the complex structure of the TJs in the CNS.[86,137]

The rate of drug penetration into the brain is estimated by the product of the BBB permeability surface area (PS) and unidirectional influx clearance from blood to the brain (CL_{in} and K_{in}).[138–140] Both parameters are measured in microliter per minute per gram brain (μL/min/g brain) or microliter per second per gram brain (μL/s/g brain).[103,141–144] Permeability measurements, conducted in situ or in cell models in vitro, provide an estimation of the unidirectional transport rate of a compound across the BBB. Moreover, in vitro, in situ, in vivo, and in silico methods are widely used in studying the permeability of the BBB.[141,145] One of the methods used for determining the CNS penetration rate depends on the determination of the drug amount in the brain following systemic administration.[12]

Eq. (4.2) can be used to determine K_{in}, where A_{brain} represents the drug amount in the brain and AUC_{plasma} represents the AUC from time zero to the time of the last measurable concentration (t).

$$K_{in} = \frac{A_{brain}}{AUC_{plasma}} \tag{4.2}$$

A relationship can be defined between PS and K_{in} based on the Renkin–Crone equation, which applies basic principles of capillary flow [Eqs. (4.3) and (4.4)].[48,140,145] The F value in the equation represents the cerebral blood flow rate or perfusion flow rate.[61] The PS product is affected by both active influx and efflux transporters and provides information about the unbound drug fraction in the brain.[24,102]

$$K_{in} = F \times \left(1 - e^{PS/F}\right) \tag{4.3}$$

Where PS is calculated using Eq. (4.4).

$$PS = -F \times \ln\left(1 - \frac{K_{in}}{F}\right) \tag{4.4}$$

Nowadays, in vitro assays, such as cell monolayers or membranes, are used for determination of the apparent permeability coefficient (P_{app}). P_{app} of a drug is independent of the design of the experiment, for this reason it is extensively used to compare permeability parameters between experiments. P_{app} is calculated by Eq. (4.5),[146–151] where dQ/dt represents the change of the drug amount per time, C_0 represents the initial drug concentration in the donor compartment, and S represents the surface area of the monolayer.

$$P_{app} = \frac{dQ}{dt} \times \frac{1}{C_0 \times S} \tag{4.5}$$

2.4 Extent of Drug Delivery to the Brain

The ratio of the total brain/plasma concentration ($K_{p,brain}$, logBB) and the ratio of unbound brain/unbound plasma concentration ($K_{p,uu,brain}$) are being used to calculate the extent of drug delivery to the brain. $K_{p,brain}$ is a commonly used parameter to evaluate brain penetration [Eq. (4.6)],[19,152] where $AUC_{tot,brain}$ and

$AUC_{tot,plasma}$ represent the area under curve for total (unbound and bound) concentrations in brain and plasma, respectively.

$$K_{p,brain} = \frac{AUC_{tot,brain}}{AUC_{tot,plasma}} \quad (4.6)$$

Factors affecting $K_{p,brain}$ are nonspecific binding of drug to plasma proteins, nonspecific binding of drug to brain tissue, delivery rate, and degree from the BBB.[9,11,14,17,18,153–155] Meanwhile, some authors assume that $K_{p,brain}$ does not reflect the BBB permeability. It is believed that the relative drug exposure in the brain is not reflected by this parameter because it is affected by the sensitivity of the drug to plasma proteins and brain tissue. Thus, the brain/blood partitioning represents an inert partitioning process of a drug into a lipid material,[156] which is caused by the relative binding affinity of compounds to the proteins and lipid contents of the brain and blood.[24]

$K_{p,uu,brain}$ partition coefficient is the ratio of unbound drug concentration in brainISF to that in the plasma [Eq. (4.7)].[154,157–160]

$$K_{p,uu,brain} = \frac{AUC_{u,brain\ ISF}}{AUC_{u,plasma}} \quad (4.7)$$

Parameters in Eq. (4.7) are given in Eqs. (4.8) and (4.9).

$$AUC_{u,brain\ ISF} = AUC_{tot,brain} \times f_{u,brain} \quad (4.8)$$

$$AUC_{u,plasma} = AUC_{tot,plasma} \times f_{u,plasma} \quad (4.9)$$

Where $AUC_{u,brainISF}$ represents AUC for unbound drug in brainISF, $AUC_{u,plasma}$ represents AUC for unbound drug in plasma, $f_{u,brain}$ represents the unbound drug fraction in brain homogenate, and $f_{u,plasma}$ represents the unbound drug fraction in the plasma. $K_{p,uu,brain}$ can also be given as a function of AUC and f_u values as shown in Eq. (4.10).[160]

$$K_{p,uu,brain} = \frac{AUC_{tot,brain}}{AUC_{tot,plasma}} \times \frac{f_{u,brain}}{f_{u,plasma}} \quad (4.10)$$

Finally, $K_{p,uu,brain}$ can be calculated from K_p, $f_{u,plasma}$, and $f_{u,brain}$ as shown in Eq. (4.11).[18,24,161–163]

$$K_{p,uu,brain} = K_{p,brain} \times \frac{f_{u,brain}}{f_{u,plasma}} \quad (4.11)$$

The most closely related parameter to a drug's pharmacodynamic (PD) profile is the $K_{p,uu,brain}$ in a condition where receptors are facing the brainISF. Furthermore, the $K_{p,uu,brain}$ value is not influenced by plasma protein binding and brain parenchymal binding, and gives a concrete value of the net result of passive and active transport across the BBB. According to Eq. (4.11), it is clear that a low $K_{p,uu,brain}$ value can be due to low nonspecific binding in plasma, a high binding in the brain, or a low $K_{p,brain}$ value. $K_{p,uu,brain}$ is the ratio of influx clearance to efflux clearance and can be expressed by Eq. (4.12). Among several methods, brain microdialysis is a direct approach to determine the free brain ($K_{p,uu,brain}$) concentration.[24,163–165]

$$K_{p,uu,brain} = \frac{C_{u,ss,brainISF}}{C_{u,ss,plasma}} = \frac{CL_{in}}{CL_{out}} \tag{4.12}$$

$$K_{p,uu,brain} = \frac{CL_{passive} + CL_{act\_uptake} - CL_{act\_efflux}}{CL_{passive} - CL_{act\_uptake} + CL_{act\_efflux} + CL_{bulk\_flow} + CL_{metabolism}} \tag{4.13}$$

Where $C_{u,ss,brainISF}$ is the unbound steady state concentration in brainISF, $C_{u,ss,plasma}$ is the unbound steady state drug concentration in plasma, and $CL_{passive}$, CL_{uptake}, and CL_{efflux} represent passive diffusional, active uptake, and efflux transport clearance at the BBB, respectively. CL_{bulk} is the bulk flow of ISF from brain to CSF and $CL_{metabolism}$ is the brain metabolic clearance at the BBB.[30,33,163,166,167]

Eq. (4.13) suggests that active efflux, brain metabolic clearance, and ISF bulk flow clearance can decrease the extent of brain penetration. For drugs with low permeability, CL_{bulk} can play an efficient role in decreasing $K_{p,uu,brain}$. The estimated bulk flow clearance is approximately 0.2–0.3 μL/min/g brain.[33,168] Additionally, $CL_{metabolism}$, can play a significant role in reducing $K_{p,uu,brain}$ as well.[33,166,167]

A $K_{p,uu,brain}$ close to unity suggests passive diffusion with the similar influx and efflux clearance rates, while a $K_{p,uu,brain}$ greater than unity suggests that influx processes are more effective than efflux transport, metabolism, and bulk flow processes. When the extent of brain penetration is lower than unity, elimination processes, such as active efflux transport, metabolism within the brain parenchyma, or bulk flow processes play a significant role.[24,163–165,169]

3 INTRABRAIN DISTRIBUTION

To determine the intrabrain drug distribution, distribution volume of unbound drug in brain ($V_{u,brain}$) and $f_{u,brain}$ parameters need to be measured. Microdialysis and brain slice techniques are used to determine the $V_{u,brain}$ value, and brain homogenate technique is used to determine the $f_{u,brain}$ value.[157,170–173] The critical parameter obtained by the microdialysis technique is $V_{u,brain}$, which is calculated from Eq. (4.14).

$$V_{u,brain} = \frac{A_{tot,brain\_incl\_blood} - V_{blood} \times C_{tot,blood}}{C_{u,brainISF}} \tag{4.14}$$

In this equation $A_{tot,brain\_incl\_blood}$ represents the amount of drug per gram brain tissue (including blood), V_{blood} represents the volume of blood brain tissue, $C_{tot,blood}$ represents the total concentration of the drug in blood, and $C_{u,brainISF}$ represents the brainISF concentration, which reflects the unbound drug.[24,173]

$A_{tot,brain\_incl\_blood}$ reflects a predetermined amount of drug that is dissolved in $V_{u,brain}$, representing an apparent volume. The inverse value of the latter is also known as the free drug fraction in the brain. The lowest value of the apparent volume is the actual volume of the brainISF, that is, 0.2 mL/g brain.[84,174] The latter is the case if the drug is only distributed in the interstitial space. A higher $V_{u,brain}$ value can be obtained if a greater part of the drug is distributed in the brain cells, in particular when it is bound to the cell content.

According to the brain slice technique, $V_{u,brain}$ can be calculated using Eq. (4.15), where A_{slice} is the amount of drug per gram of slice and C_{buffer} is the concentration of drug in the buffer.[170,172,175,176]

$$V_{u,brain} = \frac{A_{slice}}{C_{buffer}} \tag{4.15}$$

Especially in drug delivery research, the brain homogenate technique is extensively used in the determination of the $f_{u,brain}$ value. Briefly, fresh or frozen brain homogenate is diluted in phosphate-buffered saline (PBS), and subsequently equilibrated across a dialysis membrane.[9,30,177] The $f_{u,brain}$ can be calculated by the following equation:

$$\text{Undiluted\_}f_{u,brain} = \frac{1/D}{\left(\left(1/f_{u,D}\right)-1\right)+1/D} \tag{4.16}$$

Where in Eq. (4.16), D represents the dilution factor of the brain homogenate and $f_{u,D}$ represents the unbound drug fraction in diluted brain homogenate.

$$K_{p,uu,cell} = f_{u,brain} \times V_{u,brain} \tag{4.17}$$

$V_{u,brain}$ is determined by brain slice experiments, and $f_{u,brain}$ is determined by equilibrium dialysis of brain homogenates. There is a lack of direct determination technique for the intracellular unbound drug concentration. For this reason, the combination of the results of brain slice and homogenate techniques are used to estimate the intracellular distribution.[5,154,158,160,176] $K_{p,uu,cell}$ describes the ICF/ISF unbound drug concentration ratio and can be calculated using Eq. (4.18).[154,158,160,172]

$$K_{p,uu,cell} = \frac{C_{u,cell}}{C_{u,brain\ ISF}} \tag{4.18}$$

$C_{u,brainISF}$ can be calculated using Eq. (4.19).

$$C_{u,brainISF} = C_{u,plasma} \times K_{p,uu,brain} \tag{4.19}$$

Subsequently, unbound (free) drug concentration in intracellular compartments of the brain ($C_{u,cell}$) can be determined by Eq. (4.20).[178]

$$C_{u,cell} = C_{u,plasma} \times K_{p,uu,brain} \times K_{p,uu,cell} \tag{4.20}$$

Actually, the $K_{p,uu,cell}$ value reflects the $C_{u,cell}$, which would be preferred by drugs targeting intracellular CNS.[5,160,176] Besides, Liu et al. found a way to quantitate the brain entering speed of a compound by using the intrinsic brain equilibrium half-life ($t_{½\ eq,in}$).[31]

$$t_{1/2\ eq,in} = \frac{V_{brain} \times \ln 2}{PS \times f_{u,brain}} \tag{4.21}$$

In Eq. (4.21), where V_{brain} is the physiological volume of the brain tissue. The time required for reaching the brain equilibrium is determined by both the permeability across the BBB and the binding affinity to the brain tissue.[31,144,179,180]

4 CONCLUSIONS

Complex challenges have emerged in the discovery and development of effective and safe CNS drugs. To link a quantitative relation among dose, exposure, and efficacy of a drug, the related experiments with regard to the total brain concentrations, BBB permeability, and relative binding affinity to the brain tissue need to be considered and evaluated as a whole picture.

Recent findings have proved that it is important to investigate the fraction of unbound drug in the brain. To this end, it is crucial to focus on determining the free drug concentration in the brain during the development of CNS drugs. The latter is especially of importance to ensure that the target site is reached. Besides the unbound drug fraction, considering other parameters in in vivo and in vitro studies, as well as incorporating brain penetration and distribution in a suitable model, can result in the development of new compounds with optimal physicochemical, pharmacological, and PK characteristics, which will lead to increase of clinical success.

Abbreviations

$\mathbf{A_{brain}}$	Amount of drug in the brain
$\mathbf{A_{slice}}$	Amount of drug per gram of brain slice
$\mathbf{A_{tot,brain\_incl\_blood}}$	Amount of drug per gram brain tissue, including blood
ABC transporters	ATP-binding cassette transporters
ADME	Absorption, distribution, metabolism, and excretion

ATP	Adenosine triphosphate
AUC_{plasma}	Area under the curve in plasma
$AUC_{tot,brain}$	Area under the curve for total (unbound and bound) concentrations in brain
$AUC_{tot,plasma}$	Area under the curve for total (unbound and bound) concentrations in plasma
$AUC_{u,brainISF}$	Area under the curve for unbound drug in brainISF
$AUC_{u,plasma}$	Area under the curve for unbound drug in plasma
BBB	Blood–brain barrier
BCRP	Breast cancer resistance protein
BCSFB	Blood–cerebrospinal fluid barrier
BUI	Brain uptake index
C_0	Initial drug concentration in the donor compartment
C_{buffer}	Concentration of drug in the buffer
$C_{tot,blood}$	Total drug concentration in blood
$C_{u,brainISF}$	Unbound drug concentration in brainISF
$C_{u,cell}$	Unbound drug concentration in intracellular compartments of the brain
$C_{u,ss,brainISF}$	Unbound steady state drug concentration in brainISF
$C_{u,ss,plasma}$	Unbound steady state drug concentration in plasma
CL_{bulk}	Bulk flow of ISF from brain to CSF
CL_{efflux}	Efflux transport clearance at the BBB
CL_{in} (K_{in})	Unidirectional influx clearance from blood to the brain
$CL_{metabolism}$	Brain metabolic clearance at the BBB
$CL_{passive}$	Passive diffusional clearance at the BBB
CL_{uptake}	Active uptake clearance at the BBB
CNS	Central nervous system
CSF	Cerebrospinal fluid
D	Dilution factor of the brain homogenate
dQ/dt	Change of drug amount per time
ECs	Endothelial cells
F	Cerebral blood flow rate or perfusion flow rate
$f_{u,brain}$	Unbound drug fraction in brain homogenate
$f_{u,D}$	Unbound drug fraction in diluted brain homogenate
$f_{u,plasma}$	Unbound drug fraction in plasma
ISF	Brain interstitial fluid
$K_{p,brain}$ (logBB)	Ratio of the total brain/plasma concentration
$K_{p,uu,brain}$	Ratio of unbound brain/unbound plasma concentration
$K_{p,uu,cell}$	ICF/ISF unbound drug concentration ratio
Log P	Logarithm of the octanol/water partition coefficient
mAbs	Monoclonal antibodies
MRPs	Multidrug resistance proteins
P_{app}	Apparent permeability coefficient
PBS	Phosphate-buffered saline
PD	Pharmacodynamic
P-gp	P-glycoprotein
PK	Pharmacokinetics
PS	Permeability surface area product
S	Surface area of the monolayer
SLC transporters	Solute carrier transporters

$t_{1/2\,eq,in}$ Intrinsic brain equilibrium half-life
TJs Tight junctions
V_{blood} Volume of blood in brain tissue
$V_{u,brain}$ Distribution volume of unbound drug in brain

References

1. Hurko O, Ryan J. Translational research in central nervous system drug discovery. *Neurotherapeutics*. 2005;2(4):671–682.
2. Pangalos MN, Schechter LE, Hurko O. Drug development for CNS disorders: strategies for balancing risk and reducing attrition. *Nat Rev Drug Discov*. 2007;6(7):521–532.
3. Reichel A. Addressing central nervous system (CNS) penetration in drug discovery: basics and implications of the evolving new concept. *Chem Biodivers*. 2009;6(11):2030–2049.
4. World Health Organization. The World Health Report 2001: Mental Health: New Understanding, New Hope. 2001.
5. Barchet TM, Amiji MM. Challenges and opportunities in CNS delivery of therapeutics for neurodegenerative diseases. *Expert Opin Drug Deliv*. 2009;6(3):211–225.
6. Wittchen HU, Jacobi F, Rehm J, Gustavsson A, Svensson M, Jönsson B, et al. The size and burden of mental disorders and other disorders of the brain in Europe 2010. *Eur Neuropsychopharmacol*. 2011;21(9):655–679.
7. Pardridge WM. Drug and gene delivery to the brain: the vascular route. *Neuron*. 2002;36(4):555–558.
8. Reichel A. The Role of blood-brain barrier studies in the pharmaceutical industry. *Curr Drug Metab*. 2006;7(2):183–203.
9. Summerfield SG, Lucas AJ, Porter RA, Jeffrey P, Gunn RN, Read KR, et al. Toward an improved prediction of human in vivo brain penetration. *Xenobiotica*. 2008;38(12):1518–1535.
10. Summerfield S, Dong K. In vitro, in vivo and in silico models of drug distribution into the brain. *J Pharmacokinet Pharmacodyn*. 2013;40(3):301–314.
11. Liu X, Cheong J, Ding X, Deshmukh G. Use of cassette dosing approach to examine the effects of P-glycoprotein on the brain and cerebrospinal fluid concentrations in wild-type and P-glycoprotein knockout rats. *Drug Metab Dispos*. 2014;42(4):482–491.
12. Liu X, Tu M, Kelly RS, Chen C, Smith BJ. Development of a computational approach to predict blood-brain barrier permeability. *Drug Metab Dispos*. 2004;32(1):132–139.
13. Garberg P, Ball M, Borg N, Cecchelli R, Fenart L, Hurst RD, et al. In vitro models for the blood–brain barrier. *Toxicol In Vitro*. 2005;19(3):299–334.
14. Summerfield SG, Read K, Begley DJ, Obradovic T, Hidalgo IJ, Coggon S, et al. Central nervous system drug disposition: the relationship between in situ brain permeability and brain free fraction. *J Pharmacol Exp Ther*. 2007;322(1):205–213.
15. Abbott NJ, Dolman DEM, Patabendige AK. Assays to predict drug permeation across the blood-brain barrier, and distribution to brain. *Curr Drug Metab*. 2008;9(9):901–910.
16. Hammarlund-Udenaes M, Bredberg U, Friden M. Methodologies to assess brain drug delivery in lead optimization. *Curr Topics Med Chem*. 2009;9(2):148–162.
17. Hammarlund-Udenaes M. Active-site concentrations of chemicals—are they a better predictor of effect than plasma/organ/tissue concentrations?. *Basic Clin Pharmacol Toxicol*. 2010;106(3):215–220.
18. Loryan I, Sinha V, Mackie C, Van Peer A, Drinkenburg W, Vermeulen A, et al. Mechanistic understanding of brain drug disposition to optimize the selection of potential neurotherapeutics in drug discovery. *Pharm Res*. 2014;31(8):2203–2219.

19. Pardridge WM. Log(BB), PS products and in silico models of drug brain penetration. *Drug Discov Today*. 2004;9(9):392–393.
20. Hakkarainen JJ, Jalkanen AJ, Kääriäinen TM, Keski-Rahkonen P, Venäläinen T, Hokkanen J, et al. Comparison of in vitro cell models in predicting in vivo brain entry of drugs. *Int J Pharm*. 2010;402(1–2):27–36.
21. Summerfield S, Jeffrey P. Discovery DMPK: changing paradigms in the eighties, nineties and noughties. *Exp Opin Drug Discov*. 2009;4(3):207–218.
22. Reichel A, Integrated approach to optimizing CNS penetration in drug discovery: from the old to the new paradigm and assessment of drug–transporter interactions. Hammarlund-Udenaes M, de Lange ECM, Thorne RG, eds. *Drug Delivery to the Brain. AAPS Advances in the Pharmaceutical Sciences Series*, vol. 10. New York, NY: Springer; 2014:339–374.
23. Wold S. Validation of QSAR's. *Quant Struct Activity Relation*. 1991;10(3):191–193.
24. Hammarlund-Udenaes M, Fridén M, Syvänen S, Gupta A. On the rate and extent of drug delivery to the brain. *Pharm Res*. 2008;25(8):1737–1750.
25. Hammarlund-Udenaes M, Lange ECMD, Thorne RG. *Drug delivery to the brain physiological concepts, methodologies and approaches*. Verlag New York: Springer; 2014:[pp. 130–131].
26. Reichel A, Pharmacokinetics of CNS Penetration. Di L, Kerns EH, eds. *Blood-Brain Barrier in Drug Discovery: Optimizing Brain Exposure of CNS Drugs and Minimizing Brain Side Effects for Peripheral Drugs* New York: John Wiley & Sons; 2015:10–16.
27. Pardridge WM. CNS drug design based on principles of blood-brain barrier transport. *J Neurochem*. 1998;70(5):1781–1792.
28. Kreuter J. Transport of drugs across the blood-brain barrier by nanoparticles. *Curr Med Chem Central Nervous System Agent*. 2002;2(3):241–249.
29. Begley DJ. Delivery of therapeutic agents to the central nervous system: the problems and the possibilities. *Pharmacol Ther*. 2004;104(1):29–45.
30. Liu X, Chen C. Strategies to optimize brain penetration in drug discovery. *Curr Opin Drug Discov Dev*. 2005;8(4):505–512.
31. Liu X, Smith BJ, Chen C, Callegari E, Becker SL, Chen X, et al. Use of a physiologically based pharmacokinetic model to study the time to reach brain equilibrium: an experimental analysis of the role of blood-brain barrier permeability, plasma protein binding, and brain tissue binding. *J Pharmacol Exp Ther*. 2005;313(3):1254–1262.
32. Begley D, Brightman M, Structural and functional aspects of the blood-brain barrier. Prokai L, Prokai-Tatrai K, eds. *Peptide Transport and Delivery into the Central Nervous System. Progress in Drug Research*, 61. Basel: Birkhäuser; 2003:39–78.
33. Abbott NJ. Evidence for bulk flow of brain interstitial fluid: significance for physiology and pathology. *Neurochem Int*. 2004;45(4):545–552.
34. Abbott NJ. Prediction of blood–brain barrier permeation in drug discovery from in vivo, in vitro and in silico models. *Drug Discov Today*. 2004;1(4):407–416.
35. Abbott NJ, Ronnback L, Hansson E. Astrocyte-endothelial interactions at the blood-brain barrier. *Nat Rev Neurosci*. 2006;7(1):41–53.
36. Cecchelli R, Berezowski V, Lundquist S, Culot M, Renftel M, Dehouck M-P, et al. Modelling of the blood-brain barrier in drug discovery and development. *Nat Rev Drug Discov*. 2007;6(8):650–661.
37. Abbott NJ. Blood–brain barrier structure and function and the challenges for CNS drug delivery. *J Inherit Metab Dis*. 2013;36(3):437–449.
38. Bicker J, Alves G, Fortuna A, Falcão A. Blood–brain barrier models and their relevance for a successful development of CNS drug delivery systems: a review. *Eur J Pharm Biopharm*. 2014;87(3):409–432.

39. Wong AD, Ye M, Levy AF, Rothstein JD, Bergles DE, Searson PC. The blood-brain barrier: an engineering perspective. *Front Neuroeng.* 2013;6:7.
40. Reiber H, Felgenhauer K. Protein transfer at the blood cerebrospinal fluid barrier and the quantitation of the humoral immune response within the central nervous system. *Clin Chim Acta.* 1987;163(3):319–328.
41. Segal M. The choroid plexuses and the barriers between the blood and the cerebrospinal fluid. *Cell Mol Neurobiol.* 2000;20(2):183–196.
42. Pardridge WM. Drug transport across the blood-brain barrier. *J Cereb Blood Flow Metab.* 2012;32(11):1959–1972.
43. Cutler RWP, Page L, Galicich J, Watters GV. Formation and absorption of cerebrospinal fluid in man. *Brain.* 1968;91(4):707–720.
44. Davson H. The cerebrospinal fluid. In: Lajtha A, ed. *Handbook of Neurochemistry.* United States: Springer; 1969:23–48.
45. Snyder W, Cook M, Nasset E, Karhausen L, Howells GP, Tipton I. Report of the task group on reference man. *Int Comm Radiol Protect.* 1974;23:112.
46. Davies B, Morris T. Physiological parameters in laboratory animals and humans. *Pharm Res.* 1993;10(7):1093–1095.
47. Deo AK, Theil FP, Nicolas JM. Confounding parameters in preclinical assessment of blood-brain barrier permeation: an overview with emphasis on species differences and effect of disease states. *Mol Pharm.* 2013;10(5):1581–1595.
48. Crone C. The permeability of capillaries in various organs as determined by use of the 'indicator diffusion' method. *Acta Physiol Scand.* 1963;58(4):292–305.
49. Gross P, Sposito N, Pettersen S, Fenstermacher J. Differences in function and structure of the capillary endothelium in gray matter, white matter and a circumventricular organ of rat brain. *J Vascular Res.* 1986;23(6):261–270.
50. Wong A, Ye M, Levy A, Rothstein J, Bergles D, Searson PC. The blood-brain barrier: an engineering perspective. *Front Neuroeng.* 2013;6:7.
51. Bradbury M. *The Concept of a Blood-Brain Barrier.* New York, NY: Wiley; 1979.
52. Redzic Z. Molecular biology of the blood-brain and the blood-cerebrospinal fluid barriers: similarities and differences. *Fluids Barriers CNS.* 2011;8(1):3.
53. Ganesh K, Quentin RS, Mitsuhiko H, Jagan P, Mark WD. Brain uptake, pharmacokinetics, and tissue distribution in the rat of neurotoxic *N*-butylbenzenesulfonamide. *Toxicol Sci.* 2007;97(2):253–264.
54. Davson H, Segal MB. *Physiology of the CSF and Blood-Brain Barriers.* Boca Raton, FL: CRC Press; 1996.
55. Misra A, Ganesh S, Shahiwala A, Shah SP. Drug delivery to the central nervous system: a review. *J Pharm Pharm Sci.* 2003;6(2):252–273.
56. Cornford EM, Hyman S, Cornford ME, Landaw EM, Delgado-Escueta AV. Interictal seizure resections show two configurations of endothelial Glut1 glucose transporter in the human blood-brain barrier. *J Cereb Blood Flow Metab.* 1998;18(1):26–42.
57. Pardridge W. The blood-brain barrier and neurotherapeutics. *Neurotherapeutics.* 2005;2(1):1–2.
58. Reinoso RF, Telfer BA, Rowland M. Tissue water content in rats measured by desiccation. *J Pharmacol Toxicol Methods.* 1997;38(2):87–92.
59. Begley DJ. In: Begley DJ, Bradbury MW, Kreuter J, eds. *Transport to the Brain.* New York, NY: Dekker; 2000.
60. Kimelberg HK. Water homeostasis in the brain: basic concepts. *Neuroscience.* 2004;129(4):851–860.

61. Deo AK, Theil F-P, Nicolas J-M. Confounding parameters in preclinical assessment of blood–brain barrier permeation: an overview with emphasis on species differences and effect of disease states. *Mol Pharm*. 2013;10(5):1581–1595.
62. Westerhout J, Danhof M, De Lange ECM. Preclinical prediction of human brain target site concentrations: considerations in extrapolating to the clinical setting. *J Pharm Sci*. 2011;100(9):3577–3593.
63. Johanson CE, Duncan 3rd JA, Klinge PM, Brinker T, Stopa EG, Silverberg GD. Multiplicity of cerebrospinal fluid functions: new challenges in health and disease. *Cerebrospinal Fluid Res*. 2008;5:10.
64. Cserr H. Potassium exchange between cerebrospinal fluid, plasma, and brain. *Am Jof Physiol*. 1965;209(6):1219–1226.
65. Harnish PO, Samuel K. Reduced cerebrospinal fluid production in the rat and rabbit by diatrizoate ventriculocisternal perfusion. *Invest Radiol*. 1988;23(7):534–536.
66. Suzuki H, Sawada Y, Sugiyama Y, Iga T, Hanano M. Saturable transport of cimetidine from cerebrospinal fluid to blood in rats. *J Pharmacobiodyn*. 1985;8(1):73–76.
67. Stange K, Greitz D, Ingvar M, Hindmarsh T, Sollevi A. Global cerebral blood flow during infusion of adenosine in humans: assessment by magnetic resonance imaging and positron emission tomography. *Acta Physiol Scand*. 1997;160(2):117–122.
68. Eyal S, Ke B, Muzi M, Link JM, Mankoff DA, Collier AC, et al. Regional P-glycoprotein activity and inhibition at the human blood–brain barrier as imaged by positron emission tomography. *Clin Pharmacol Ther*. 2010;87(5):579–585.
69. Harashima H, Sawada Y, Sugiyama Y, Iga T, Hanano M. Analysis of nonlinear tissue distribution of quinidine in rats by physiologically based pharmacokinetics. *J Pharmacokin Biopharm*. 1985;13(4):425–440.
70. Linde R, Hasselbalch SG, Topp S, Paulson OB, Madsen PL. Global cerebral blood flow and metabolism during acute hyperketonemia in the awake and anesthetized rat. *J Cereb Blood Flow Metab*. 2006;26(2):170–180.
71. Todd MM, Weeks JB, Warner DS. Microwave fixation for the determination of cerebral blood volume in rats. *J Cereb Blood Flow Metab*. 1993;13(2):328–336.
72. Shockley RP, LaManna JC. Determination of rat cerebral cortical blood volume changes by capillary mean transit time analysis during hypoxia, hypercapnia and hyperventilation. *Brain Res*. 1988;454(1):170–178.
73. Pardridge WM, Fierer G. Blood-brain barrier transport of butanol and water relative to *N*-isopropyl-*p*-iodoamphetamine as the internal reference. *J Cereb Blood Flow Metab*. 1985;5(2):275–281.
74. Begley DJ, Regina A, Roux F, Rollinson C, Abbott J, Khan EU, The role of brain extracellular fluid production and efflux mechanisms in drug transport to the brain. Begley DJ, Bradbury MW, Kreuter J, eds. *The Blood-Brain Barrier and Drug Delivery to the CNS* New York: Marcel Dekker Inc; 2000:93–108.
75. Cserr HF, Cooper DN, Suri PK, Patlak CS. Efflux of radiolabeled polyethylene glycols and albumin from rat brain. *Am J Physiol*. 1981;240(4):F319–F328.
76. Nilsson C, Stahlberg F, Thomsen C, Henriksen O, Herning M, Owman C. Circadian variation in human cerebrospinal fluid production measured by magnetic resonance imaging. *Am J Physiol*. 1992;262(1):R20–R24.
77. Szentistvanyi I, Patlak CS, Ellis RA, Cserr HF. Drainage of interstitial fluid from different regions of rat brain. *Am J Physiol*. 1984;246(6):F835–F844.
78. Johansson E, Mansson S, Wirestam R, Svensson J, Petersson JS, Golman K, et al. Cerebral perfusion assessment by bolus tracking using hyperpolarized 13C. *Magn Reson Med*. 2004;51(3):464–472.

79. De Lange EC. The mastermind approach to CNS drug therapy: translational prediction of human brain distribution, target site kinetics, and therapeutic effects. *Fluids Barriers CNS*. 2013;10(1):12.

80. Westerhout J, Ploeger B, Smeets J, Danhof M, de Lange EC. Physiologically based pharmacokinetic modeling to investigate regional brain distribution kinetics in rats. *AAPS J*. 2012;14(3):543–553.

81. Kohn MI, Tanna NK, Herman GT, Resnick SM, Mozley PD, Gur RE, et al. Analysis of brain and cerebrospinal fluid volumes with MR imaging. Part I. Methods, reliability, and validation. *Radiology*. 1991;178(1):115–122.

82. Bass NH, Lundborg P. Postnatal development of bulk flow in the cerebrospinal fluid system of the albino rat: c. *Brain Res*. 1973;52:323–332.

83. Segal MB, Fluid compartments of the central nervous system. Zheng W, Chodobski A, eds. *The Blood-Cerebrospinal Fluid Barrier* Boca Raton: CRC Press; 2005:83–100.

84. Nicholson C, Phillips JM. Ion diffusion modified by tortuosity and volume fraction in the extracellular microenvironment of the rat cerebellum. *J Physiol*. 1981;321(1):225–257.

85. Woodward D, Reed D, Woodbury D. Extracellular space of rat cerebral cortex. *Am J Physiol*. 1967;212(2):367–370.

86. Phillipson M, Kaur J, Colarusso P, Ballantyne CM, Kubes P. Endothelial domes encapsulate adherent neutrophils and minimize increases in vascular permeability in paracellular and transcellular emigration. *PLoS One*. 2008;3(2):1–8.

87. Vorbrodt AW, Dobrogowska DH. Molecular anatomy of intercellular junctions in brain endothelial and epithelial barriers: electron microscopist's view. *Brain Res Rev*. 2003;42(3):221–242.

88. Bhowmik A, Khan R, Ghosh MK. Blood brain barrier: a challenge for effectual therapy of brain tumors. *BioMed Res Int*. 2015;2015:20.

89. Crone C, Christensen O. Electrical resistance of a capillary endothelium. *J Gen Physiol*. 1981;77(4):349–371.

90. Crone C, Olesen SP. Electrical resistance of brain microvascular endothelium. *Brain Res*. 1982;241(1):49–55.

91. Smith QR, Rapoport SI. Cerebrovascular permeability coefficients to sodium, potassium, and chloride. *J Neurochem*. 1986;46(6):1732–1742.

92. Hawkins BT, Egleton RD. Fluorescence imaging of blood–brain barrier disruption. *J Neurosci Methods*. 2006;151(2):262–267.

93. Fischer H, Gottschlich R, Seelig A. Blood-brain barrier permeation: molecular parameters governing passive diffusion. *J Membrane Biol*. 1998;165(3):201–211.

94. Crone C. The permeability of brain capillaries to non-electrolytes. *Acta Physiol Scand*. 1965;64(4):407–417.

95. Ferguson R, Woodbury D. Penetration of ^{14}C-inulin and ^{14}C-sucrose into brain, cerebrospinal fluid, and skeletal muscle of developing rats. *Exp Brain Res*. 1969;7(3):181–194.

96. Ohno K, Pettigrew KD, Rapoport SI. Lower limits of cerebrovascular permeability to nonelectrolytes in the conscious rat. *Am J Physiol*. 1978;235(3):H299–H307.

97. Hitchcock SA, Pennington LD. Structure–brain exposure relationships. *J Med Chem*. 2006;49(26):7559–7583.

98. Kreuter J. Nanoparticulate systems for brain delivery of drugs. *Adv Drug Deliv Rev*. 2001;47(1):65–81.

99. Gabathuler R. Approaches to transport therapeutic drugs across the blood–brain barrier to treat brain diseases. *Neurobiol Dis*. 2010;37(1):48–57.

100. Lipinski CA, Lombardo F, Dominy BW, Feeney PJ. Experimental and computational approaches to estimate solubility and permeability in drug discovery and development settings. *Adv Drug Deliv Rev*. 2012;64(suppl):4–17.

101. Chen Y, Liu L. Modern methods for delivery of drugs across the blood–brain barrier. *Adv Drug Deliv Rev*. 2012;64(7):640–665.

102. Nagpal K, Singh SK, Mishra DN. Drug targeting to brain: a systematic approach to study the factors, parameters and approaches for prediction of permeability of drugs across BBB. *Exp Opin Drug Deliv*. 2013;10(7):927–955.

103. Gratton JA, Abraham MH, Bradbury MW, Chadha HS. Molecular factors influencing drug transfer across the blood-brain barrier. *J Pharm Pharmacol*. 1997;49(12):1211–1216.

104. Kelder J, Grootenhuis PJ, Bayada D, Delbressine LC, Ploemen J-P. Polar molecular surface as a dominating determinant for oral absorption and brain penetration of drugs. *Pharm Res*. 1999;16(10):1514–1519.

105. Lipinski CA. Drug-like properties and the causes of poor solubility and poor permeability. *J Pharmacol Toxicol Methods*. 2000;44(1):235–249.

106. Abraham MH. The factors that influence permeation across the blood–brain barrier. *Eur J Med Chem*. 2004;39(3):235–240.

107. Fu X-C, Wang G-P, Shan H-L, Liang W-Q, Gao J-Q. Predicting blood–brain barrier penetration from molecular weight and number of polar atoms. *Eur J Pharm Biopharm*. 2008;70(2):462–466.

108. Pardridge W. The blood-brain barrier: bottleneck in brain drug development. *Neurotherapeutics*. 2005;2(1):3–14.

109. Seelig A. The role of size and charge for blood–brain barrier permeation of drugs and fatty acids. *J Mol Neurosci*. 2007;33(1):32–41.

110. Rapoport SI, Levitan H. Neurotoxicity of X-ray contrast media. *Am J Roentgenol*. 1974;122(1):186–193.

111. Wang R, Fu Y, Lai L. A new atom-additive method for calculating partition coefficients. *J Chem Info Comp Sci*. 1997;37(3):615–621.

112. Norinder U, Sjöberg P, Österberg T. Theoretical calculation and prediction of brain–blood partitioning of organic solutes using MolSurf parametrization and PLS statistics. *J Pharm Sci*. 1998;87(8):952–959.

113. Young RC, Mitchell RC, Brown TH, Ganellin CR, Griffiths R, Jones M, et al. Development of a new physicochemical model for brain penetration and its application to the design of centrally acting H2 receptor histamine antagonists. *J Med Chem*. 1988;31(3):656–671.

114. Mukherjee S, Ghosh RN, Maxfield FR. Endocytosis. *Physiol Rev*. 1997;77(3):759–803.

115. Abbott NJ, Romero IA. Transporting therapeutics across the blood-brain barrier. *Mol Med Today*. 1996;2(3):106–113.

116. Bickel U, Yoshikawa T, Pardridge WM. Delivery of peptides and proteins through the blood–brain barrier. *Adv Drug Deliv Rev*. 2001;46(1–3):247–279.

117. Smith MW, Gumbleton M. Endocytosis at the blood–brain barrier: from basic understanding to drug delivery strategies. *J Drug Target*. 2006;14(4):191–214.

118. Pardridge WM. Blood–brain barrier delivery. *Drug Discov Today*. 2007;12(1–2):54–61.

119. Lajoie P, Nabi IR, Lipid rafts, caveolae, and their endocytosis. Jeon KW, ed. *International Review of Cell and Molecular Biology*, vol. 282. New York: Academic Press; 2010:135–163.

120. Higgins CF, Gottesman MM. Is the multidrug transporter a flippase?. *Trends Biochem Sci*. 1992;17(1):18–21.

121. Matheny CJ, Lamb MW, Brouwer KLR, Pollack GM. Pharmacokinetic and pharmacodynamic implications of P-glycoprotein modulation. *Pharmacotherapy*. 2001;21(7):778–796.

122. Kimura Y, Kodan A, Matsuo M, Ueda K. Cholesterol fill-in model: mechanism for substrate recognition by ABC proteins. *J Bioenerg Biomembr*. 2007;39(5–6):447–452.

123. Thiebaut F, Tsuruo T, Hamada H, Gottesman MM, Pastan I, Willingham MC. Immunohistochemical localization in normal tissues of different epitopes in the multidrug transport protein P170: evidence for localization in brain capillaries and crossreactivity of one antibody with a muscle protein. *J Histochem Cytochem*. 1989;37(2):159–164.

124. Tsuji A, Terasaki T, Takabatake Y, Tenda Y, Tamai I, Yamashima T, et al. P-glycoprotein as the drug efflux pump in primary cultured bovine brain capillary endothelial cells. *Life Sci*. 1992;51(18):1427–1437.

125. Löscher W, Potschka H. Role of drug efflux transporters in the brain for drug disposition and treatment of brain diseases. *Progr Neurobiol*. 2005;76(1):22–76.

126. Jardetzky O. Simple allosteric model for membrane pumps. *Nature*. 1966;211(5052):969–970.

127. Saier MH. A functional-phylogenetic classification system for transmembrane solute transporters. *Microbiol Mol Biol Rev*. 2000;64(2):354–411.

128. Forrest LR, Krämer R, Ziegler C. The structural basis of secondary active transport mechanisms. *Biochim Biophys Acta*. 2011;1807(2):167–188.

129. Schlessinger A, Khuri N, Giacomini KM, Sali A. Molecular modeling and ligand docking for solute carrier (SLC) transporters. *Curr Topics Med Chem*. 2013;13(7):843–856.

130. Hammarlund-Udenaes M, Paalzow L, de Lange EM. Drug equilibration across the blood-brain barrier—pharmacokinetic considerations based on the microdialysis method. *Pharm Res*. 1997;14(2):128–134.

131. Syvänen S, Xie R, Sahin S, Hammarlund-Udenaes M. Pharmacokinetic consequences of active drug efflux at the blood–brain barrier. *Pharm Res*. 2006;23(4):705–717.

132. Bulat M, Živković B. Neurochemical study of the cerebrospinal fluid. In: Marks N, Rodnight R, eds. *Research Methods in Neurochemistry*. United States: Springer; 1978:57–89.

133. Hammarlund-Udenaes M, Friden M, Syvanen S, Gupta A. On the rate and extent of drug delivery to the brain. *Pharm Res*. 2008;25(8):1737–1750.

134. Oldendorf WH. Measurement of brain uptake of radiolabeled substances using a tritiated water internal standard. *Brain Res*. 1970;24(2):372–376.

135. Oldendorf WH. Lipid solubility and drug penetration of the blood brain barrier. *Exp Biol Med*. 1974;147(3):813–816.

136. Pardridge WM. Kinetics of competitive inhibition of neutral amino acid transport across the blood-brain barrier. *J Neurochem*. 1977;28(1):103–108.

137. Serlin Y, Shelef I, Knyazer B, Friedman A. Anatomy and physiology of the blood–brain barrier. *Semin Cell Dev Biol*. 2015;38:2–6.

138. Ohno K, Pettigrew K, Rapoport S. Lower limits of cerebrovascular permeability to nonelectrolytes in the conscious rat. *Am J Physiol*. 1978;235(3):H299–H307.

139. Patlak CS, Blasberg RG, Fenstermacher JD. Graphical evaluation of blood-to-brain transfer constants from multiple-time uptake data. *J Cereb Blood Flow Metab*. 1983;3(1):1–7.

140. Takasato Y, Rapoport SI, Smith QR. An in situ brain perfusion technique to study cerebrovascular transport in the rat. *Am J Physiol*. 1984;247(3):H484–H493.

141. Smith Q, A Review of blood-brain barrier transport techniques. Nag S, ed. *The Blood-Brain Barrier: Biology and Research Protocols. Methods in Molecular Medicine* New York City, USA: Humana Press; 2003:193–208.

142. Zhao R, Kalvass JC, Pollack G. Assessment of blood–brain barrier permeability using the in situ mouse brain perfusion technique. *Pharm Res*. 2009;26(7):1657–1664.

143. Dagenais C, Avdeef A, Tsinman O, Dudley A, Beliveau R. P-glycoprotein deficient mouse in situ blood-brain barrier permeability and its prediction using an in combo PAMPA model. *Eur J Pharm Sci*. 2009;38(2):121–137.

144. Hammarlund-Udenaes M, Pharmacokinetic concepts in brain drug delivery. Hammarlund-Udenaes M, de Lange ECM, Thorne RG, eds. *Drug Delivery to the Brain. AAPS Advances in the Pharmaceutical Sciences Series*, vol. 10. New York, NY: Springer; 2014:127–161.

145. Smith Q, Samala R, In Situ and in vivo animal models. Hammarlund-Udenaes M, de Lange ECM, Thorne RG, eds. *Drug Delivery to the Brain. AAPS Advances in the Pharmaceutical Sciences Series*, vol. 10. New York, NY: Springer; 2014:199–211.

146. Poulin P, Theil F-P. Prediction of pharmacokinetics prior to in vivo studies. 1. Mechanism-based prediction of volume of distribution. *J Pharm Sci*. 2002;91(1):129–156.

147. Poulin P, Theil F-P. Prediction of pharmacokinetics prior to in vivo studies. II. Generic physiologically based pharmacokinetic models of drug disposition. *J Pharm Sci*. 2002;91(5):1358–1370.

148. Rodgers T, Leahy D, Rowland M. Physiologically based pharmacokinetic modeling 1: predicting the tissue distribution of moderate-to-strong bases. *J Pharm Sci*. 2005;94(6):1259–1276.

149. Rodgers T, Rowland M. Physiologically based pharmacokinetic modelling 2: predicting the tissue distribution of acids, very weak bases, neutrals and zwitterions. *J Pharm Sci*. 2006;95(6):1238–1257.

150. Krämer SD, Lombardi D, Primorac A, Thomae AV, Wunderli-Allenspach H. Lipid-bilayer permeation of drug-like compounds. *Chem Biodivers*. 2009;6(11):1900–1916.

151. Quignot N. Modeling bioavailability to organs protected by biological barriers. *In Silico Pharmacol*. 2013;1(1):1–9.

152. Rowland M, Tozer TN. *Clinical Pharmacokinetics and Pharmacodynamics: Concepts and Applications*. Baltimore, United States: Wolters Kluwer/Lippincott Williams & Wilkins;2011; 2011:159–176.

153. Kalvass JC, Olson ER, Cassidy MP, Selley DE, Pollack GM. Pharmacokinetics and pharmacodynamics of seven opioids in P-glycoprotein-competent mice: assessment of unbound brain EC50,u and correlation of in vitro, preclinical, and clinical data. *J Pharmacol Exp Ther*. 2007;323(1):346–355.

154. Fridén M, Gupta A, Antonsson M, Bredberg U, Hammarlund-Udenaes M. In vitro methods for estimating unbound drug concentrations in the brain interstitial and intracellular fluids. *Drug Metabo Dispos*. 2007;35(9):1711–1719.

155. Watson J, Wright S, Lucas A, Clarke KL, Viggers J, Cheetham S, et al. Receptor occupancy and brain free fraction. *Drug Metab Dispos*. 2009;37(4):753–760.

156. Van de Waterbeemd H, Smith D, Jones B. Lipophilicity in PK design: methyl, ethyl, futile. *J Comp Aided Mol Design*. 2001;15(3):273–286.

157. Cory Kalvass J, Maurer TS. Influence of nonspecific brain and plasma binding on CNS exposure: implications for rational drug discovery. *Biopharm Drug Dispos*. 2002;23(8):327–338.

158. Maurer TS, DeBartolo DB, Tess DA, Scott DO. Relationship between exposure and nonspecific binding of thirty-three central nervous system drugs in mice. *Drug Metab Dispos*. 2005;33(1):175–181.

159. Summerfield SG, Jeffrey P. In vitro prediction of brain penetration—a case for free thinking?. *Exp Opin Drug Discov*. 2006;1(6):595–607.

160. Fridén M, Bergström F, Wan H, Rehngren M, Ahlin G, Hammarlund-Udenaes M, et al. Measurement of unbound drug exposure in brain: modeling of pH partitioning explains diverging results between the brain slice and brain homogenate methods. *Drug Metab Dispos*. 2011;39(3):353–362.

161. Doran A, Obach RS, Smith BJ, Hosea NA, Becker S, Callegari E, et al. The impact of P-glycoprotein on the disposition of drugs targeted for indications of the central nervous system: evaluation using the MDR1A/1B knockout mouse model. *Drug Metab Dispos*. 2005;33(1):165–174.

162. Kalvass JC, Maurer TS, Pollack GM. Use of plasma and brain unbound fractions to assess the extent of brain distribution of 34 drugs: comparison of unbound concentration ratios to in vivo P-glycoprotein efflux ratios. *Drug Metab Dispos*. 2007;35(4):660–666.

163. Fridén M, Winiwarter S, Jerndal G, Bengtsson O, Wan H, Bredberg U, et al. Structure–brain exposure relationships in rat and human using a novel data set of unbound drug concentrations in brain interstitial and cerebrospinal fluids. *J Med Chem*. 2009;52(20):6233–6243.

164. Boström E, Simonsson USH, Hammarlund-Udenaes M. In vivo blood-brain barrier transport of oxycodone in the rat: indications for active influx and implications for pharmacokinetics/pharmacodynamics. *Drug Metab Dispos*. 2006;34(9):1624–1631.

165. Gupta A, Chatelain P, Massingham R, Jonsson EN, Hammarlund-Udenaes M. Brain distribution of cetirizine enantiomers: comparison of three different tissue-to-partition coefficients: K_p, $K_{p,u}$, and $K_{p,uu}$. *Drug Metab Dispos*. 2006;34(2):318–323.

166. Cserr HF, Cooper DN, Milhorat TH. Flow of cerebral interstitial fluid as indicated by the removal of extracellular markers from rat caudate nucleus. *Exp Eye Res*. 1977;25(suppl 1):461–473.

167. Rosenberg GA, Kyner WT, Estrada E. Bulk flow of brain interstitial fluid under normal and hyperosmolar conditions. *Am J Physiol*. 1980;238(1):F42–F49.

168. Cserr HF, Patlak CS, Secretion and bulk flow of interstitial fluid. Bradbury MB, ed. *Physiology and Pharmacology of the Blood-Brain Barrier. Handbook of Experimental Pharmacology*, 103. Berlin Heidelberg: Springer; 1992:245–261.

169. Jeffrey P, Summerfield SG. Challenges for blood-brain barrier (BBB) screening. *Xenobiotica*. 2007;37(10–11):1135–1151.

170. Kakee A, Terasaki T, Sugiyama Y. Brain efflux index as a novel method of analyzing efflux transport at the blood-brain barrier. *J Pharmacol Exp Ther*. 1996;277(3):1550–1559.

171. Wang Y, Welty D. The simultaneous estimation of the influx and efflux blood-brain barrier permeabilities of gabapentin using a microdialysis-pharmacokinetic approach. *Pharm Res*. 1996;13(3):398–403.

172. Fridén M, Ducrozet F, Middleton B, Antonsson M, Bredberg U, Hammarlund-Udenaes M. Development of a high-throughput brain slice method for studying drug distribution in the central nervous system. *Drug Metab Dispos*. 2009;37(6):1226–1233.

173. Fridén M, Ljungqvist H, Middleton B, Bredberg U, Hammarlund-Udenaes M. Improved measurement of drug exposure in the brain using drug-specific correction for residual blood. *J Cereb Blood Flow Metab*. 2010;30(1):150–161.

174. Nicholson C, Syková E. Extracellular space structure revealed by diffusion analysis. *Trends Neurosci*. 1998;21(5):207–215.

175. Becker S, Liu X. Evaluation of the utility of brain slice methods to study brain penetration. *Drug Metab Dispos*. 2006;34(5):855–861.

176. Loryan I, Hammarlund-Udenaes M, Drug discovery methods for studying brain drug delivery and distribution. Hammarlund-Udenaes M, de Lange ECM, Thorne RG, eds. *Drug Delivery to the Brain. AAPS Advances in the Pharmaceutical Sciences Series*, vol. 10. New York, NY: Springer; 2014:271–316.

177. Di L, Umland JP, Chang G, Huang Y, Lin Z, Scott DO, et al. Species independence in brain tissue binding using brain homogenates. *Drug Metab Dispos*. 2011;39(7):1270–1277.

178. Liu X, Van Natta K, Yeo H, Vilenski O, Weller PE, Worboys PD, et al. Unbound drug concentration in brain homogenate and cerebral spinal fluid at steady state as a surrogate for unbound concentration in brain interstitial fluid. *Drug Metab Dispos*. 2009;37(4):787–793.

179. Kielbasa W, Stratford RE. Exploratory translational modeling approach in drug development to predict human brain pharmacokinetics and pharmacologically relevant clinical doses. *Drug Metab Dispos*. 2012;40(5):877–883.

180. Ball K, Bouzom F, Scherrmann J-M, Walther B, Declèves X. Development of a physiologically based pharmacokinetic model for the rat central nervous system and determination of an in vitro–in vivo scaling methodology for the blood–brain barrier permeability of two transporter substrates, morphine and oxycodone. *J Pharm Sci*. 2012;101(11):4277–4292.

Nose-to-Brain Drug Delivery

CHAPTER 5

Nasal Physiology and Drug Transport

Emine Sekerdag, MSc
Koç University, Istanbul, Turkey

CONTENTS

1 INTRODUCTION

Nowadays, noninvasive drug targeting to the brain is a very important, but difficult task for researchers in the field of neurological and psychiatric disorders. The blood–brain barrier (BBB) is a major obstacle for many lipophilic (MW > 600 Da) and most of the hydrophobic drugs that need to reach the brain and/or central nervous system (CNS) to show effectiveness.[1,2] Many studies have been performed to overcome and/or manipulate the BBB to meet the required conditions for drug delivery to the brain. Recently, a different approach to reach the brain that bypasses BBB is of interest, namely the nose-to-brain route.

The nose-to-brain route is known from the beginning of the last century. A discovery had been taken place at that time: the olfactory neurons, a direct connection between the outside air and the brain. For this discovery, researchers injected a dye into the ventricles of monkeys and showed presence of the dye in the nasal mucosa. From this experiment it was concluded that the cerebrospinal fluid (CSF) had found its way to the nasal mucosa by drainage via the olfactory neurons.[3,4] Considering this route of CSF to the nasal mucosa, it was believed that the opposite pathway, from nasal mucosa to the CSF, would be available as well. Several studies demonstrated the opposite pathway by nasal application of several dyes,[5–7] viruses, and metals.[8–11] This new discovery was found to be the key for circumventing the BBB and a new avenue was opened to investigate drug transport into the brain. However, researchers had many difficulties with the experimental setup due to the intrinsic sensitive properties of the brain tissue and CSF.

This chapter will focus on the anatomy and physiology of the nasal cavity and will discuss the different transport pathways after nasal drug administration.

Nanotechnology Methods for Neurological Diseases and Brain Tumors. http://dx.doi.org/10.1016/B978-0-12-803796-6.00005-8

2 ANATOMY OF THE NOSE

The nose is divided into two symmetrical nasal cavities by the nasal septum. Each halve has a surface area of approximately 75 cm^2 and is further determined by three areas: the vestibular, respiratory, and olfactory regions. The vestibular region is located at the entrance of the nasal cavity and is involved in filtering particles present in inhaled air. Moreover, the vestibular region is not involved in absorption functions. The respiratory epithelium is highly vascularized and located on the dorsal and ventral sections of the nasal cavity. With a surface area of approximately 160 cm^2 in humans, the respiratory region consists of ciliated and nonciliated columnar, goblet, and basal cells. The goblet cells produce mucus, covering the epithelium of the respiratory region, which is cleared away by the cilia. This mechanism is called the mucociliary clearance. The frequency of movement of this cilia is 1000 beats/min and can thereby move the cilia with 5 mm/min.[12]

In general, the main function of the respiratory region is drug absorption. The third region, the olfactory region, is located at the top of the nose and just below the cribriform plate of the ethmoid bone (Fig. 5.1).[13,14] This olfactory region, acts as a bridge between the cranial and nasal cavity, has a surface area of approximately 10 cm^2 (approximately 8% of the total nasal surface area), and contains olfactory receptor cells, sustentacular cells, and basal cells (Fig. 5.2). Each cell has a different function. The olfactory receptor cells are bipolar neurons that connect the olfactory bulb with the nasal cavity. These bipolar neurons have dendrites at one end terminating in the olfactory epithelium and axons on the other end extending from the soma in the epithelium of the olfactory region, which pass through the cribriform plate and terminate into the olfactory bulb. Once in the olfactory bulb, these axons connect with the synapses of juxtaglomerular neurons, tufted cells, and mitral cells in the

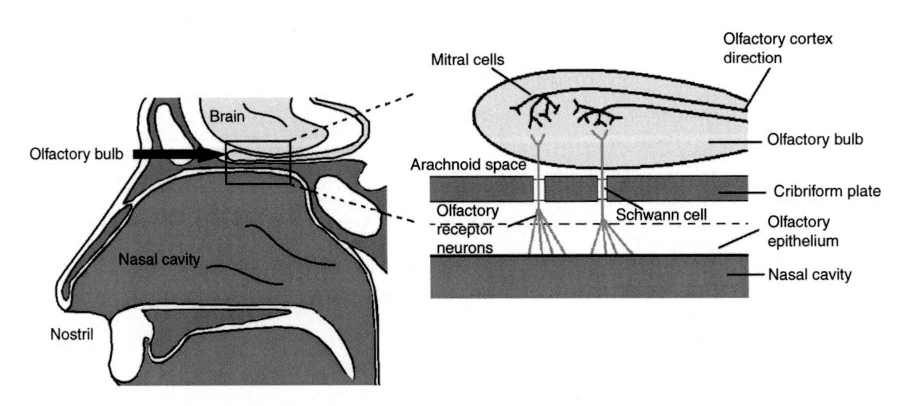

FIGURE 5.1 A schematic illustration of the human nasal cavity and the olfactory region.

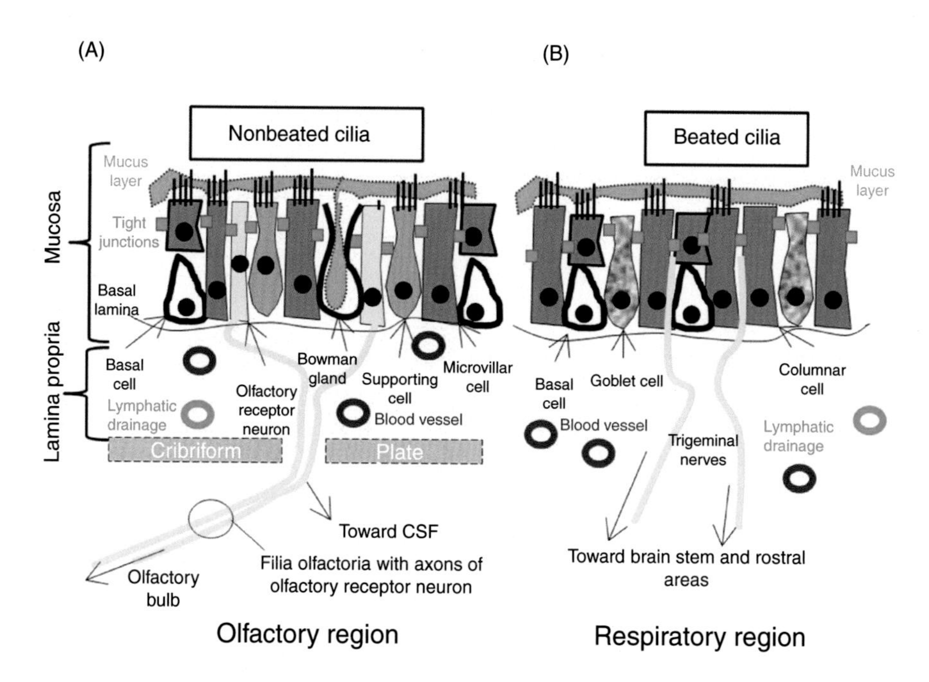

FIGURE 5.2 (A) A microscopic view of the olfactory mucosa, which represents ≤10% of the nasal cavity. (B) A microscopic view of the respiratory mucosa, which represents 80%–90% of the nasal cavity. *Image adapted from Van Woensel M et al. Formulations for intranasal delivery of pharmacological agents to combat brain disease: a new opportunity to tackle GBM?* Cancers. *2013;5(3):1020–1048.*[16]

glomeruli.[15] The latter explains why the olfactory bulb forms a potential transportation route of drugs to the brain and CSF.

Another type of cell located in the olfactory region is the sustentacular cell. These cells are columnar in shape and have microvilli at the apical side.[17] The sustentacular cells, alternated by olfactory receptor neurons in the epithelium, provide mechanical support by forming an entanglement with the cilia of the olfactory receptor neurons. A third cell type located in the olfactory region is the basal cell. These cells are small in size and have a conical shape. These cells undergo continuous mitotic activity and therefore have the ability to differentiate into olfactory receptor neurons in physiological conditions (turnover), as well as after injured conditions.[13,14,18] The continuous replacement of receptor neurons causes a "leaky" environment in the nasal epithelium, which improves the nasal drug delivery to the brain.[19]

Other structures found in the nasal cavity are Bowman's glands, blood vessels, and lymphatic vessels.[20] Furthermore, the lumen and epithelial barrier in the nasal cavity contain the enzymes exopeptidases (mono- and diamino peptidases) and endopeptidases (serine and cysteine). Exopeptidases can cleave peptides at their N- and C-terminals where endopeptidases can attack the internal bonds of peptides.[21] In addition, enzyme isoforms of CYT P450 (CYP1A, CYP2A, and CYP2E), carboxylesterases, S-transferases, and glutathione are also found in the human nasal cavity.[22]

3 NASAL TRANSPORT MECHANISMS

3.1 Mucus Layer

The first step of the absorption mechanism for a compound is passing through the mucus layer of the nasal epithelium. This mucus layer contains 95% water, 2% mucin (which is a protein that has the affinity to bind to the entering compounds and therefore hinders the transit), 1% salt, 1% albumin, lysozymes, lactoferrin, immunoglobulins, and lipids.[23] The pH of the nasal mucus is in the range of 5.5–6.5.[24] Furthermore, the epithelium of the olfactory region lacks the dynein arm (which is necessary for motion) and therefore is not involved in mucociliary clearance, in contrast to the respiratory region.[25] After diffusion through the mucus layer, compounds may be transferred through one or more transport routes with different characteristics (Fig. 5.3).

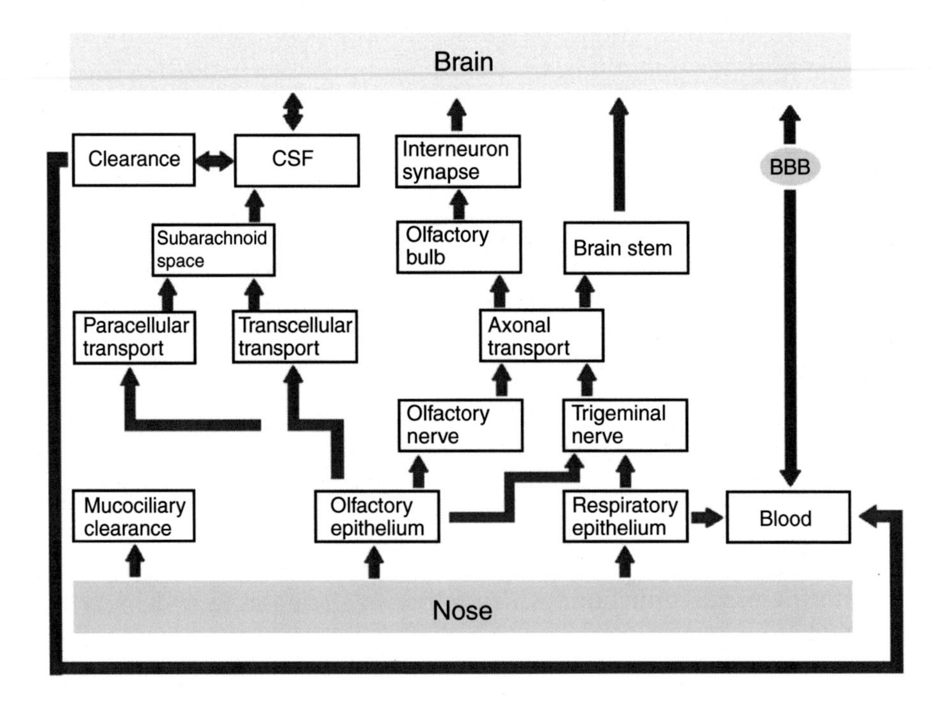

FIGURE 5.3 An overview of possible fate of drugs after nasal administration.
BBB, Blood–brain barrier; *CSF*, cerebrospinal fluid.

Table 5.1 Olfactory Pathways and Involved Transport Mechanisms

Pathways	Transport	Compounds
Paracellular route	• Rapid transport route through tight junctions (<2 h)	Low MW and/or hydrophilic
Transcellular route	• Passive diffusion • Receptor-mediated endocytosis • Fluid-phase endocytosis (hours to several days)	Large MW or lipophilic
Neuronal route	• Intercellular axonal	Dependent on the axon diameter (in humans: axon diameter is 100–700 nm)

MW, Molecular weight.

3.2 Olfactory Pathway

The olfactory pathway is a very important and potential transport route in drug delivery to the brain and/or CSF. In this pathway the drug molecules are transported to the brain parenchyma and/or CSF by the following routes: (1) paracellular route, (2) transcellular route, and (3) neuronal route[13,26,27] (Fig. 5.3 and Table 5.1).

The integrity of the nasal epithelium, including the tight junctions, adherent junctions, desmosomes, and gap junctions between the epithelial cells, decides which compounds can enter by the paracellular transport.[28] For example, the hydrophilic gap between two epithelial cells is approximately 8 Å.[29] On the other hand, researchers believe that transcellular transport can allow the transit of compounds larger than 20 nm with several transport mechanisms depending on the size of the compound.[30] After paracellular and transcellular transport, the drugs can enter the subarachnoid space and CSF, respectively. From the CSF, drug compounds can enter the brain through the CSF barrier or can be cleared away through the CSF turnover. The latter, can lead to reuptake of the drug compound from the blood into the brain through BBB pathway, depending on the nature of the compound. However for the latter, compounds need to be hydrophilic and smaller than at least 200 nm to be able to pass through the BBB, which is a great challenge for most therapeutics agents.

The neuronal pathway is believed to be the determining step of the nose-to-brain route. In this pathway the compounds are intercellularly transported by the axons, which are ensheathed by Schwann cells, to the olfactory bulb and neuron synapses of the brain. The neuronal transport is mainly controlled by the diameter of the axon. Also, perineural channels (ionic reservoir) of 10–15 nm and mesaxon pores (transit extracellular fluid) are believed to play a role in this transport mechanism.

3.3 Trigeminal Pathway

Several studies have shown that one more pathway is involved in the nose-to-brain drug delivery, namely the trigeminal pathway.[26,27] The trigeminal nerve is the largest cranial nerve and is involved in sensation and motor functions. Researchers have shown that trigeminal nerves, normally linked with the respiratory region of the nose, might also be partially branched below the olfactory epithelium. After compounds diffuse trough the nasal mucosa, they may reach the branches of the trigeminal nerves in both the respiratory and olfactory regions, and therefore can be transported by the axonal route via the brain stem.

4 FACTORS AFFECTING NOSE-TO-BRAIN TRANSPORT

Despite the existence of several routes for drug delivery to the brain, after compounds reached the brain, the drug can still be transferred back to the microcirculation by P-glycoproteins (efflux transporters of the BBB)[13,22,31,32] or eliminated from the CSF back to the blood circulation without reaching its target.[33,34]

On the other hand, several studies reported that physicochemical and physiological conditions and formulation characteristics of drugs influence intranasal drug delivery to the brain and/or CSF.[22,35] Examples of these factors are listed in Table 5.2.

The most important factors affecting the nose-to-brain route are poor bioavailability, mucociliary clearance, and enzymatic degradation. The bioavailability is poor for polar drugs (~10%) and macromolecules (~1%), which is primarily caused by low membrane permeability across the olfactory mucosa.[32] Also, tight junctions play an important role in limiting the transition of large molecules.[37,38] The mucociliary clearance, that is, the rapid clearance of the drug formulation from the nasal cavity, leads to a reduced retention (contact) time of the drug at the absorption site and therefore limits drug absorption through the olfactory mucosa. Studies revealed that the half-life time for clearance of nonbioadhesive liquid or powder formulations in the nasal cavity is between

Table 5.2 List of Several Factors Influencing Intranasal Drug Delivery[36]

Physicochemical Factors	Physiological Factors	Formulation Factors
• Molecular weight • Particle size • Chemical form • Charge • Polymorphism • Solubility and dissolution rate	• Disposition • Mucociliary clearance • Degradation • Disease condition • Temperature	• Drug concentration • Drug dose • pH • Viscosity • Buffer capacity • Administration volume

Presented factors belong to the drug compound and/or formulation.

15 and 30 min.[35,39] Furthermore, therapeutic proteins and peptides can be degraded by enzymes in the lumen of the nasal cavity or during transition through the epithelial membrane.[40] Subsequently, the latter leads to poor bioavailability. In addition, the nasal mucosa consisting of mucosa secretases and degradation enzymes contributes to the pseudofirst pass effect, which forms an enzyme barrier in the nasal cavity. This enzyme barrier reduces the amount of therapeutic proteins and peptides during nasal drug delivery.

5 PRINCIPLES OF DRUG ADMINISTRATION

Nasal drug administration has many advantages, including being noninvasive, bypassing the systemic circulation and its side effects, and bypassing direct clearance by the liver and the kidneys. Also, patient compliance, self-administration, and the possible application for chronic diseases play important roles in investigating and considering this route in the treatment of neurological disorders.

Besides the advantages, there are several issues that need to be taken into consideration during nasal drug administration regarding the experimental setup. The translation from human to laboratory animals needs to be kept in mind when designing an experiment. For example, the volume of the nasal cavity is 0.032, 0.26, and 25 cm^3 in mice, rats, and humans, respectively.[41] Also, the CSF turnover is very important, which is 1.5 h in mice and 5 h in humans. Another difference between rodent and human nasal cavities is the ratio of the olfactory region which needs to be translated with caution. The olfactory region in rats is 50% of the total nasal cavity, while in humans this is only 8%.

Potential drug candidates can be dissolved or dispersed in liquid solutions for intranasal delivery in maximal volumes of 24 μL in mice, 40–100 μL in rats, and 400 μL in humans. Hence, these formulations can also be administered in powder forms. Moreover, Charlton et al. has demonstrated the superiority of drops over sprays in nasal drug delivery.[42] In the meantime, healthcare professionals came up with potential devices, which are being implemented in nasal drug delivery to improve the drug administration.[43] Additionally, due to the deep location of the olfactory region in the nasal cavity and primary factors, such as inhalation, clearance, and gravity, the head position and retention time in this position are extremely important factors that influence loss of administered formulation during the nose-to-brain drug delivery.[44] Hereupon, in animal studies it was found that the supine position is mostly preferred for nasal drug administration.[45]

6 CONCLUSIONS

The nose-to-brain route is a potential route for delivering therapeutic agents to the brain in a noninvasive way bypassing the BBB hurdle and without systemic side effects. Although the mechanisms are not yet fully understood and validated completely, many studies revealed possible pathways in which drugs reach the brain. The axonal transport of drug compounds to the CSF and the brain are opening new perspectives to the treatment complexities concerning these disorders. The olfactory region is therefore the target zone in nasal drug delivery and many studies have been performed to investigate potential treatment candidates. In addition to many limitations in drug administration and hurdles in achieving desired drug concentrations in the brain, there are still many challenges for utilizing the nose-to-brain route before it becomes clinically relevant in the field of neurological diseases.

Abbreviations

BBB Blood–brain barrier
CNS Central nervous system
CSF Cerebrospinal fluid

References

1. Pardridge WM. Brain drug delivery and blood–brain barrier transport. *Drug Deliv.* 1993;1(2):83–101.
2. Scherrmann J-M. Drug delivery to brain via the blood–brain barrier. *Vasc Pharmacol.* 2002;38(6):349–354.
3. Jackson RT, Tigges J, Arnold W. Subarachnoid space of the CNS, nasal mucosa, and lymphatic system. *Arch Otolaryngol.* 1979;105(4):180–184.
4. Yoffey J. Passage of fluid and other substances through the nasal mucosa. *J LaryngolOtol.* 1958;72(05):377–384.
5. Arnold W, Vosteen K-H. Zur sekretorischen aktivitatder interdentalzellen des limbus spiralis. *Acta Oto-Laryngol.* 1973;75(2–6):192–202.
6. Faber WM. The nasal mucosa and the subarachnoid space. *Am J Anat.* 1937;62(1):121–148.
7. Wang H-W, Su W-Y, Wang J-Y. Retrograde axonal transport of true blue dye by the peripheral autonomic nerves in canine nasal mucosa. *Arch Oto-Rhino-Laryngol.* 1987;244(5):295–299.
8. Martinez J, et al. Experimental *Naegleria* meningoencephalitis in mice. Penetration of the olfactory mucosal epithelium by *Naegleria* and pathologic changes produced: a light and electron microscope study. *Lab Invest.* 1973;29(2):121–133.
9. Perlman S, Evans G, Afifi A. Effect of olfactory bulb ablation on spread of a neurotropic coronavirus into the mouse brain. *J Exp Med.* 1990;172(4):1127–1132.
10. Sunderman FW. Nasal toxicity, carcinogenicity, and olfactory uptake of metals. *Ann Clin Lab Sci.* 2001;31(1):3–24.
11. Tjälve H, Henriksson J. Uptake of metals in the brain via olfactory pathways. *Neurotoxicology.* 1998;20(2–3):181–195.
12. Schipper NG, Verhoef JC, Merkus FW. The nasal mucociliary clearance: relevance to nasal drug delivery. *Pharm Res.* 1991;8(7):807–814.
13. Illum L. Transport of drugs from the nasal cavity to the central nervous system. *Eur J Pharm Sci.* 2000;11(1):1–18.

14. Mistry A, Stolnik S, Illum L. Nanoparticles for direct nose-to-brain delivery of drugs. *Int J Pharm*. 2009;379(1):146–157.
15. Shipley MT, Ennis M. Functional organization of olfactory system. *J Neurobiol*. 1996;30(1): 123–176.
16. Van Woensel M, et al. Formulations for intranasal delivery of pharmacological agents to combat brain disease: a new opportunity to tackle GBM?. *Cancers*. 2013;5(3):1020–1048.
17. Morrison EE, Costanzo RM. Morphology of olfactory epithelium in humans and other vertebrates. *Microsc Res Tech*. 1992;23(1):49–61.
18. Illum L. Nasal drug delivery, possibilities, problems and solutions. *J Control Release*. 2003;87(1):187–198.
19. Caggiano M, Kauer JS, Hunter DD. Globose basal cells are neuronal progenitors in the olfactory epithelium: a lineage analysis using a replication-incompetent retrovirus. *Neuron*. 1994;13(2):339–352.
20. Mathison S, Nagilla R, Kompella UB. Nasal route for direct delivery of solutes to the central nervous system: fact or fiction?. *J Drug Target*. 1998;5(6):415–441.
21. Jadhav KR, et al. Nasal drug delivery system-factors affecting and applications. *Curr Drug Ther*. 2007;2(1):27–38.
22. Vyas TK, et al. Intranasal drug delivery for brain targeting. *Curr Drug Deliv*. 2005;2(2):165–175.
23. Kaliner M, et al. Human respiratory mucus. *J Allergy Clin Immunol*. 1984;73(3):318–323.
24. Khutoryanskiy VV. Advances in mucoadhesion and mucoadhesive polymers. *Macromol Biosci*. 2011;11(6):748–764.
25. Moran DT, et al. The fine structure of the olfactory mucosa in man. *J Neurocytol*. 1982;11(5): 721–746.
26. Thorne R, et al. Delivery of interferon-β to the monkey nervous system following intranasal administration. *Neuroscience*. 2008;152(3):785–797.
27. Thorne R, et al. Delivery of insulin-like growth factor-I to the rat brain and spinal cord along olfactory and trigeminal pathways following intranasal administration. *Neuroscience*. 2004;127(2):481–496.
28. Altner H, Altner-Kolnberger I. Freeze-fracture and tracer experiments on the permeability of the zonulae occludentes in the olfactory mucosa of vertebrates. *Cell Tissue Res*. 1974;154(1):51–59.
29. Van Itallie CM, Anderson JM. Claudins and epithelial paracellular transport. *Annu Rev Physiol*. 2006;68:403–429.
30. Conner SD, Schmid SL. Regulated portals of entry into the cell. *Nature*. 2003;422(6927): 37–44.
31. Begley DJ. Delivery of therapeutic agents to the central nervous system: the problems and the possibilities. *Pharmacol Ther*. 2004;104(1):29–45.
32. Illum L. Nasal drug delivery: new developments and strategies. *Drug Discov Today*. 2002;7(23):1184–1189.
33. Graff CL, Pollack GM. P-glycoprotein attenuates brain uptake of substrates after nasal instillation. *Pharm Res*. 2003;20(8):1225–1230.
34. Miller DS. Regulation of P-glycoprotein and other ABC drug transporters at the blood–brain barrier. *Trends Pharm Sci*. 2010;31(6):246–254.
35. Wu H, Hu K, Jiang X. *From nose to brain: understanding transport capacity and transport rate of drugs*. 2008;5(10):1159–1168.
36. Bonthagarala B, et al. Nasal drug delivery: a potential route for brain targetting. *Int J Adv Sci Res*. 2015;1(2):65–70.

37. Madara JL, Dharmsathaphorn K. Occluding junction structure-function relationships in a cultured epithelial monolayer. *J Cell Biol*. 1985;101(6):2124–2133.

38. McMartin C, et al. Analysis of structural requirements for the absorption of drugs and macromolecules from the nasal cavity. *J Pharm Sci*. 1987;76(7):535–540.

39. Soane R, et al. Evaluation of the clearance characteristics of bioadhesive systems in humans. *Int J Pharm*. 1999;178(1):55–65.

40. Minn A, et al. Drug transport into the mammalian brain: the nasal pathway and its specific metabolic barrier. *J Drug Target*. 2002;10(4):285–296.

41. Jafek BW. Ultrastructure of human nasal mucosa. *Laryngoscope*. 1983;93(12):1576–1599.

42. Charlton S, et al. Distribution and clearance of bioadhesive formulations from the olfactory region in man: effect of polymer type and nasal delivery device. *Eur J Pharm Sci*. 2007;30(3):295–302.

43. Craft S, et al. Intranasal insulin therapy for Alzheimer disease and amnestic mild cognitive impairment: a pilot clinical trial. *Arch Neurol*. 2012;69(1):29–38.

44. Hardy J, Lee S, Wilson C. Intranasal drug delivery by spray and drops. *J Pharm Pharmacol*. 1985;37(5):294–297.

45. van Den Berg MP, et al. Serial cerebrospinal fluid sampling in a rat model to study drug uptake from the nasal cavity. *J Neurosci Methods*. 2002;116(1):99–107.

CHAPTER 6

Challenges of the Nose-to-Brain Route

Emine Sekerdag, MSc
Koç University, Istanbul, Turkey

1 INTRODUCTION

The blood–brain barrier (BBB), blocks the free entrance of many therapeutic drugs into the brain, which are used in the treatment of neurological and psychiatric disorders. One approach in bypassing the BBB and reaching the brain with an effective drug concentration is the nose-to-brain route. This method is especially practical compared to currently used conventional and invasive treatment methods.[1,2] Besides, those invasive methods are inconvenient, expensive, and have a high risk on side effects.

A variety of therapeutic agents have been used for intranasal drug delivery for the treatment of neurological disorders. Neurotrophins,[3,4] neuropeptides,[5] cytokines,[6] polynucleotides,[7] and small molecules[8] are several examples in this context. In general, drug compounds in nanomolar range are preferred for intranasal delivery.[1]

Intranasal administration is performed by applying the drug in the nasal cavity, deep enough to reach the olfactory region, so that drugs can be transported by several mechanisms to the brain and/or cerebrospinal fluid (CSF). These pathways are described in Chapter 5 and are the olfactory epithelium route, the olfactory nerve route,[9] and the trigeminal nerve route.[10,11] The olfactory epithelium route is further divided into two pathways, the paracellular and transcellular transport routes. Dependent on the nasal environment and drug/formulation characteristics, one or more of the mentioned absorption/transport mechanisms will take place. Besides the transport routes, the experimental study design and pre- and post-treatment procedures need to be examined carefully to enable translation of the obtained data to a possible clinical level.

This chapter presents a general view on the current status of the nose-to-brain route in the scientific world, the faced challenges with regard to drug transport to the brain, and their analyses.

CONTENTS

Nanotechnology Methods for Neurological Diseases and Brain Tumors. http://dx.doi.org/10.1016/B978-0-12-803796-6.00006-X

2 NOSE-TO-BRAIN RESEARCH

To obtain relevant outcomes when performing nose-to-brain drug delivery research, the study design needs to meet certain criteria. When administering intranasally, the control group should be treated via the most relevant administration route to enable correct conclusions about the obtained data. Furthermore, certain sampling and analyzing techniques are necessary to achieve a complete picture of the nose-to-brain route.

First, pharmacokinetic and pharmacodynamic analyses are of great value to understand the drug effects and its transport route. Therefore, to relate pharmacokinetic effects to drug plasma levels after different administration routes, studies should compare intranasal administration with intravenous administration. Many studies were performed with this approach, but studies including placebo or no treatment are also available. In general, drug distribution in the brain and/or the CSF is determined by the CSF/plasma ratios. The actual pathway of drug transport can be retrieved from CSF/plasma ratios compared to the ratios obtained from the intravenous route. Moreover, in experimental models CSF is mainly sampled by puncturing the cisterna magna and collecting 75–150 μL in rats.[12,13] It is considered that if the CSF/plasma ratios after intranasal drug administration are significantly higher than the intravenous route, the drug is transported directly from the nose to the brain and/or CSF.[14] If the ratios are similar[15,16] or nasally applied drug CSF/plasma ratios are lower than the ratios for intravenous application, the drug is considered as not transported by the olfactory receptor neurons, but transported via the blood circulation.

Second, dissected or sliced brain tissues need to be analyzed with several techniques to prove the presence of the drug at the target site, as well as to confirm pharmacokinetic and pharmacodynamic studies. These techniques are mostly performed with radioactive labeling studies. However, high-performance liquid chromatography (HPLC) or radioimmunoassay studies could also be performed after the extraction of the drug of interest. Also, staining techniques are very useful in this approach; however, staining techniques require microscopic evaluation, which also requires the use of the most relevant control group (intravenous application). The detection of the drug of interest in the olfactory tract can indicate direct nose-to-brain transport, but it may be misleading us, as the necessary pharmacological active drug concentration in the brain should also be measured to conclude whether the drug is effective.

Although, both CSF sampling and brain tissue analysis provide a better understanding of the drug distribution in the central nervous system (CNS), these studies need a large number of laboratory animals and are therefore very difficult to carry out. Additionally, the relative area for nasal drug delivery in rodents is very high compared to humans, making clinical translation difficult.

3 OVERCOMING BARRIERS IN NASAL DRUG DELIVERY

There are difficulties to actually demonstrate nose-to-brain drug delivery, and there are also several hurdles for the delivery mechanism itself. Many potential drug candidates fail to pass through the nasal mucosa/epithelium because of the mucociliary clearance, which is present in the nasal cavity, and drugs therefore lack the ability to even reach the olfactory tract. Besides, several studies have shown that proteins, peptides, and DNA molecules reach the brain in very low amounts due to the unfavorable nasal environment.[17–19] Also, many formulation-dependent characteristics are involved, and current research is now focused on bypassing the mucociliary clearance issue, achieving a rapid onset, improving the absorption and permeability of these compounds, increasing bioavailability, and/or developing other techniques to enable drugs to reach the olfactory tract, subsequently allowing the delivery of the drug to the brain.

Several methods that mimic preferred conditions for the nose-to-brain route are available, with the aim to enhance brain uptake by increasing the permeability, prolonging the retention time on the nasal epithelium and therefore also on the olfactory region, and increasing bioavailability. These conditions can be achieved by inducing physiological changes in the nasal environment or by incorporating the drug into specific transport enhancing solutions or nanocarrier systems.

3.1 Physiological Possibilities

First, it should be noted that when administering a drug in the nasal cavity, the drug formulation should be applied in less-ciliated areas in the nasal cavity, which means deep in the nasal cavity and in the olfactory region as close as possible. Second, bioadhesive excipients can be used in the drug formulation to overcome the rapid mucociliary clearance in the nasal cavity.[20,21] Third, the enzymatic degradation of proteins and peptides in the nasal cavity can be bypassed by coadministering enzyme inhibitors[22] or by creating conditions of enzyme saturation.

On the other hand, a possible way to improve transnasal absorption is by increasing the permeability of the drug through the nasal mucosa via enhancers. These enhancers alter the permeability of the epithelial membrane or tight junctions, or work by inhibiting enzymatic degradation. However, the structural organization of the nasal epithelium should be preserved to avoid side effects. For example, using cationic compounds, such as chitosan, as enhancers may cause minimal or no damage to the nasal mucosa.[23]

3.2 Surface Modifications of Drugs

Besides the issues from the perspective of the nasal cavity, the surface of the drug compound can be manipulated as well. The preferred physicochemical and biological characteristics of the drug surface can be mimicked by a prodrug approach that allows and/or increases drug absorption.[24] In the work of Kao et al. it was reported that the intranasal administration of a prodrug form of levodopa into rats resulted in higher levodopa levels in the CSF compared to the original drug form. The latter was even higher through intranasal administration when compared to its equivalent intravenous dose.[25] Despite these effective results, the prodrug approach (structure change) may cause alterations in the biological function of the drug. For this reason, this method needs to be investigated carefully when considered.

On the other hand, the mucus layer in the nasal cavity contains the saccharides *N*-acetylglucosamine and L-fucose. The drug of interest or drug delivery system could be engineered in such a way that lectins could be incorporated on their surface. Lectins, wheat germ agglutinin (WGA), and *Ulex europaeus* agglutinin (UEA 1) bind to *N*-acetylglucosamine and L-fucose. Once bound, these induce cellular signals and endocytosis and lead to facilitated absorption and subsequently to transport into the brain.[26–28]

In addition to the surface manipulation of the drug compound, brain homing peptides (grafted to the surface of molecules) can also be used, which specifically target the CNS.[29,30]

3.3 Nanoparticle Formulations

Another very interesting and potential drug delivery approach is the use of drug delivery systems, such as nanocarriers. Drug delivery systems can be very effective in prolonging the retention time of the drug in the nasal mucosa, as well as maintaining a high drug concentration at the location.[20,31] One of these drug delivery system approaches is the use of nanoparticle formulation.

The drug can be incorporated into nanoparticles before nasal application to maintain the physiological stability and function of the drug, achieve a high/desired concentration of drug at the target site, increase the half-life time of the active compound, and ensure a better pharmacological effect after the nose-to-brain route. In this way, gel formulations, emulsions, micelles, and liposomes can also be developed to improve nasal drug delivery to the brain.

Nanoparticles usually have a size in the range of 1–1000 nm and consist of compounds that can adsorb, covalently bind, or entrap the desired drug molecule. Regarding the drug delivery mechanisms of nanoparticles to cross the BBB it is considered that nanoparticles can: (1) open tight junctions between endothelial cells and mediate the transition of the incorporated drug, (2) allow direct transport through transcytosis and/or endocytosis mechanisms, (3) cause a concentration gradient across membranes and thereby improve drug transport, (4) cause membrane fluidization with surfactants on their surface, and (5) inhibit P-glycoprotein efflux due to the presence of coated agents on their surface.[32]

Depending on their construction, nanoparticles protect the drug from degradation and clearance and improve permeability through membranes.[33] As mentioned before, drugs can be functionalized with lectins on their surface. Similarly, nanoparticles can also be functionalized by conjugating lectins to their surface.[28,34] An example can be found in a study of Gao et al. where poly(ethylene glycol)–poly(lactic acid) (PEG–PLA) nanoparticles were modified with WGA on their surface and were administered intranasally to rats. A facilitated uptake of an encapsulated fluorescence tracer in the CNS was demonstrated.[28]

A list of several drugs for the treatment of neurological diseases incorporated in particle formulations or enhancing solutions can be found in Table 6.1. Most of these drugs are marketed drugs or potential candidates used for other administration routes rather than intranasal application. However, these drugs were not able to reach the brain with an effective concentration, had poor bioavailability, and/or caused a slow onset that resulted in a need for an alternative method. Therefore, these drugs were formulated with drug delivery systems or enhanced, and intranasal administration resulted in significant improvements in treatments of the subsequent neurological disorder.

4 IN VITRO AND IN VIVO MODELS FOR NASAL DRUG DELIVERY

Several in vitro and in vivo models are available for mimicking neurological disease conditions for the nose-to-brain route. For in vitro models, excised olfactory epithelium,[71] excised mucosal tissue, or cell culture monolayers can be used.[72,73] In vivo studies have been performed in different laboratory animals (rats, mice, rabbits, and monkeys) but particularly in rats,[74] and demonstrated the direct transport of low-MW drugs, peptides, and proteins from the nasal cavity to the CNS.

Many in vivo studies have revealed that intranasally applied low-MW drugs directly enter the brain and/or CSF.[16,75,76] In case of large-MW drugs, several studies have reported successful nose-to-brain (and/or CNS) delivery of proteins and peptides in different species.[77–80]

Besides neurological disorders, brain tumors have the attention of the nose-to brain method as well. Studies targeting glioblastoma multiforme with the nose-to-brain route are reviewed intensively by Van Woensel et al.[81] In these studies the efficacy of pharmacological agents of intranasal administration over intravenous application has been demonstrated in the treatment of glioblastoma multiforme in both animal models,[82–85] as well as in a clinical study.[86]

Table 6.1 List of Several Drugs Incorporated in Different Drug Delivery Systems to Improve Nose-to-Brain Transport in the Treatment of Neurological Disorders

Formulation Type	Active Drug(s)	Targeted Neurological Disorder(s)
Nanoparticle	Venlafaxine[35–38]	Depression/bipolar disorder
	Bromocriptine[39,40]	Parkinson's disease
	Estradiol[41]	Alzheimer's disease
	Risperidone[42,43]	Psychotic disorder/schizophrenia
	Olanzapine[44]	Psychotic disorder/schizophrenia
	Ropinirole HCL[45]	Parkinson's disease
	Thymoquinone[46]	Alzheimer's disease
	Rivastigmine[47]	Alzheimer's disease
	Buspirone HCL[48]	Attention deficit hyperactivity
	NAPVSIPQ[49]	Alzheimer's disease
	Basic fibroblast growth factor[50]	Alzheimer's disease
Liposome	Rivastigmine[51]	Alzheimer's disease
Emulsion	Zonisamide[52]	Epilepsy
	Clonazepam[53]	Status epilepticus
	Growth differentiation factor 5[54]	Parkinson's disease
	Olanzapine[55]	Psychotic disorder/schizophrenia
	Diazepam[56]	Status epilepticus
	Buspirone HCL[57]	Attention deficit hyperactivity
	Rivastigmine[58]	Alzheimer's disease
Gel	Ropinirole[59]	Parkinson's disease
Micelle	Zolmitriptan[60]	Migraine
Solution	Insulin[61–64]	Alzheimer's disease/memory loss
	Iduronidase[65]	Lysosomal storage disease
	Melanocortin[66]	Obesity/cognitive disorder
	Mesenchymal stem cells[67]	Parkinson's disease
	Deferoxamine[68]	Ischemic stroke/intracerebral hemorrhage
	Erythropoietin[69,70]	Ischemic stroke

5 CONCLUSIONS

The nose-to-brain drug delivery route is gaining interest in the pharmaceutical, nanoscience, neuroscience, and neurooncology fields. Due to the need of alternative routes to deliver the current marketed drugs and potential treatment drugs in a more effective but noninvasive manner to the brain, this route is very promising. Although, further studies are needed to better understand the transport mechanism, the ability of bypassing the BBB, the ease of self-administration, the low cost profile, and patient compliance make the nose-to-brain route attractive and it is further endeavored to investigate the possibilities of translating research more to the clinical level. This chapter underlined the fact that the nose-to-brain route has many advantages, but also limiting steps which can be improved by the application of nanotechnology-based techniques. For the latter, a multidisciplinary approach is important to enable progress in this field.

Abbreviations

BBB	Blood–brain barrier
CNS	Central nervous system
CSF	Cerebrospinal fluid
HPLC	High-performance liquid chromatography
PEG–PLA	Poly(ethylene glycol)–poly(lactic acid)
UEA	*Ulex europaeus* agglutinin
WGA	Wheat germ agglutinin

References

1. Dhanda DS, Frey W, Leopold D, Kompella UB. Approaches for drug deposition in the human olfactory epithelium. *Drug Deliv.* 2005;5(4).
2. William H, Frey I. Bypassing the blood–brain barrier to deliver therapeutic agents to the brain and spinal cord. *Drug Deliv Technol.* 2002;2(5).
3. Chen X, Fawcett J, Rahman Y, Ala T, Frey I. Delivery of nerve growth factor to the brain via the olfactory pathway. *J Alzheimers Dis.* 1998;1(1):35–44.
4. Thorne R, Frey 2nd W. Delivery of neurotrophic factors to the central nervous system. *Clin Pharmacokinet.* 2001;40(12):907–946.
5. Hanson LR, Martinez P, Taheri S, Kamsheh L, Mignot E, Frey W. Intranasal administration of hypocretin 1 (orexin A) bypasses the blood-brain barrier and targets the brain: a new strategy for the treatment of narcolepsy. *Drug Deliv Technol.* 2004;4:65–71.
6. Yu Y-P, Xu Q-Q, Zhang Q, Zhang W-P, Zhang L-H, Wei E-Q. Intranasal recombinant human erythropoietin protects rats against focal cerebral ischemia. *Neurosci Lett.* 2005;387(1): 5–10.
7. Han I-K, Kim MY, Byun H-M, Hwang TS, Kim JM, Hwang KW, et al. Enhanced brain targeting efficiency of intranasally administered plasmid DNA: an alternative route for brain gene therapy. *J Mol Med.* 2007;85(1):75–83.
8. Barakat N, Omar S, Ahmed A. Carbamazepine uptake into rat brain following intra-olfactory transport. *J Pharm Pharmacol.* 2006;58(1):63–72.
9. Frey WH, Liu J, Chen X, Thorne RG, Fawcett JR, Ala TA, et al. Delivery of 125I-NGF to the brain via the olfactory route. *Drug Deliv.* 1997;4(2):87–92.
10. Ross TM, Martinez PM, Renner JC, Thorne RG, Hanson LR, Frey Ii WH. Intranasal administration of interferon beta bypasses the blood–brain barrier to target the central nervous system and cervical lymph nodes: a non-invasive treatment strategy for multiple sclerosis. *J Neuroimmunol.* 2004;151(1–2):66–77.
11. Thorne RG, Pronk GJ, Padmanabhan V, Frey Ii WH. Delivery of insulin-like growth factor-I to the rat brain and spinal cord along olfactory and trigeminal pathways following intranasal administration. *Neuroscience.* 2004;127(2):481–496.
12. Chou RC, Levy G. Effect of heparin or salicylate infusion on serum protein binding and on concentrations of phenytoin in serum, brain and cerebrospinal fluid of rats. *J Pharmacol Exp Ther.* 1981;219(1):42–48.
13. Waynforth H, Flecknell P. Cisternal puncture (and intracisternal injection). Experimental Techniques in the Rat. London: Academic Press; 1980:[pp. 59–61].
14. Chow HHS, Anavy N, Villalobos A. Direct nose–brain transport of benzoylecgonine following intranasal administration in rats. *J Pharm Sci.* 2001;90(11):1729–1735.

15. Hussain MA, Rakestraw D, Rowe S, Aungst BJ. Nasal administration of a cognition enhancer provides improved bioavailability but not enhanced brain delivery. *J Pharm Sci*. 1990;79(9):771–772.
16. Sherry Chow HH, Chen Z, Matsuura GT. Direct transport of cocaine from the nasal cavity to the brain following intranasal cocaine administration in rats. *J Pharm Sci*. 1999;88(8):754–758.
17. Chen XD, Xie GZ, Rahman MS. Application of the distribution factor concept in correlating thermal conductivity data for fruits and vegetables. *Int J Food Prop*. 1998;1(1):35–44.
18. Dufes C, Olivier JC, Gaillard F, Gaillard A, Couet W, Muller JM. Brain delivery of vasoactive intestinal peptide (VIP) following nasal administration to rats. *Int J Pharm*. 2003;255(1–2):87–97.
19. Kim TW, Chung H, Kwon IC, Sung HC, Jeong SY. In vivo gene transfer to the mouse nasal cavity mucosa using a stable cationic lipid emulsion. *Mol Cells*. 2000;10(2):142–147.
20. Kumar M, Misra A, Babbar A, Mishra A, Mishra P, Pathak K. Intranasal nanoemulsion based brain targeting drug delivery system of risperidone. *Int J Pharm*. 2008;358(1):285–291.
21. Vyas TK, Babbar A, Sharma R, Singh S, Misra A. Preliminary brain-targeting studies on intranasal mucoadhesive microemulsions of sumatriptan. *AAPS PharmSciTech*. 2006;7(1):E49–E57.
22. Graff CL, Pollack GM. P-Glycoprotein attenuates brain uptake of substrates after nasal instillation. *Pharm Res*. 2003;20(8):1225–1230.
23. Illum L, Farraj NF, Davis SS. Chitosan as a novel nasal delivery system for peptide drugs. *Pharm Res*. 1994;11(8):1186–1189.
24. Pezron I, Mitra AK, Duvvuri S, Tirucherai GS. Prodrug strategies in nasal drug delivery. *Exp Opin Ther Patents*. 2002;12(3):331–340.
25. Kao HD, Traboulsi A, Itoh S, Dittert L, Hussain A. Enhancement of the systemic and CNS specific delivery of L-dopa by the nasal administration of its water soluble prodrugs. *Pharm Res*. 2000;17(8):978–984.
26. Gabor F, Bogner E, Weissenboeck A, Wirth M. The lectin–cell interaction and its implications to intestinal lectin-mediated drug delivery. *Adv Drug Deliv Rev*. 2004;56(4):459–480.
27. Gao X, Wu B, Zhang Q, Chen J, Zhu J, Zhang W, et al. Brain delivery of vasoactive intestinal peptide enhanced with the nanoparticles conjugated with wheat germ agglutinin following intranasal administration. *J Control Release*. 2007;121(3):156–167.
28. Gao X, Tao W, Lu W, Zhang Q, Zhang Y, Jiang X, et al. Lectin-conjugated PEG–PLA nanoparticles: preparation and brain delivery after intranasal administration. *Biomaterials*. 2006;27(18):3482–3490.
29. Frenkel D, Solomon B. Filamentous phage as vector-mediated antibody delivery to the brain. *Proc Natl Acad Sci USA*. 2002;99(8):5675–5679.
30. Janda KJ. Delivery of active proteins to the central nervous system using phage vectors. Google Patents. 2007.
31. Dalpiaz A, Gavini E, Colombo G, Russo P, Bortolotti F, Ferraro L, et al. Brain uptake of an anti-ischemic agent by nasal administration of microparticles. *J Pharm Sci*. 2008;97(11):4889–4903.
32. Kreuter J, Gelperina S. Use of nanoparticles for cerebral cancer. *Tumori*. 2007;94(2):271–277.
33. Vila A, Sánchez A, Tobío M, Calvo P, Alonso MJ. Design of biodegradable particles for protein delivery. *J Control Release*. 2002;78(1–3):15–24.
34. Bies C, Lehr CM, Woodley JF. Lectin-mediated drug targeting: history and applications. *Adv Drug Deliv Rev*. 2004;56(4):425–435.
35. Haque S, Md S, Fazil M, Kumar M, Sahni JK, Ali J, et al. Venlafaxine loaded chitosan NPs for brain targeting: pharmacokinetic and pharmacodynamic evaluation. *Carbohydr Polym*. 2012;89(1):72–79.

36. Alam MI, Baboota S, Ahuja A, Ali M, Ali J, Sahni JK. Intranasal administration of nanostructured lipid carriers containing CNS acting drug: Pharmacodynamic studies and estimation in blood and brain. *J Psych Res*. 2012;46(9):1133–1138.

37. Dange SM, Kamble MS, Bhalerao KK, Chaudhari PD, Bhosale AV, Nanjwade BK, et al. Formulation and evaluation of venlafaxine nanostructured lipid carriers. *J Bionanosci*. 2014;8(2): 81–89.

38. Haque S, Md S, Sahni JK, Ali J, Baboota S. Development and evaluation of brain targeted intranasal alginate nanoparticles for treatment of depression. *J Psych Res*. 2014;48(1):1–12.

39. Md S, Haque S, Fazil M, Kumar M, Baboota S, Sahni JK, et al. Optimised nanoformulation of bromocriptine for direct nose-to-brain delivery: biodistribution, pharmacokinetic and dopamine estimation by ultra-HPLC/mass spectrometry method. *Exp Opin Drug Deliv*. 2014;11(6):827–842.

40. Md S, Khan RA, Mustafa G, Chuttani K, Baboota S, Sahni JK, et al. Bromocriptine loaded chitosan nanoparticles intended for direct nose to brain delivery: pharmacodynamic, pharmacokinetic and scintigraphy study in mice model. *Eur J Pharm Sci*. 2013;48(3):393–405.

41. Wang X, Chi N, Tang X. Preparation of estradiol chitosan nanoparticles for improving nasal absorption and brain targeting. *Eur J Pharm Biopharm*. 2008;70(3):735–740.

42. Kumar M, Pathak K, Misra A. Formulation and characterization of nanoemulsion-based drug delivery system of risperidone. *Drug Dev Ind Pharm*. 2009;35(4):387–395.

43. Patel S, Chavhan S, Soni H, Babbar AK, Mathur R, Mishra AK, et al. Brain targeting of risperidone-loaded solid lipid nanoparticles by intranasal route. *J Drug Target*. 2011;19(6):468–474.

44. Seju U, Kumar A, Sawant KK. Development and evaluation of olanzapine-loaded PLGA nanoparticles for nose-to-brain delivery: in vitro and in vivo studies. *Acta Biomater*. 2011;7(12):4169–4176.

45. Jafarieh O, Md S, Ali M, Baboota S, Sahni JK, Kumari B, et al. Design, characterization, and evaluation of intranasal delivery of ropinirole-loaded mucoadhesive nanoparticles for brain targeting. *Drug Dev Ind Pharm*. 2015;41(10):1674–1681.

46. Alam S, Khan ZI, Mustafa G, Kumar M, Islam F, Bhatnagar A, et al. Development and evaluation of thymoquinone-encapsulated chitosan nanoparticles for nose-to-brain targeting: a pharmacoscintigraphic study. *Int J Nanomed*. 2012;7:5705–5718.

47. Fazil M, Md S, Haque S, Kumar M, Baboota S, Sahni JK, et al. Development and evaluation of rivastigmine loaded chitosan nanoparticles for brain targeting. *Eur J Pharm Sci*. 2012;47(1): 6–15.

48. Khan MS, Patil K, Yeole P, Gaikwad R. Brain targeting studies on buspirone hydrochloride after intranasal administration of mucoadhesive formulation in rats. *J Pharm Pharmacol*. 2009;61(5):669–675.

49. Liu Z, Jiang M, Kang T, Miao D, Gu G, Song Q, et al. Lactoferrin-modified PEG-co-PCL nanoparticles for enhanced brain delivery of NAP peptide following intranasal administration. *Biomaterials*. 2013;34(15):3870–3881.

50. Zhang C, Chen J, Feng C, Shao X, Liu Q, Zhang Q, et al. Intranasal nanoparticles of basic fibroblast growth factor for brain delivery to treat Alzheimer's disease. *Int J Pharm*. 2014;461(1): 192–202.

51. Yang Z-Z, Zhang Y-Q, Wang Z-Z, Wu K, Lou J-N, Qi X-R. Enhanced brain distribution and pharmacodynamics of rivastigmine by liposomes following intranasal administration. *Int J Pharm*. 2013;452(1–2):344–354.

52. Shahiwala A, Dash D. Preparation and evaluation of microemulsion based formulations for rapid-onset intranasal delivery of zonisamide. *Adv Sci Lett*. 2010;3(4):442–446.

53. Vyas TK, Babbar A, Sharma R, Singh S, Misra A. Intranasal mucoadhesive microemulsions of clonazepam: preliminary studies on brain targeting. *J Pharm Sci*. 2006;95(3):570–580.

54. Hanson LR, Fine JM, Hoekman JD, Nguyen TM, Burns RB, Martinez PM, et al. Intranasal delivery of growth differentiation factor 5 to the central nervous system. *Drug Deliv.* 2012;19(3):149–154.
55. Kumar M, Misra A, Mishra A, Mishra P, Pathak K. Mucoadhesive nanoemulsion-based intranasal drug delivery system of olanzapine for brain targeting. *J Drug Target.* 2008;16(10):806–814.
56. Li L, Nandi I, Kim KH. Development of an ethyl laurate-based microemulsion for rapid-onset intranasal delivery of diazepam. *Int J Pharm.* 2002;237(1):77–85.
57. Bshara H, Osman R, Mansour S, El-Shamy AE-HA. Chitosan and cyclodextrin in intranasal microemulsion for improved brain buspirone hydrochloride pharmacokinetics in rats. *Carbohydr Polym.* 2014;99:297–305.
58. Shah BM, Misra M, Shishoo CJ, Padh H. Nose to brain microemulsion-based drug delivery system of rivastigmine: formulation and ex-vivo characterization. *Drug Deliv.* 2015;22(7): 918–930.
59. Khan S, Patil K, Bobade N, Yeole P, Gaikwad R. Formulation of intranasal mucoadhesive temperature-mediated in situ gel containing ropinirole and evaluation of brain targeting efficiency in rats. *J Drug Target.* 2010;18(3):223–234.
60. Jain R, Nabar S, Dandekar P, Patravale V. Micellar nanocarriers: potential nose-to-brain delivery of zolmitriptan as novel migraine therapy. *Pharm Res.* 2010;27(4):655–664.
61. Benedict C, Hallschmid M, Hatke A, Schultes B, Fehm HL, Born J, et al. Intranasal insulin improves memory in humans. *Psychoneuroendocrinology.* 2004;29(10):1326–1334.
62. Benedict C, Hallschmid M, Schmitz K, Schultes B, Ratter F, Fehm HL, et al. Intranasal insulin improves memory in humans: superiority of insulin aspart. *Neuropsychopharmacology.* 2007;32(1):239–243.
63. Freiherr J, Hallschmid II M, Frey WH, Brünner YF, Chapman CD, Hölscher C, et al. Intranasal insulin as a treatment for Alzheimer's disease: a review of basic research and clinical evidence. *CNS Drugs.* 2013;27(7):505–514.
64. Reger MA, Watson GS, Green PS, Baker LD, Cholerton B, Fishel MA, et al. Intranasal insulin administration dose-dependently modulates verbal memory and plasma β-amyloid in memory-impaired older adults. *J Alzheimer Dis.* 2008;13(3):323.
65. Wolf DA, Hanson LR, Aronovich EL, Nan Z, Low WC, Frey WH, et al. Lysosomal enzyme can bypass the blood–brain barrier and reach the CNS following intranasal administration. *Mol Genet Metab.* 2012;106(1):131–134.
66. Hallschmid M, Benedict C, Schultes B, Perras B, Fehm H-L, Kern W, et al. Towards the therapeutic use of intranasal neuropeptide administration in metabolic and cognitive disorders. *Regul Pept.* 2008;149(1):79–83.
67. Danielyan L, Schäfer R, von Ameln-Mayerhofer A, Bernhard F, Verleysdonk S, Buadze M, et al. Therapeutic efficacy of intranasally delivered mesenchymal stem cells in a rat model of Parkinson disease. *Rejuvenation Res.* 2011;14(1):3–16.
68. Hanson LR, Roeytenberg A, Martinez PM, Coppes VG, Sweet DC, Rao RJ, et al. Intranasal deferoxamine provides increased brain exposure and significant protection in rat ischemic stroke. *J Pharmacol Exp Ther.* 2009;330(3):679–686.
69. Fletcher L, Kohli S, Sprague SM, Scranton RA, Lipton SA, Parra A, et al. Intranasal delivery of erythropoietin plus insulin-like growth factor-I for acute neuroprotection in stroke: laboratory investigation. *J Neurosurg.* 2009;111(1):164–170.
70. García-Rodríguez JC, Sosa Teste I. The nasal route as a potential pathway for delivery of erythropoietin in the treatment of acute ischemic stroke in humans. *Sci World J.* 2009;9:970–981.
71. Mistry A, Stolnik S, Illum L. Nose-to-brain delivery: investigation of the transport of nanoparticles with different surface characteristics and sizes in excised porcine olfactory epithelium. *Mol Pharm.* 2015;12(8):2755–2766.

72. Schmidt MC, Peter H, Lang SR, Ditzinger G, Merkle HP. In vitro cell models to study nasal mucosal permeability and metabolism. *Adv Drug Deliv Rev*. 1998;29(1):51–79.

73. Schmidt MC, Simmen D, Hilbe M, Boderke P, Ditzinger G, Sandow J, et al. Validation of excised bovine nasal mucosa as in vitro model to study drug transport and metabolic pathways in nasal epithelium. *J Pharm Sci*. 2000;89(3):396–407.

74. Illum L. Transport of drugs from the nasal cavity to the central nervous system. *Eur J Pharm Sci*. 2000;11(1):1–18.

75. David G, Puri C, Kumar TA. Bioavailability of progesterone enhanced by intranasal spraying. *Experientia*. 1981;37(5):533–534.

76. Sakane T, Akizuki M, Yoshida M, Yamashita S, Nadai T, Hashida M, et al. Transport of cephalexin to the cerebrospinal fluid directly from the nasal cavity. *J Pharm Pharmacol*. 1991;43(6): 449–451.

77. Alcalá-Barraza SR, Lee MS, Hanson LR, McDonald AA, Frey WH, McLoon LK. Intranasal delivery of neurotrophic factors BDNF, CNTF, EPO, and NT-4 to the CNS. *J Drug Target*. 2010;18(3):179–190.

78. Liu X-F, Fawcett JR, Thorne RG, DeFor TA, Frey WH. Intranasal administration of insulin-like growth factor-I bypasses the blood–brain barrier and protects against focal cerebral ischemic damage. *J Neurol Sci*. 2001;187(1):91–97.

79. Migliore MM, Vyas TK, Campbell RB, Amiji MM, Waszczak BL. Brain delivery of proteins by the intranasal route of administration: a comparison of cationic liposomes versus aqueous solution formulations. *J Pharm Sci*. 2010;99(4):1745–1761.

80. Vajdy M, O'Hagan DT. Microparticles for intranasal immunization. *Adv Drug Deliv Rev*. 2001;51(1):127–141.

81. Van Woensel M, Wauthoz N, Rosière R, Amighi K, Mathieu V, Lefranc F, et al. Formulations for intranasal delivery of pharmacological agents to combat brain disease: a new opportunity to tackle GBM?. *Cancers*. 2013;5(3):1020–1048.

82. Hashizume R, Ozawa T, Gryaznov SM, Bollen AW, Lamborn KR, Frey WH, et al. New therapeutic approach for brain tumors: Intranasal delivery of telomerase inhibitor GRN163. *Neurooncology*. 2008;10(2):112–120.

83. Özduman K, Wollmann G, Piepmeier JM, Van den Pol AN. Systemic vesicular stomatitis virus selectively destroys multifocal glioma and metastatic carcinoma in brain. *J Neurosci*. 2008;28(8):1882–1893.

84. Reitz M, Demestre M, Sedlacik J, Meissner H, Fiehler J, Kim SU, et al. Intranasal delivery of neural stem/progenitor cells: a noninvasive passage to target intracerebral glioma. *Stem Cells Transl Med*. 2012;1(12):866–873.

85. Shingaki T, Inoue D, Furubayashi T, Sakane T, Katsumi H, Yamamoto A, et al. Transnasal delivery of methotrexate to brain tumors in rats: a new strategy for brain tumor chemotherapy. *Mol Pharm*. 2010;7(5):1561–1568.

86. Da Fonseca CO, Simao M, Lins IR, Caetano RO, Futuro D, Quirico-Santos T. Efficacy of monoterpene perillyl alcohol upon survival rate of patients with recurrent glioblastoma. *J Cancer Res Clin Oncol*. 2011;137(2):287–293.

Nanoscience in Targeted Brain Drug Delivery

CHAPTER 7

Nanoscience in Targeted Brain Drug Delivery

Meltem Çetin, PhD*, Eren Aytekin, MD, Burçin Yavuz, PhD**, Sibel Bozdağ-Pehlivan, PhD****

**Ataturk University, Erzurum, Turkey; **Hacettepe University, Ankara, Turkey*

CONTENTS

1 IMPORTANCE AND APPLICATION OF NANOTECHNOLOGY-BASED BRAIN DRUG DELIVERY SYSTEMS

The successful treatments for neurological diseases (NDs) [such as Alzheimer's disease (AD) and Parkinson's disease (PD) etc.] have improved with time; however, drug delivery to the brain still remains limited due to the restrictive mechanism of the blood–brain barrier (BBB), which is a critical problem to overcome.[1,2] The BBB, formed principally from capillary endothelial cells without fenestrations, is the major interface between brain and the systemic circulation and protects the brain from potentially harmful substances and microorganisms, and maintains the homeostasis of the central nervous system (CNS). There are several transport routes (such as transcellular transport and carrier- and specific receptor–mediated systems) for solute molecules to cross the BBB (Fig. 7.1). Small molecules with molecular weight of <400 Da, with a Log octanol/water partition coefficient between 0.5 and 6.0, neutral or significantly uncharged small molecules at physiological pH 7.4, or small lipid-soluble molecules cross the BBB by passive diffusion. Besides, specific and selective transporters [e.g., Na^+-independent transporter (such as glucose transporter GLUT1 for D-glucose); Na^+-independent neutral amino acid transporter (system L) for amino acids (such as L-tyrosine, L-tryptophan, and L-histidine); CAT1 (SCL7A2) transporter for the basic amino acids (such as L-lysine and L-arginine)] supply the CNS with glucose, free fatty acids, amino acids, vitamins, minerals, and electrolytes. In spite of all these transport mechanisms, 98% of small molecules and nearly all large molecules, such as recombinant proteins or gene-based medicines, cannot cross the BBB. Therefore, the development of noninvasive strategies, including exploration of nanotechnology-based drug delivery systems (NBDDS) (such as various types of nanoparticles (NPs), nanocapsules, liposomes, dendrimers, carbon nanotubes, polymeric

Nanotechnology Methods for Neurological Diseases and Brain Tumors. http://dx.doi.org/10.1016/B978-0-12-803796-6.00007-1

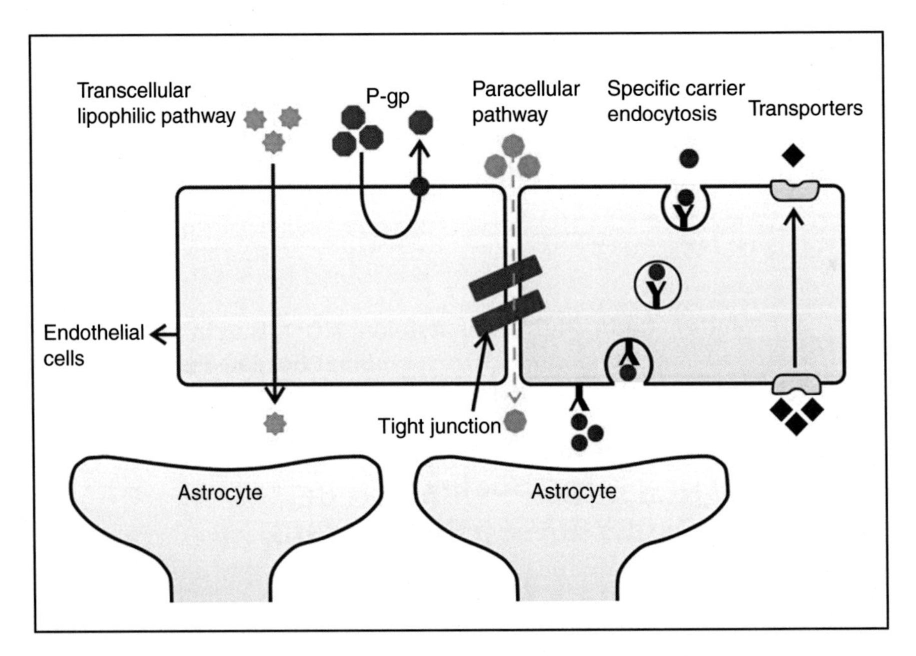

FIGURE 7.1 A schematic diagram of the transport mechanisms across the blood–brain barrier (BBB).
P-gp, P-glycoprotein.

micelles, etc.) using advanced formulation design strategies to enhance therapeutics delivery across the BBB, is a crucial requirement.[1,3–5]

Nanotechnology combines the knowledge from chemistry, physics, and other related disciplines and thus is a multidisciplinary field of theoretical and experimental area of applied science and technology.[6] The expression "nano" is derived from *nanos*, which is an ancient Greek word that means dwarf/extremely small.[7] The idea of nanotechnology was first introduced by Richard Feynman in 1959. However, the term nanotechnology was first used by Professor Norio Taniguchi (Tokyo Science University) in 1974 and defined as "Nano-technology is the production technology to obtain extra high accuracy and ultrafine dimensions, i.e. the preciseness and fineness of the order of 1 nm (nanometer), 10^{-9} m in length" in a paper on ion-sputter machining.[8,9] Nanotechnology has a significant impact on almost all industrial areas, including the pharmaceutical technology field.[10,11]

1.1 Nanoparticles

NPs are defined as submicronic (1–1000 nm) colloidal systems[12] (Fig. 7.2). NPs have been investigated as delivery systems for a wide number of drugs and have several advantages compared to the traditional dosage forms, such as

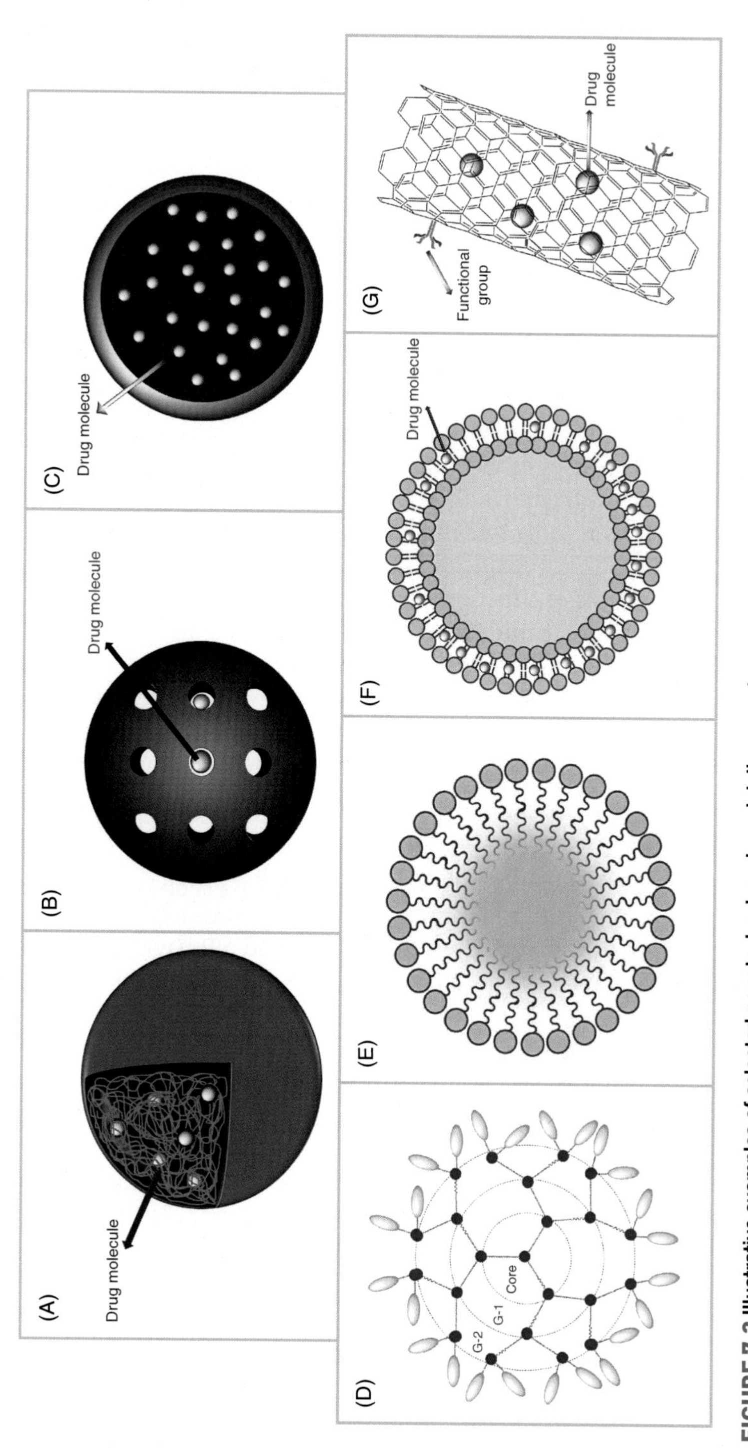

FIGURE 7.2 Illustrative examples of selected nanotechnology-based delivery systems.
(A) Polymeric nanoparticles (NPs), (B) hollow silica NPs (C) nanocapsules, (D) dendrimers, (E) polymeric micelles, (F) liposomes, and (G) carbon nanotubes.

reduction of drug's side effects and repeated doses, increase in patient comfort and compliance, control of drug release or the site of drug release, and increase in the relative bioavailability of drugs by the effective protection of drugs against degradation (e.g., enzymatic).[13–15] Different nanoparticulate drug delivery systems (such as polymeric NPs, metal NPs, molecular targeted NPs, etc.) have been developed over the past 3 decades.[16] Nanosized drug delivery systems have to be biocompatible and nontoxic. However, their undesirable effects strongly depend on their hydrodynamic size (smaller particles have a greater surface area).[17]

1.1.1 Metallic Nanoparticles

Metallic NPs are composed of inert and biocompatible metals, such as gold and iron oxide. However, a significant fraction of these particles after repeated administration accumulates in the body and can lead to toxicity.[16] In addition, magnetic NPs based on the usage three elements (iron, cobalt, and nickel) are promising as theranostic tools, which may be used for the combination of drug delivery and diagnosis. Iron oxide is FDA approved, and magnetic materials and magnetic properties of superparamagnetic iron oxide nanoparticles (SPIONs) have the capacity for multifunctional applications due to their unique biocompatible and magnetic properties.[18–20] Furthermore, metallic NPs can be used for different purposes, such as inhibition of amyloid β (Aβ) fibrillations, determination of acetylcholinesterase (AChE), drug delivery, visualization of Aβ plaques in the diagnosis, and treatment of neurodegenerative diseases.[21–23]

The protein fibrillation process is the main cause of neurodegenerative diseases (e.g., AD and PD). Gold NPs are strong inhibitors for Aβ fibrillations. These NPs have a promising potential in inhibiting the progress of the disease through control of the fibrillation kinetic process.[24] Mirsadeghi et al.[24] investigated the effects of differently sized and shaped gold NPs (with diameters of 20–30 nm for gold nanospheres and with aspect ratios of 4 and 20 for gold nanorods) on the Aβ fibrillation process in the presence and absence of protein sources, such as human plasma and fetal bovine serum. The surfaces of NPs were covered by biomolecules (such as protein), and the available surface area and composition of NPs have an important role in inhibiting or enhancing the fibrillation process. This study showed that the inhibitory effects of gold NPs on the Aβ fibrillation process were reduced in the presence of protein sources compared to bare NPs. These proteins covered the NPs surface, regardless of the size and shape of NPs. Besides, the types and amount of protein sources had noticeable effects on the fibrillation process.

In another study, SPIONs inhibited the fibrillation process, and the antifibrillation effects of NPs were strongly dependent on their sizes and surface charges.[25]

Nerve growth factor (NGF) has a key role in development of the peripheral nervous system and in the viability of peripheral sympathetic neurons. It also promotes and maintains the phenotype of cholinergic neurons in CNS. There are alterations in NGF and its receptor levels in patients with AD. NGF may be useful to improve cholinergic function and the memory of these patients and thus, NGF has a potential therapeutic activity in the treatment of AD and PD.[26] However, NGF is not able to significantly penetrate the BBB due to its size and polarity. Besides, NGF has rapid degradation, leading to a short biological half-life.[27] Therefore, Marcus et al.[27] prepared NGF-conjugated iron oxide NPs. After conjugation to iron oxide NPs, the stability of NGF was significantly improved and the half-life of NGF extended due to slower degradation. Also, NGF-conjugated NPs significantly elevated neurite outgrowth, enhanced the complexity of the neuronal branching trees, and induced cell differentiation in PC12 cells (a rat pheochromocytoma cell line used for investigating neuronal differentiation) as compared to free NGF.

1.1.2 Silica Nanoparticles

Silica has been used in artificial implants (because of the osteogenic properties of its composites), drug delivery systems, and as an enhancer of biocompatibility of different drug delivery systems (biopolymers, micelles, and magnetic NPs).[28] Silica-based delivery systems are classified as xerogels and mesoporous silica NPs [such as MCM-41 (Mobil Composition of Matter) and SBA-15 (Santa Barbara University mesoporous silica material)]. Mesoporous silica has a honeycomb-like porous structure with hundreds of mesopores, and is useful to encapsulate relatively large amounts of both small and large molecules due to its high surface area (>900 m^2/g) and large pore volume (>0.9 cm^3/g). Furthermore, mesoporous silica NPs have good chemical and thermal stabilities, and have been used for the delivery of several drugs, such as anticancer drugs, heart disease drugs, and antibiotics, due to their excellent biocompatibility.[17,28,29]

Quercetin is a flavonoid that has been extensively studied in the past 3 decades due to its potent antioxidant and radical-scavenging properties. It is considered as a new neuroprotectant, as it decreased levels of reactive oxygen species. However, it has been showed that quercetin improved memory and spatial learning in D-galactose–treated aging in mice by increasing brain antioxidant capacity.[30,31] Nday et al.[32] synthesized quercetin as metal-chelating flavonoid-loaded silica NPs, with surface modified with cetyltrimethylammonium bromide (CTAB) or poly(ethylene glycol) (PEG) by hydrolysis and polycondensation of alkoxide precursors. In addition, the effect of the NPs against Cu(II)-induced oxidative stress in neuronal and glial cultures was investigated. Quercetin-loaded NPs (<150 μM) showed a protective antioxidant effect against Cu(II)-mediated oxidative stress, and the specific action of PEG-modified NPs on oxidative stress was more pronounced compared to that of CTAB-modified NPs.

On the other hand, some studies reported that nanosilica materials showed in vitro and in vivo toxicity and certain hazards.[33–37] Silica NPs induce oxidative stress in cells by producing reactive oxygen species, activating antioxidant enzymes (e.g., superoxide dismutase and heme oxygenase 1), decreasing glutathione levels, and increasing production of malondialdehyde. The effects of silica NPs on cells depend on their size and the cell type.[17,33] Furthermore, the potential neurotoxicity of silica NPs is still unclear. Silica NPs used increasingly in imaging/diagnosis and drug delivery to the CNS have potential side effects on the brain. In the study of Yang et al.,[34] the cellular uptake of silica NPs in human SK-N-SH and mouse neuro2a (N2a) neuroblastoma cells was investigated. The following results were found in this study: (1) silica NPs were widely localized in the cytoplasm of the cells; (2) application of silica NPs decreased cell density, induced cellular apoptosis, significantly decreased cell viability, and increased levels of intracellular reactive oxygen species (in dose-dependent manner) in both cell lines; and (3) increased deposit of intracellular Aβ(1–42) and phosphorylation of tau at Ser262 and Ser396. As a result, they reported that the application of silica NPs may induce neurotoxicity and contribute to the risk of developing AD.[34]

1.1.3 Polymeric Nanoparticles

Polymeric NPs are prepared using synthetic polymers [e.g., polyglycolides (PGA), polylactides (PLA), poly(lactide-*co*-glycolide) (PLGA), poly-e-caprolactone (PCL), polyacrylamide, and polyacrylate] or natural polymers (e.g., chitosan, albumin, and gelatin). Especially, the most popular FDA-approved biodegradable aliphatic polyesters, such as PLGA, PCL, and PLA, have significantly been used in the field of CNS drug delivery and are well known for their biocompatibility, biodegradability, and nontoxic properties. Thus, these polymers are suitable matrices for modified release in drug delivery systems. PLGA is a diblock copolymer and is degraded relatively faster than PLA, which delivers pharmaceutical agents over months, and can ensure a show release over a period of 2–6 weeks.[17,38,39]

PLGA NPs were widely studied to deliver varied therapeutic agents, such as dopamine, bacoside-A, growth factors, peptide inhibitor, curcumin, huperzine A (HA), and selegiline, to the brain to treat neurodegenerative disease.[40–46]

Pahuja et al.[40] used PLGA for the delivery of dopamine to the brain for the treatment of PD. The efficiency of the dopamine-loaded NPs was evaluated in 6-OHDA–induced Parkinsonian rats. The particle size of dopamine-loaded PLGA NPs and FITC-linked PLGA NPs prepared, using the double emulsion solvent evaporation method, were about 120 and 157 nm with a narrow range, respectively. Their delivery to striatum crossing BBB after intravenous (IV) injection was carried out successfully, and dopamine was released slowly from dopamine-loaded NPs after internalization in the brain. They found that the NPs notably reversed neurochemical and neurobehavioral abnormalities in rats with PD, reduced dopamine autoxidation–related toxicity, and therefore did not cause production of excess reactive oxygen species, dopaminergic neuron degeneration, and abnormal cardiovascular alterations.

Surface modification of NPs with surfactants (e.g., polysorbates) is used to enhance drug delivery to the brain. In this strategy, plasma apolipoprotein E or B (apo-E or -B) can be adsorbed on the polysorabate-coated NPs. Apo-E contributes to the transport of lipids into the brain by the low-density lipoprotein (LDL) receptors (LDLRs) [e.g., low-density lipoprotein receptor–related protein (LRP-1) and very low–density lipoprotein receptor (VLDLR)] present on the BBB. Thus, the NPs, mimicking the LDL, are taken up into the cerebral endothelium by an endocytotic mechanism followed by transcytosis.[41,47,48] Jose at al.[41] prepared bacoside-A (as neuroprotective drug)–loaded, surface-modified PLGA NPs using O/W emulsion–solvent evaporation method to treat neurodegenerative disease. These NPs had an encapsulation efficiency value of 57%. The surface charge of drug loaded NPs was negative (−19 mV), which has an important effect on physical stability of NPs. Tween 80 was used to modify the surface of NPs, so that they could cross BBB. They found that the NPs were successful in delivering a 10-fold bacoside-A dose into the brain compared to the solution of pure bacoside-A after an intraperitoneal injection to adult albino Wistar rats. The surface modification of NPs with polysorbate 80 is a highly effective strategy to deliver drugs across the BBB and to treat CNS diseases.

Peptide inhibitor (PGQ9, with amino acid sequence: KKQQQQQQQQP-GQQQQPQQQQPGQQQQQQQQPGQQQQQQQQKK)–loaded PLGA NPs were prepared by nanoprecipitation method to inhibit polyglutamine aggregation as a therapeutic approach for Huntington's disease (HD), which is caused by a CAG trinucleotide repeat expansion located in the first exon of the HD gene and is a progressive neurodegenerative disorder.[43,49] The mean size of PGQ9-loaded PLGA NPs was 172 nm, with negative zeta potential (about −30 mV), and a good encapsulation efficiency (78%).[43] The size of NPs is an important factor (<200 nm), which enhances the in vivo half-life of NPs without the activation of the reticuloendothelial system and splenic filtration.[43,50] The inhibitory effect of NPs was investigated with the aggregation-prone Q35P10 peptide, which represents the N-terminal part of Huntington protein, and it was found that the NPs stopped the elongation phase of Q35P10 aggregation. It was reported that inhibitor-loaded NPs were considered to be a significant approach in the treatment of HD.[43]

Age-related neurodegenerative diseases, commonly defined by the progressive loss of neurons, are expected to increase by 12% in 2030 due to increasing life expectancy and increased proportion of aged population. Therefore, induction of neurogenesis, a process where new neurons are generated from neural stem cells that declines with age, can be a potential approach focusing on replacing lost neurons during the progression of several neurodegenerative disease, such as AD, HD, PD, etc. Two neurogenic brain regions, that is, the subventricular zone (SVZ) of the lateral ventricle and the subgranular zone (SGZ) of the hippocampus are important for neurogenesis.[44,51] Tiwari et al.[44] developed curcumin-loaded PLGA NPs to induce adult neurogenesis in AD. They reported that the PLGA NPs containing curcumin significantly induced neuronal differentiation by enhancing nuclear translocation of β-catenin and increasing phosphorylation of glycogen synthase kinase-3β. Further, the NPs enhanced neural stem cell proliferation by increasing reelin and Pax6 (a gene heavily expressed in neural stem cells) expressions in the hippocampus in vitro and in vivo (in the hippocampus and SVZ). These NPs were internalized by the neural stem cells in culture and reached the brain. In addition, curcumin-loaded NPs reversed Aβ-mediated inhibitory effects on neurogenesis and memory learning in rats with AD through the activation of the canonical Wnt/β-catenin pathway involved in the regulation of adult hippocampal neurogenesis, the development of neuroepithelium in hippocampus, and self-renewal of neural stem cells.

PEG, a potential hydrophilic cover for hydrophobic NPs, is used to obtain long-circulating NPs by reducing the clearance of NPs from the blood due to successful RES escape. Therefore, PEGylation provides great opportunities for drug targeting.[52,53] PEGylated NPs cross the BBB by interacting with microvascular transport proteins (e.g., apo-E) and use the LDLR-mediated pathway. Several PEGylated drug delivery systems (liposome, micelles, and NPs) permeate the BBB and accumulate in the brain.[52]

Zhang et al.[45] prepared surface-modified NPs containing HA as a model drug to increase HA penetration into the brain. HA is a strongly selective and reversible inhibitor of AChE and has been approved to be used to improve memory deficit in AD and elderly patients in China.[45] First, HA-loaded PEG–PLGA NPs were prepared by emulsion–solvent evaporation method, and later, the surface of NPs was modified with aprotinin (Apr). Apr, a protease inhibitor, inhibits the activity of AChE and has a high affinity toward LRP, which is a cell surface receptor that is highly expressed in vessel endothelial cells of the BBB. Furthermore, recently Apr has been used as a promising targeting ligand for delivery to the brain with efficient BBB transport. Besides, borneol solution in PEG 400, which was intragastrically coadministrated with IV-applied HA-loaded Apr NPs, was evaluated for efficient brain delivery in rats with AD. Borneol, usually orally administered, has a low molecular weight and is a lipophilic compound that can be rapidly absorbed via the gastrointestinal tract. Besides, borneol promotes the passing of NPs into brain by loosening the intercellular tight junction in the BBB and enhancing epithelial junction permeability. Further, it accelerates the transport of drugs by increasing the volume and number of pinocytosis vesicles in BBB cells and also inhibits the activity of P-glycoprotein

(P-gp). It was also found that both conjugation of Apr and coadministration of borneol significantly enhanced the delivery of NPs to the brain and that surface-modified NPs containing HA improved memory in rats with AD. Borneol can be a promising promoter for the targeting of nanocarriers to the brain.[45]

Li et al.[54] developed a targeting drug delivery system using a phage-displayed peptide (TGN; TGNYKALHPHNGC) as a targeting motif. The NAP (NAPVSIPQ)-loaded PEG–PLGA NPs were prepared, conjugated with TGN, and evaluated in vivo after IV administration in mice. NAP, an active fragment of activity-dependent neuroprotective protein, which is important for brain development, showed significant neuroprotective effects on neurodegenerative disease and has potential for AD therapy. However, it is not stable due to enzymatic degradation and has a short half-life. The NPs protected NAP from enzymatic degradation, and the delivery of NAP-loaded NPs to the brain was enhanced significantly after conjugation with TGN.

Angiopep-2–conjugated PEG–PCL NPs were also developed for brain targeting. Angiopep-2 is a ligand peptide for LRP receptor and is composed of 19 amino acids. Thus, the delivery of Angiopep-conjugated NPs to the brain was carried out through a LRP receptor–mediated transcytosis process. Angiopep-2–conjugated NPs are able to cross the BBB.[55]

Chitosan, a biocompatible and biodegradable polysaccharide polymer, is widely used in the pharmaceutical field. Its properties make chitosan a good candidate for the design of drug delivery systems, especially nano/microparticles. Furthermore, it improves the mucosal and transmucosal delivery of drugs mostly due to its mucoadhesive and absorption-enhancing properties, which are closely related to its positive charge. Chitosan has potential for transmucosal drug delivery, as it temporarily widens tight junctions between epithelial cells and facilitates the transport of poorly absorbable drugs through well-organized epithelial barriers. It has further been used in the preparation of brain drug delivery systems of different types of drugs/agents, such as gallic acid, dopamine, ropinirole, and rivastigmine.[47,56–59]

Chitosan NPs containing gallic acid, which is an antioxidant, were prepared by the modified ionotropic gelation method and subsequently coated with Tween 80, and the antiamnesic activity of these NPs was evaluated in scopolamine (SC)-induced amnesic mice. Gallic acid–loaded and Tween 80–coated NPs reversed SC-induced amnesic activity by reducing AChE activity, as compared to pure gallic acid.[47] Tween 80–coated chitosan NPs have also been designed by spontaneous emulsification method, so that rivastigmine could be used for the treatment of AD. Moreover, coating with Tween 80 altered the biodistribution of NPs.[58]

Chitosan NPs can be used to efficiently deliver small interfering RNAs (siRNAs) in neuronal cells. RNA interference technology is a therapeutic method for the reduction of pathological molecules in neurons.[60] TAT (RKKRRQRRR)-conjugated PEG–chitosan NPs containing siRNA have been fabricated using TAT as a cell-penetrating oligopeptide and PEG as a linker between chitosan and peptide. TAT-coated NPs are able to cross the BBB and localize in neurons. TAT destabilizes the lipid bilayer of the cell membrane through an energy-independent pathway, and arginine and lysine amino acids in TAT can translocate across biological membrane easily due to its strong cell adherence. These NPs may provide a promising carrier to deliver siRNA to neuronal cells with minimal toxicity and thereby treat neurodegenerative diseases.[60]

1.1.4 Lipid Nanoparticles

Recently, lipid NPs [e.g., solid lipid nanoparticles (SLNs) and nanostructured lipid carriers (NLCs)] have been utilized for the delivery of drugs to the brain. SLNs are colloidal dispersions having a mean particle size of 50–1000 nm and consist of a solid core. SLNs were prepared using solid lipids, being considered as safe (generally regarded as safe or GRAS), and useful biocompatible and biodegradable materials for drug delivery. Lipids, such as triacylglycerols, fatty acids, fatty alcohols, waxes, and cationic lipids, are commonly used in their solid phase at room temperature. SLNs have significant advantages in terms of the optimization of drug release, long shelf life, low chronic toxicity, site-specific targeting, protection of incorporated labile drugs, excellent physical stability, and various application routes (e.g., parenteral, dermal, ocular, oral, and pulmonary). There are some limitations for these systems, such as the low loading capacity, the risk of drug leakage after polymorphic transitions during storage, and the requirement of a high amount of water (>70%) for their composition. Both hydrophilic and hydrophobic drugs are encapsulated within SLNs. Drugs are dispersed in the lipid core or adsorbed on the surface, if a drug is molecularly dispersed within the lipid core or the drug concentration is relatively close to or at its saturation solubility in the lipid melt, then prolonged or/and controlled release profiles are achieved.[38,61–65] SLNs have great potential to overcome BBB and increase the concentration of drugs in the brain for the treatment of CNS disorders.[66]

The surface charges of SLNs affect their biodistribution. Positively charged and drug-loaded tripalmitin NPs showed higher blood concentrations, prolonged residence time, and a higher accumulation in the brain and bone after IV administration in mice compared to negatively charged SLN and pure drug.[67] In another study, it was showed that SLN could cross the BBB and accumulate in the brain parenchyma after IV administration to rats.[68]

SLNs containing superparamagnetic iron oxide were also prepared for use as a nuclear magnetic resonance contrast agent for CNS. These NPs were able to overcome BBB after IV administration to rats.[69]

Laserra et al.[70] developed lipoyl memantine–loaded SLNs to enhance their solubility and absorption through the gastrointestinal tract. Lipoyl memantine is a new potential anti-Alzheimer codrug synthesized from memantine (a noncompetitive NMDA receptor antagonist that is approved by FDA for the treatment of AD) and (*R*)-α-lipoic acid (a natural neuroprotective agent). It inhibits Aβ(1–42) aggregation and scavenges free radicals. The authors prepared and optimized the formulation (170 nm), which had a zeta potential of −33.8 mV, and an encapsulation efficiency of 88%. Cytotoxicity and oxidative damage tests were also carried out to evaluate the suitability of SLNs for brain delivery of lipoyl memantine. It was found that there were no significant changes in cell viability during 24 h for the lipoyl memantine–loaded SLNs (for 10 and 100 μM) and SLNs can be useful for the treatment of neurodegenerative disease, such as AD.

SLNs can be also applied as transdermal delivery systems. Therefore, a HA-loaded SLN–based gel formulation was designed to evaluate for the treatment of AD. In vitro permeability study showed that HA was released from SLN-based gel and permeated through the abdominal skin of rat. The study also determined that the formulation does not irritate the skin. The effects of HA-loaded SLN–based gel on learning and memory processes in mice with transient memory deficit were evaluated. The deficit was produced using SC and was especially evident in prolonging transfer latency. As a result, it was stated that reduced transfer latency over the period of 3 days, as well as significant improvement in cognitive function in mice, was obtained.[71]

NLC-containing ubiquinone (coenzyme Q10) was formulated by Nanjwade et al.[72] by the use of solid lipids and spatially incompatible liquid lipids. The addition of a liquid lipid into the solid lipid enhanced ubiquinone encapsulation efficiency and also improved drug release properties from ubiquinone-loaded NLC. In the in vivo study, ubiquinone-loaded NLC showed more antioxidant activity compared to ubiquinone solution and thus, it might be used to increase the antioxidant enzyme activity and reduce the oxidative stress in AD, PD, etc.[72]

Other examples concerning NP formulations used for brain drug targeted delivery are presented in Table 7.1.

1.2 Liposomes

Liposomes are vesicles that are majorly formed by phospholipids with spherical self-closed shape.[84,85] Liposome membranes are composed of two lipid layers that is collectively called a bilayer membrane.[86] Size of liposomes can vary in a wide range of 50–1000 nm.[87] Also liposomes may contain more than one bilayer membrane.[88] Thus classification of liposomes can be made based on size and number of bilayer membranes, which is summarized in Table 7.2.[89–91]

Table 7.1 Examples of NP Formulations Used for Brain Drug Targeted Delivery

NP Types	Materials/ Polymers	Therapeutics/ Active Agents	Key Achievements	Major Outcomes	References
Metallic NPs					
	Gold	None	Provides conductive matrix for the electrochemical detection of neurotransmitters, such as dopamine, L-DOPA, adrenaline, and noradrenaline	Quantitative analysis of neurotransmitters and quantitative assay of tyrosinase, which may be important for clinical diagnostics	[22]
	Gold	POMD and POMD-peptide	Provide one-step synthesis and could act as efficient vehicles for drug delivery across the BBB	Inhibits Aβ aggregation against AD	[73]
	Iron oxide	HO and HI peptides	Improved permeation across BBB	Early diagnosis of AD by molecular image	[74]
	Super paramagnetic iron oxide	DDNP	High binding affinities toward Aβ	Early diagnosis of AD	[75]
Silica NPs					
	Silica	Clioquinol as metal chelator	H_2O_2-responsive mesoporous silica NPs functionalized with phenylboronic acid released metal chelator in the case of increased levels of H_2O_2	Efficient inhibition of Aβ aggregation in the presence of H_2O_2	[76]
Polymeric NPs					
	PLGA	GDNF/VEGF	Combined administration of VEGF NPs and GDNF NPs (at lower doses of VEGF and GDNF)	Restoration of damaged brain areas in the treatment of PD	[42]
	PLGA–PEG	Selegiline (a monoamine oxidase-B inhibitor)	Increasing incubation time (until 6 h) of selegiline-loaded PEG-PLGA NPs with Aβ (1–40) and Aβ (1–42) and also increasing selegiline concentration resulted in increasing of the destabilizing effect of NPs on Aβ-fibrils	In vitro destabilizing effect on Aβ fibrils against AD	[77]
	Poly(alkyl cyanoacrylate)–PEG	Selegiline	Selegiline-decorated PEG-poly(alkyl cyanoacrylate) NPs had a lower interaction with Aβ-fibril (1–42) compared to the nonfunctionalized nanoparticles, because, higher selegiline content caused a decrease in the linear PEG chain density at the surface of the NPs	Higher linear PEG density at the surface of the NPs results in higher binding activity towards Aβ-fibril (1–42)	[46]

Poly(alkyl cyanoacrylate)–PEG	—	Prolonged circulation time in blood following IV application, selective adsorption of apo-E	Improve AD condition by blood peptide clearance	[78]
Chitosan	Ropinirole	Due to mucoadhesive properties of chitosan, increased nasal residence time of the NPs following nasal application and increased drug levels in the brain	Treatment of PD	[59]
PLGA–Chitosan	Anti-amyloid antibody IgG 4.1	Improved the BBB crossing ability and targeting of cerebrovascular Aβ proteins	Diagnosis and treatment of AD and cerebral amyloid angiopathy	[79]
Lipid NPs				
Solid lipids–Tween 80	Piperine	Effective drug delivery across the BBB	Therapeutic effects in AD by decreasing the amyloidal and tangle contents	[80]
Solid lipids	Sesamol	Increased drug bioavailability and brain penetration	Significant reduction in tumor necrosis factor-α	[81]
	Curcumin	Significantly increased the bioavailability of curcumin	Normalization of the brain microstructural elements and significant improvement in induced brain damage	[82]
	Quercetin	Improved permeation across BBB	Better memory retention in aluminium-induced neurotoxicity	[31]
	Bromocriptine	Prolonged drug release for 48 h	Rapid onset and prolonged activity compared to free drug against PD	[83]
Phospholipid-based gelatin	bFGF	Suitable system for nose-to-brain drug delivery	Evident therapeutic effect by stimulation of dopaminergic function for survival synapses on hemiparkinsonian rats	[65]

Aβ, β-Amyloid plaques (Aβ); AD, Alzheimer's disease; apo-E, apolipoprotein E; BBB, blood–brain barrier; bFGF, basic fibroblast growth factor; DDNP, 1,1-dicyano-2-[6-(dimethylamino)-naphthalene-2-yl] propene carboxyl derivative; GDNF, glial cell line–derived neurotrophic factor; IV, intravenous; NP, nanoparticle; PD, Parkinson's disease; PEG, poly(ethylene glycol); peptide HO, C-IPLPFYN-C; peptide HI, C-FRHMTEQC; PLGA, poly(lactic-co-glycolic) acid; POMD, polyoxometalate with Wells–Dawson structure; VEGF, vascular endothelial growth factor.

Table 7.2 Classification of Liposomes

Type of Vesicles	Estimated Diameter Size (nm)	Number of Bilayers
Small unilamellar	20–100	1
Large unilamellar	>100	1
Giant unilamellar	>1000	1
Oligolamellar	100–500	5
Multilamellar	>500	5–25
Multivesicular	>1000	Contains several vesicles

Data adapted from Laouini A, Jaafar-Maalej C, Limayem-Blouza I, Sfar S, Charcosset C, Fessi H. Preparation, characterization and applications of liposomes: state of the art. J Colloid Sci Biotechnol. *2012;1(2):147–168*[89]*; Tyagi S, Sharma PK, Malviya R. Advancement and patents on liposomal drug delivery.* Global J Pharmacol. *2015;9(2):166–173*[90]*; Tikshdeep C, Sonia A, Bharat P, Abhishek C. Liposome drug delivery: a review.* Int J Pharm Chem Sci. *2012;1(3):754–764.*[91]

Liposomes consist of both a hydrophilic core and a hydrophobic bilayer membrane, and this amphiphilic structure allows them to contain hydrophilic, as well as hydrophobic molecules.[92] Liposomes offer several advantages, such as enhancing pharmacokinetic properties; biodistribution and bioavailability of drug molecules, by protecting them from enzymatic degradation and macrophages; as well as increasing blood circulation time.[89,93] In addition, liposomes provide targeted selectivity, enhanced stability, reduced toxicity, and bind to site-specific ligands.[94,95]

Liposomes have been used as brain drug delivery carriers for a long time. They can be used to decrease effective drug dose due to their drug-entrapping feature and enhanced drug penetration through BBB.[96] Furthermore, employing surface modification by antibody binding to liposomes can allow coupling with receptor molecules on endothelial cells of the BBB.[96]

Results from studies of Di Stefano et al. indicated that dopamine-loaded liposomes prolonged dopamine release[97] and also demonstrated that liposomes showed better results when compared with free drug in the treatment of PD.[98] In addition, L-DOPA dimeric prodrugs encapsulated in liposomes showed basal dopamine levels 3 times higher than L-DOPA itself.[98]

Rivastigmine concentration and exposure are enhanced by liposomal formulations compared to nonliposomal formulations applied for drug delivery to the brain for the treatment of AD.[99,100] Furthermore, liposomes offer hope due to their targeting ability to Aβ, which is responsible for the amyloid plaques in AD patient brains.[101,102]

1.3 Dendrimers

Dendrimers can be defined as nanosized globular molecules with a well-defined unique 3D architecture. Low polydispersity, multivalency, and modifiable composition based on intended purposes are some of the favorable characteristics of dendrimers.[103] Dendrimers consist of three different domains which are: core, branching layer, and corona. Branching layers composed of repeating units are called generations, and they surround the core, while terminal reactive functional groups are at the outer periphery, forming the corona.[104]

Dendrimers are one of the favored drug delivery systems due to their tunable properties, such as size, shape, flexibility, functional groups, and number of generations.[105] Dendrimers can be used in drug delivery with two different approaches. First, when drug molecules are entrapped in dendrimer cavities, solubility of these hydrophobic molecules increases. Second, drug molecules may be conjugated to functional groups on the outer surface, and the release rate of drugs can be controlled by the feature linkers that are used during conjugation.[106]

Dendrimers can prevent the fibrillation of Aβ peptides by prohibiting their aggregation.[107,108] Wasiak et al. showed that cationic phosphorus dendrimers can modify aggregation of Aβ peptides and MAP-Tau proteins, which are the two key conducive agents for AD. Also, it was demonstrated that the dendrimers can reduce the toxicity of aggregated Aβ peptides.[109] In another study, maltose-decorated poly(propyleneimine) dendrimer formulations were effective on the fibril formation process of amyloid peptides.[110]

Fibrillar aggregation of α-synuclein, which is related with PD, can be inhibited by dendrimer formulations. Besides, PAMAM dendrimers may decompose aggregated fibrils depending on their concentration and generation.[111] The effect of generation was demonstrated by the comparison of PAMAM G4 and PAMAM 3.5 in the study by Milowska et al., and it was found that G4 generation inhibits α-synuclein fibrillation, while G3.5 does not influence fibrillation of α-synuclein.[112]

1.4 Polymeric Micelles

Micelles are colloidal particles with a core–shell architecture that is generally around 5–100 nm. Formation of micelles is established by amphiphilic unimers, which are usually block copolymers with hydrophilic and hydrophobic groups. These amphiphilic unimers aggregate by self-assembly at critical micelle concentrations and form micelles with hydrophobic core and hydrophilic corona.[113]

Surface functionality of micelles represented by hydrophilic block (for which PEG is generally favored due to its biocompatible), steric protector nature, and good stealth properties is FDA approved. Different from hydrophilic block, hydrophobic blocks can be chosen and modified from a wide variety of alternatives to obtain suitable formulations for drug molecules with high drug loading capacity and good compatibility.[114,115]

Polymeric micelles are able to enhance bioavailability of drug molecules by protecting the loaded drug, releasing the loaded drug at targeted sites, prolonging the residence time of loaded drug, and inhibiting efflux pumps.[116]

Due to these advantages, polymeric micelles can be used for CNS drug delivery. Kabanov et al. showed that haloperidol delivery to brain was increased with the use of polymeric micelle formulation formed by Pluronic copolymers with conjugation of antibody against insulin.[117]

Lu et al. developed polymeric micelle formulation loaded with resveratrol, which ensures protection from Aβ peptide toxicity. By using this system, protective feature of resveratrol was obtained without long-term cytotoxicity.[118]

1.5 Carbon Nanotubes

Carbon nanotubes (CNTs) can be defined as allotropes of carbons in tubular shape, which looks like graphene sheet rolls.[119] The diameter of CNTs varies from 1 to 100 nm, while the length of CNTs may be up to millimeter scale.[120] CNTs can be classified based on the number of layers: single-walled carbon nanotubes (SWCNTs) and multiple-walled carbon nanotubes (MWCNTs).[121] In addition to this classification, SWCNTs can be categorized as armchair carbon nanotubes, zigzag carbon nanotubes, and chiral carbon nanotubes by their crystallographic configurations.[122] Each type of CNTs has advantages, disadvantages, and unique properties due to their distinctive characteristics.[123]

CNTs have large specific surface areas on which a wide range of molecular moieties can be attached. Also CNTs can be functionalized to enhance their physicochemical properties and permeation abilities through cell membranes.[124]

Studies have shown that the delivery of drugs and genes to brain can be achieved with functionalized CNTs.[125] Thus, CNTs provide a new drug delivery approach to CNS diseases.

Delivery of acetyl choline (ACh), which is one of the drug molecules that is used in treatment of AD, to mice brain was achieved with SWCNTs by Yang et al. SWCNTs were used in this study due to their permeation ability into the brain and their absorption ability. Once this system entered the neuron's lysosome, ACh molecules were released and exhibited their neurotransmitter role.[126]

The ability to inhibit Aβ fibrillation of CNTs has been shown by Li et al. Average β-sheet content, which is strongly related to AD, ranged from 44.5% to 7.9% due to the presence of SWCNTs.[127] In another study, Xue et al. showed that SWCNTs are able to restore defective autophagy, which can cause neurodegenerative disorder development, such as AD.[128]

In addition to the feasible drug delivery and unique properties, in study by Vitale et al., CNTs were used as microelectrodes for electrical stimulations that

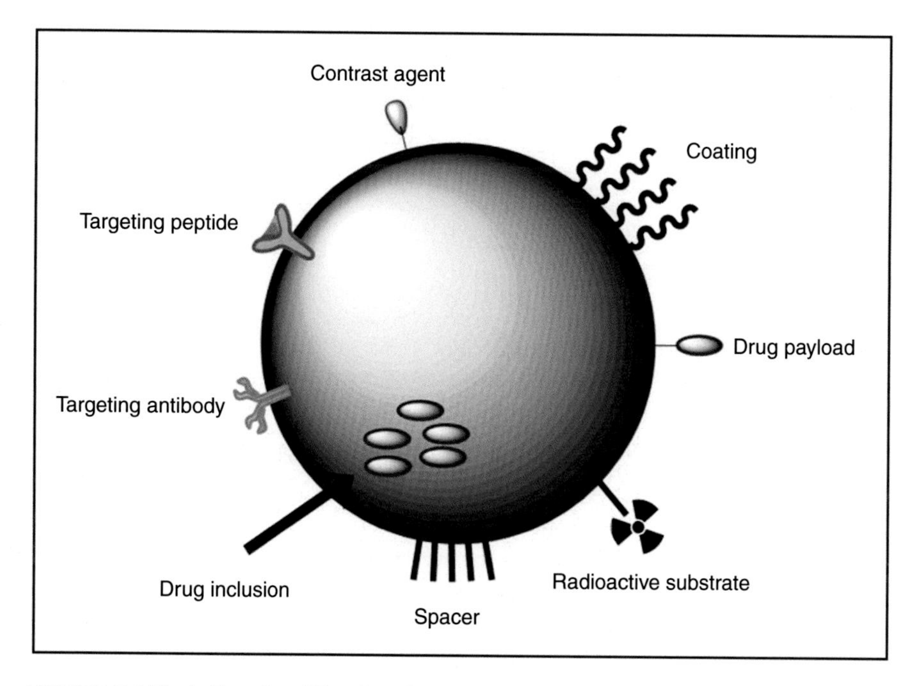

FIGURE 7.3 Illustration of multifunctionalized NPs.

were used in the treatment of neurological disorders, such as PD, due to their biological stability, mechanical strength, and electrical conductivity. This study showed that CNTs are promising candidates, in terms of their use as microelectrodes.[129]

NBDDS can be used solve potential problems of CNS-targeted therapies, such as passing the BBB. Besides, multifunctional properties, such as bioactivity, targeting, imaging capabilities, and gene delivery, can be accomplished concurrently with these systems (Fig. 7.3).

2 NEUROTOXICITY

Various chemicals have been reported to be more toxic in the nanoparticulate form than their microsized particles.[130] Although studies have shown that combustion-derived NPs, such as diesel exhaust particles and particulate matters, are neurotoxic, there is no certain information about manufactured NPs due to their aggregation properties.[131]

It was reported that oxidative stress induction by NP surface functionality plays an important role in the common toxicity mechanism of NPs, as oxidative stress triggers inflammation in the NP-depositing organs.[132,133] Potential pathways of NP neurotoxicity have been presented in Fig. 7.4.

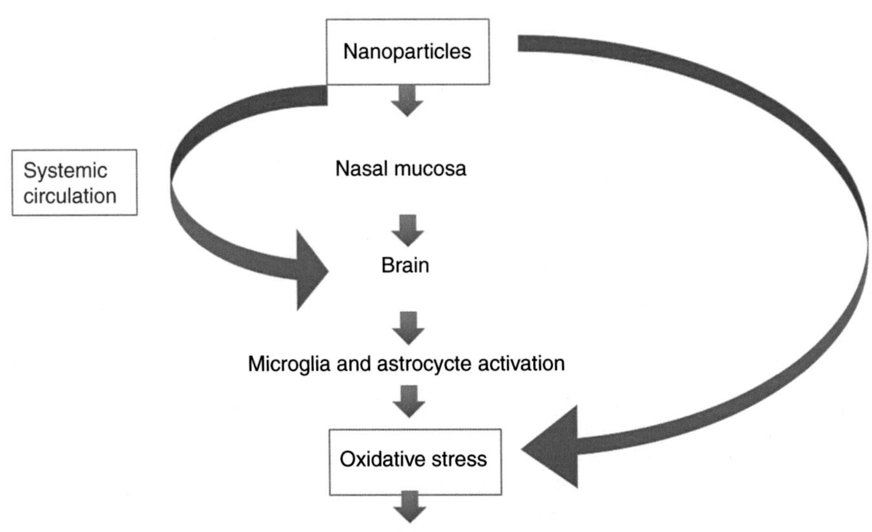

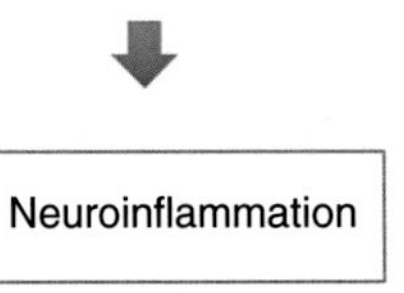

FIGURE 7.4 Potential pathways of NP-related neurotoxicity. *Data adapted from Win-Shwe TT, Fujimaki H. Nanoparticles and neurotoxicity.* Int J Mol Sci. *2011;12(9):6267–6280.*[134]

NP toxicity is known to be related to several parameters, such as chemical composition, crystal structure, purity, size, shape, and surface charge, which may result in a different distribution, accumulation, and transport of the nanomaterials.[135,136] NPs, similar to reactive oxygen species, are reportedly associated with neurodegenerative disorders, including PD, AD, and HD.[137] Based on studies performed using PCl2 cells, which are commonly employed for NP neurotoxicity studies, decreased PC12 cell viability, disturbed cell cycle, and nonuniform membrane borders were reported in the presence of carbon nanotubes,[138] anionic magnetic NPs,[139] and manganese oxide NPs.[140] Furthermore, several studies have reported potential NP toxicity in neuronal and glial cells based on in vitro nanotoxicological studies. However in vitro studies have some limitations and in vivo studies are required to assess organ sensitivity, pharmacokinetic factors, and interactions between cells.[134]

It was reported that following intranasal instillation of 14-nm carbon black NPs to mice, levels of neurotransmitters and proinflammatory cytokines were altered, and extracellular glutamate and glycine levels and IL-1β mRNAs were upregulated.[141] Another study showed that aluminum oxide NPs induced apoptosis and impaired spatial learning behavior.[142] Silica NPs, which are commonly used in drug delivery, diagnosis, and imaging for CNS, have been investigated in terms of neurotoxicity by Wu et al.[143] It was reported that silica NPs penetrated into the brain, reduced cell viability, disturbed cell cycle, triggered oxidative stress, induced lactate dehydrogenase levels and apoptosis, and activated the p53-mediated signaling pathway, following intranasal application.[143]

Numerous in vivo and in vitro studies have been performed to study the toxic effects of nanosized materials; however, there is still a lack of the understanding of the potential safety issues regarding nanomaterials. Also batch-to-batch inconsistency is a major problem in NP production, which results in unreliable results. Brain-targeted NBDDS should be carefully designed and evaluated before these systems will be eligible for clinical translation. Thus sensitive and specific methods are needed to quantify nanomaterials, and appropriate dose–response studies should be performed in neurotoxicological studies.[144]

3 CLINICAL CONSIDERATIONS

Nanotechnology-based drug delivery and diagnosis are promising innovations; however, unknown characteristics of nanosized materials raise fears for safety of the new system. As important as animal studies and laboratory experiments are, they cannot replace clinical studies, which offer knowledge about efficacy and safety about these new nanosystems. It is important to understand the stability of these nanotechnology products before clinical applications. Even if the interventions must be tested in vitro and in vivo prior to human studies, it is inevitable that human subjects may be exposed to some level of risks.[145] As the risk of failure in clinical trials is too expensive, standardized nanosafety research is required to develop more reliable systems and eliminate potential problematic candidates.[144]

Considering these risks, FDA requires preclinical studies involving animals, human cells, or tissues before initiating a clinical trial for a nanomedicinal product. Following FDA's evaluation of these results, they allow for Phase I studies with a small group of subjects[25–100] to determine the maximum tolerable dose for the product. These studies are followed by Phase II trials with 100–500 subjects, which evaluate the efficacy and safety of the developed nanomedicine. FDA allows conducting Phase III trials (500–3000 subjects) for drugs that past Phase II studies, and it is determined whether the application will be approved or not based on the results of Phase II.[146]

Since the first nanodrug Doxil has been approved by FDA, clinical studies for NBDDS have increased. As of January 2012, it was reported that 149 of the 789 ongoing clinical trials involved nanotechnology-based systems and 72% of these trials were found to be cancer treatment related.[147–150]

Table 7.3 Clinical Studies for Nanotechnology-Based Drug Delivery Systems (NBDDS) for Treatment of Neurological Diseases (NDs)

NBDDS	Diseases	Administration Routes	Main Conclusions
Magnetic iron oxide NP	Glioblastoma	Intratumoral	Safe and effective Longer overall survival[152]
Liposomal doxorubicin	Glioblastoma	Intravenous	No meaningful improvement[153]
	Brain metastases	Intravenous	Well tolerated Encouraging activity in brain metastases[154]
Liposomal cytarabine	Primary CNS tumors	Lumbar injection	Well tolerated and efficacious[155]
	Brain tumors	Intratechal	Well tolerated[156,157] Should be used only with dexamethasone
	Embryonal neoplasms	Intratechal	Improvement in response Low toxicity (for otherwise fatal patients)[158]
Cationic IL-12 liposomes	Malignant glioma	Intratumoral	Terminated due to the preclinical neurotoxicity reports[159]
Cationic IFN-β liposomes	Malignant glioma	Intratumoral	Feasible and safe[160]

CNS, Central nervous system; NP, nanoparticle.

Several NBDDS for brain drug delivery have been reported to be successful based on preclinical studies. Unfortunately there are very few clinical trials and these studies are mainly for CNS tumors or malignancies[151] as presented in Table 7.3.

4 ETHICAL AND REGULATORY ISSUES

The main ethical issue about nanotechnology is the conflict of risks and benefits associated with clinical implementation. Do these new systems overstrain the existing ethical and legal regulations, and is there any need for new regulations? When it comes to nanomedicinal concerns, laboratory and animal experiments, documentation, safety report, and long-term observation are some of the regulations that are required to reduce the risk to subjects.[145,161] The key point of the ethical issues about nanotechnology is that it is not possible to obtain a stable assessment, as every new development might result in benefits, as well as potential problems and health risks. Thus it seems unlikely to have a generic regulatory approach for environmental, health, and safety regulations.[162]

It has been claimed in a recent study[163] that there is no need for a special regulation of nanomedicine clinical trials and that current guidelines seem adequate. However, there are heightened concerns for identifying the protocols for novel technologies, such as gene therapy.[164] Also current guidelines and official regulatory approaches mainly focus on patients and clinical trial participants, whereas nanomedicine trails might affect other people, such as researchers, laboratory, and factory staff and their family members.[165]

FDA is one of the most important regulatory agencies, which ensures the safety and efficacy of drugs, medical devices, vaccines, veterinary products, and tobacco products. So far FDA has chosen to regulate nanomedicines and nanoproducts solely based on the current regulations; however, it's believed that a guidance or regulation for nanotechnology products is in order. As of 2012, FDA started to take an approach that determined whether the product involved use of nanotechnology or contained nanomaterials.[166]

Various nanoproducts have been approved and marketed; however, there is a growing concern that these products might be unsafe for human consumption due to unexpected toxicity effects based on their increased reactivity and tissue accumulation.[167,168] Considering these concerns, nanoproducts should be handled case by case rather than applying the current regulations, until new regulations will be available.[169]

5 CONCLUSIONS

NDs are major health diseases that cause high morbidity worldwide. For these burdens, NBDDS may serve as a revolutionary step in the nanoneuroscience field, enabling the delivery of therapeutic drugs across the BBB and promoting functional regeneration of damaged target sites, as well as maintaining neuroprotection. Although, the use of NBDDS is challenging in integrated therapies and other diagnostic methods, they have to be further improved for their physical, chemical, and biological characteristics. Besides the possible toxicities and immunological reactions, as well as the lack of scalability and hurdles with batch-to-batch variation, there is a still room for improvements and safer strategies. However, multidisciplanary approaches, as well as the translation to clinical applications, are critical and encouraged in this regard.

Abbreviations

Aβ Amyloid β
ACh Acetyl choline
AChE Acetyl cholinesterase
AD Alzheimer's disease
Apr Aprotinin
BBB Blood–brain barrier
CNS Central nervous system
CNTs Carbon nanotubes
CTAB Cetyltrimethylammonium bromide

GRAS	Generally regarded as safe
HA	Huperzine A
HD	Huntington's disease
IV	Intravenous
LDL	Low-density lipoprotein
LDLR	Low-density lipoprotein receptor
LRP	Low-density lipoprotein receptor–related protein
MCM	Mobil Composition of Matter
MWCNTs	Multiple-walled carbon nanotubes
N2a	Neuro2a
NBDDS	Nanotechnology-based drug delivery system
NDs	Neurological diseases
NGF	Nerve growth factor
NLCs	Nanostructured lipid carriers
NPs	Nanoparticles
PCL	Poly-e-caprolactone
PD	Parkinson's disease
PEG	Poly(ethylene glycol)
PGA	Polyglycolides
PLA	Polylactides
PLGA	Poly(lactide-*co*-glycolide)
SBA	Santa Barbara University mesoporous silica material
SC	Scopolamine
SGZ	Subgranular zone
siRNA	Small interfering RNA
SLNs	Solid lipid nanoparticles
SPIONs	Superparamagnetic iron oxide nanoparticles
SVZ	Subventricular zone
SWCNTs	Single-walled carbon nanotubes
VLDLR	Very low–density lipoprotein receptor

References

1. Shilo M, Motiei M, Hana P, Popovtzer R. Transport of nanoparticles through the blood–brain barrier for imaging and therapeutic applications. *Nanoscale*. 2014;6(4):2146–2152.
2. Garbayo E, Ansorena E, Blanco-Prieto MJ. Brain drug delivery systems for neurodegenerative disorders. *Curr Pharm Biotechnol*. 2012;13(12):2388–2402.
3. Koffie RM, Farrar CT, Saidi L-J, William CM, Hyman BT, Spires-Jones TL. Nanoparticles enhance brain delivery of blood–brain barrier-impermeable probes for in vivo optical and magnetic resonance imaging. *Proc Natl Acad Sci USA*. 2011;108(46):18837–18842.
4. Chen Y, Liu L. Modern methods for delivery of drugs across the blood–brain barrier. *Adv Drug Deliv Rev*. 2012;64(7):640–665.
5. Stamatovic SM, Sladojevic N, Keep RF, Andjelkovic AV. Blood-brain barrier permeability: from bench to bedside. Management Of Epilepsy–Research, Results and Treatment. InTech; 2011.
6. Kim KY. Research training and academic disciplines at the convergence of nanotechnology and biomedicine in the United States. *Nat Biotechnol*. 2007;25(3):359.

7. Brune H, Ernst H, Grunwald A, Grünwald W, Hofmann H, Krug H, et al. *Nanotechnology: Assessment and Perspectives*. Berlin: Springer Science and Business Media; 2006.
8. Pierotti MA, Lombardo C, Rosano C. Nanotechnology: going small for a giant leap in cancer diagnostics and therapeutics. *Tumori*. 2007;94(2):191–196.
9. Taniguchi N. On the basic concept of nano-technology. *Proceedings of the International Conference on Production Engineering*. Tokyo: Japan Society of Precision Engineering; 1974.
10. Gupta A, Arora A, Menakshi A, Sehgal A, Sehgal R. Nanotechnology and its applications in drug delivery: a review. *WebMed Central*. 2012;3(1):WMC002867.
11. Wen MM, El-Salamouni NS, El-Refaie WM, Hazzah HA, Ali MM, Tosi G, Farid RM, Blanco-Prieto MJ, Billa N, Hanafy AS. Nanotechnology-based drug delivery systems for Alzheimer's disease management: technical, industrial, and clinical challenges. *J Control Release*. 2017;245:95–107.
12. Brigger I, Dubernet C, Couvreur P. Nanoparticles in cancer therapy and diagnosis. *Adv Drug Delivery Rev*. 2002;54(5):631–651.
13. Varde NK, Pack DW. Microspheres for controlled release drug delivery. *Exp Opin Biol Ther*. 2004;4(1):35–51.
14. Kumar M. Nano and microparticles as controlled drug delivery devices. *J Pharm Pharm Sci*. 2000;3(2):234–258.
15. Madhav NS, Kala S. Review on microparticulate drug delivery system. *Int J PharmTech Res*. 2011;3(3):1242–1254.
16. Wang AZ, Langer R, Farokhzad OC. Nanoparticle delivery of cancer drugs. *Annu Rev Med*. 2012;63:185–198.
17. Wilczewska AZ, Niemirowicz K, Markiewicz KH, Car H. Nanoparticles as drug delivery systems. *Pharmacol Rep*. 2012;64(5):1020–1037.
18. Busquets MA, Sabaté R, Estelrich J. Potential applications of magnetic particles to detect and treat Alzheimer's disease. *Nanoscale Research Lett*. 2014;9(1):538.
19. Dürr S, Janko C, Lyer S, Tripal P, Schwarz M, Zaloga J, et al. Magnetic nanoparticles for cancer therapy. *Nanotechnol Rev*. 2013;2(4):395–409.
20. Amiri H, Saeidi K, Borhani P, Manafirad A, Ghavami M, Zerbi V. Alzheimer's disease: pathophysiology and applications of magnetic nanoparticles as MRI theranostic agents. *ACS Chem Neurosci*. 2013;4(11):1417–1429.
21. Ahmad MZ, Akhter S, Jain GK, Rahman M, Pathan SA, Ahmad FJ, et al. Metallic nanoparticles: technology overview and drug delivery applications in oncology. *Exp Opin Drug Deliv*. 2010;7(8):927–942.
22. Baron R, Zayats M, Willner I, Dopamine-. L-DOPA-, adrenaline-, and noradrenaline-induced growth of Au nanoparticles: assays for the detection of neurotransmitters and of tyrosinase activity. *Anal Chem*. 2005;77(6):1566–1571.
23. Zhou Y, Dong H, Liu L, Xu M. Simple Colorimetric Detection of amyloid β-peptide (1-40) based on aggregation of gold nanoparticles in the presence of copper ions. *Small*. 2015;11(18):2144–2149.
24. Mirsadeghi S, Dinarvand R, Ghahremani MH, Hormozi-Nezhad MR, Mahmoudi Z, Hajipour MJ, et al. Protein corona composition of gold nanoparticles/nanorods affects amyloid beta fibrillation process. *Nanoscale*. 2015;7(11):5004–5013.
25. Mahmoudi M, Serpooshan V. Large protein absorptions from small changes on the surface of nanoparticles. *J Phys Chem C*. 2011;115(37):18275–18283.
26. Williams BJ, Eriksdotter-Jonhagen M, Granholm A-C. Nerve growth factor in treatment and pathogenesis of Alzheimer's disease. *Progr Neurobiol*. 2006;80(3):114–128.

27. Marcus M, Skaat H, Alon N, Margel S, Shefi O. NGF-conjugated iron oxide nanoparticles promote differentiation and outgrowth of PC12 cells. *Nanoscale*. 2015;7(3):1058–1066.

28. Slowing II, Vivero-Escoto JL, Wu C-W, Lin VS-Y. Mesoporous silica nanoparticles as controlled release drug delivery and gene transfection carriers. *Adv Drug Deliv Rev*. 2008;60(11):1278–1288.

29. Jang S-F, Liu W-H, Song W-S, Chiang K-L, Ma H-I, Kao C-L, et al. Nanomedicine-based neuroprotective strategies in patient specific-iPSC and personalized medicine. *Int J Mol Sci*. 2014;15(3):3904–3925.

30. Sun SW, Yu HQ, Zhang H, Zheng YL, Wang JJ, Luo L. Quercetin attenuates spontaneous behavior and spatial memory impairment in D-galactose–treated mice by increasing brain antioxidant capacity. *Nutr Res*. 2007;27(3):169–175.

31. Dhawan S, Kapil R, Singh B. Formulation development and systematic optimization of solid lipid nanoparticles of quercetin for improved brain delivery. *J Pharm Pharmacol*. 2011;63(3):342–351.

32. Nday CM, Halevas E, Jackson GE, Salifoglou A. Quercetin encapsulation in modified silica nanoparticles: potential use against Cu (II)-induced oxidative stress in neurodegeneration. *J Inorg Biochem*. 2015;145:51–64.

33. Kim IY, Joachim E, Choi H, Kim K. Toxicity of silica nanoparticles depends on size, dose, and cell type. *Nanomedicine*. 2015;11(6):1407–1416.

34. Yang X, He Ce, Li J, Chen H, Ma Q, Sui X, et al. Uptake of silica nanoparticles: neurotoxicity and Alzheimer-like pathology in human SK-N-SH and mouse neuro2a neuroblastoma cells. *Toxicol Lett*. 2014;229(1):240–249.

35. Fubini B, Hubbard A. Reactive oxygen species (ROS) and reactive nitrogen species (RNS) generation by silica in inflammation and fibrosis. *Free Radical Biol Med*. 2003;34(12):1507–1516.

36. Lin W, Huang Y-w, Zhou X-D, Ma Y. In vitro toxicity of silica nanoparticles in human lung cancer cells. *Toxicol App Pharmacol*. 2006;217(3):252–259.

37. Cho M, Cho W-S, Choi M, Kim SJ, Han BS, Kim SH, et al. The impact of size on tissue distribution and elimination by single intravenous injection of silica nanoparticles. *Toxicol Lett*. 2009;189(3):177–183.

38. Goyal K, Koul V, Singh Y, Anand A. Targeted drug delivery to central nervous system (CNS) for the treatment of neurodegenerative disorders: trends and advances. *Cent Nerv Syst Agents Med Chem*. 2014;14(1):43–59.

39. Mansour HM, Sohn M, Al-Ghananeem A, DeLuca PP. Materials for pharmaceutical dosage forms: molecular pharmaceutics and controlled release drug delivery aspects. *Int J Mol Sci*. 2010;11(9):3298–3322.

40. Pahuja R, Seth K, Shukla A, Shukla RK, Bhatnagar P, Chauhan LKS, et al. Trans-blood brain barrier delivery of dopamine-loaded nanoparticles reverses functional deficits in parkinsonian rats. *ACS Nano*. 2015;9(5):4850–4871.

41. Jose S, Sowmya S, Cinu T, Aleykutty N, Thomas S, Souto E. Surface modified PLGA nanoparticles for brain targeting of Bacoside-A. *Eur J Pharm Sci*. 2014;63:29–35.

42. Herrán E, Requejo C, Ruiz-Ortega JA, Aristieta A, Igartua M, Bengoetxea H, et al. Increased antiparkinson efficacy of the combined administration of VEGF-and GDNF-loaded nanospheres in a partial lesion model of Parkinson's disease. *Int J Nanomed*. 2014;9:2677.

43. Joshi AS, Thakur AK. Biodegradable delivery system containing a peptide inhibitor of polyglutamine aggregation: a step toward therapeutic development in Huntington's disease. *J Pept Sci*. 2014;20(8):630–639.

44. Tiwari SK, Agarwal S, Seth B, Yadav A, Nair S, Bhatnagar P, et al. Curcumin-loaded nanoparticles potently induce adult neurogenesis and reverse cognitive deficits in Alzheimer's disease model via canonical Wnt/β-catenin pathway. *ACS Nano*. 2013;8(1):76–103.

45. Zhang L, Han L, Qin J, Lu W, Wang J. The use of borneol as an enhancer for targeting aprotinin-conjugated PEG-PLGA nanoparticles to the brain. *Pharm Res*. 2013;30(10):2560–2572.

46. Le Droumaguet B, Souguir H, Brambilla D, Verpillot R, Nicolas J, Taverna M, et al. Selegiline-functionalized, PEGylated poly (alkyl cyanoacrylate) nanoparticles: Investigation of interaction with amyloid-β peptide and surface reorganization. *Int J Pharm*. 2011;416(2):453–460.

47. Nagpal K, Singh S, Mishra D. Nanoparticle mediated brain targeted delivery of gallic acid: in vivo behavioral and biochemical studies for protection against scopolamine-induced amnesia. *Drug Deliv*. 2013;20(3–4):112–119.

48. Wohlfart S, Gelperina S, Kreuter J. Transport of drugs across the blood–brain barrier by nanoparticles. *J Control Release*. 2012;161(2):264–273.

49. Möncke-Buchner E, Reich S, Mücke M, Reuter M, Messer W, Wanker EE, et al. Counting CAG repeats in the Huntington's disease gene by restriction endonuclease EcoP15I cleavage. *Nucleic Acids Res*. 2002;30(16):e83.

50. Kulkarni SA, Feng S-S. Effects of particle size and surface modification on cellular uptake and biodistribution of polymeric nanoparticles for drug delivery. *Pharm Res*. 2013;30(10):2512–2522.

51. Latchney SE, Eisch AJ. Therapeutic application of neural stem cells and adult neurogenesis for neurodegenerative disorders: regeneration and beyond. *Eur J Neurodegener Dis*. 2012;1(3):335.

52. Grover A, Hirani A, Sutariya V. Nanoparticle-based brain targeted delivery systems. *J Biomol Res Ther*. 2013;2:1.

53. Olivier J-C. Drug transport to brain with targeted nanoparticles. *NeuroRx*. 2005;2(1):108–119.

54. Li J, Zhang C, Li J, Fan L, Jiang X, Chen J, et al. Brain delivery of NAP with PEG-PLGA nanoparticles modified with phage display peptides. *Pharm Res*. 2013;30(7):1813–1823.

55. Xin H, Sha X, Jiang X, Chen L, Law K, Gu J, et al. The brain targeting mechanism of Angiopep-conjugated poly (ethylene glycol)-*co*-poly (ε-caprolactone) nanoparticles. *Biomaterials*. 2012;33(5):1673–1681.

56. Rodrigues S, Dionísio M, López CR, Grenha A. Biocompatibility of chitosan carriers with application in drug delivery. *J Funct Biomater*. 2012;3(3):615–641.

57. Trapani A, De Giglio E, Cafagna D, Denora N, Agrimi G, Cassano T, et al. Characterization and evaluation of chitosan nanoparticles for dopamine brain delivery. *Int J Pharm*. 2011;419(1):296–307.

58. Wilson B, Samanta MK, Muthu MS, Vinothapooshan G. Design and evaluation of chitosan nanoparticles as novel drug carrier for the delivery of rivastigmine to treat Alzheimer's disease. *Ther Deliv*. 2011;2(5):599–609.

59. Jafarieh O, Md S, Ali M, Baboota S, Sahni J, Kumari B, et al. Design, characterization, and evaluation of intranasal delivery of ropinirole-loaded mucoadhesive nanoparticles for brain targeting. *Drug Dev Ind Pharm*. 2015;41(10):1674–1681.

60. Malhotra M, Tomaro-Duchesneau C, Prakash S. Synthesis of TAT peptide-tagged PEGylated chitosan nanoparticles for siRNA delivery targeting neurodegenerative diseases. *Biomaterials*. 2013;34(4):1270–1280.

61. Md S, Haque S, Sahni JK, Baboota S, Ali J. New non-oral drug delivery systems for Parkinson's disease treatment. *Exp Opin Drug Deliv*. 2011;8(3):359–374.

62. Blasi P, Giovagnoli S, Schoubben A, Ricci M, Rossi C. Solid lipid nanoparticles for targeted brain drug delivery. *Adv Drug Deliv Rev*. 2007;59(6):454–477.

63. Martins SM, Sarmento B, Nunes C, Lúcio M, Reis S, Ferreira DC. Brain targeting effect of camptothecin-loaded solid lipid nanoparticles in rat after intravenous administration. *Eur J Pharm Biopharm*. 2013;85(3):488–502.

64. Uchegbu IF, Schätzlein AG, Cheng WP, Lalatsa A. *Fundamentals of Pharmaceutical Nanoscience*. New York, NY: Springer; 2013.
65. Zhao Y-Z, Li X, Lu C-T, Lin M, Chen L-J, Xiang Q, et al. Gelatin nanostructured lipid carriers-mediated intranasal delivery of basic fibroblast growth factor enhances functional recovery in hemiparkinsonian rats. *Nanomedicine*. 2014;10(4):755–764.
66. Masserini M. Nanoparticles for brain drug delivery. *ISRN Biochem*. 2013;2013:238428.
67. Reddy LH, Sharma RK, Chuttani K, Mishra AK, Murthy RR. Etoposide-incorporated tripalmitin nanoparticles with different surface charge: formulation, characterization, radiolabeling, and biodistribution studies. *AAPS J*. 2004;6(3):55–64.
68. Podio V, Zara GP, Carazzone M, Cavalli R, Gasco MR. Biodistribution of stealth and non-stealth solid lipid nanospheres after intravenous administration to rats. *J Pharm Pharmacol*. 2000;52(9):1057–1063.
69. Peira E, Marzola P, Podio V, Aime S, Sbarbati A, Gasco MR. In vitro and in vivo study of solid lipid nanoparticles loaded with superparamagnetic iron oxide. *J Drug Target*. 2003;11(1):19–24.
70. Laserra S, Basit A, Sozio P, Marinelli L, Fornasari E, Cacciatore I, et al. Solid lipid nanoparticles loaded with lipoyl–memantine codrug: Preparation and characterization. *Int J Pharm*. 2015;485(1):183–191.
71. Patel PA, Patil SC, Kalaria DR, Kalia YN, Patravale VB. Comparative in vitro and in vivo evaluation of lipid based nanocarriers of Huperzine A. *Int J Pharm*. 2013;446(1):16–23.
72. Nanjwade BK, Kadam VT, Manvi F. Formulation and characterization of nanostructured lipid carrier of ubiquinone (coenzyme Q10). *J Biomed Nanotechnol*. 2013;9(3):450–460.
73. Gao N, Sun H, Dong K, Ren J, Qu X. Gold-nanoparticle-based multifunctional amyloid-β inhibitor against Alzheimer's disease. *Chemistry*. 2015;21(2):829–835.
74. Ansciaux E, Burtea C, Laurent S, Crombez D, Nonclercq D, Vander Elst L, et al. In vitro and in vivo characterization of several functionalized ultrasmall particles of iron oxide, vectorized against amyloid plaques and potentially able to cross the blood–brain barrier: toward earlier diagnosis of Alzheimer's disease by molecular imaging. *Contrast Media Mol Imaging*. 2015;10(3):211–224.
75. Zhang D, Fa H-B, Zhou J-T, Li S, Diao X-W, Yin W. The detection of β-amyloid plaques in an Alzheimer's disease rat model with DDNP-SPIO. *Clin Radiol*. 2015;70(1):74–80.
76. Geng J, Li M, Wu L, Chen C, Qu X. Mesoporous silica nanoparticle-based H_2O_2 responsive controlled-release system used for Alzheimer's disease treatment. *Adv Healthcare Mater*. 2012;1(3):332–336.
77. Baysal I, Yabanoglu-Ciftci S, Tunc-Sarisozen Y, Ulubayram K, Ucar G. Interaction of selegiline-loaded PLGA-b-PEG nanoparticles with beta-amyloid fibrils. *J Neural Transm*. 2013;120(6):903–910.
78. Brambilla D, Verpillot R, Le Droumaguet B, Nicolas J, Taverna M, Kóňa J, et al. PEGylated nanoparticles bind to and alter amyloid-beta peptide conformation: toward engineering of functional nanomedicines for Alzheimer's disease. *ACS Nano*. 2012;6(7):5897–5908.
79. Jaruszewski KM, Ramakrishnan S, Poduslo JF, Kandimalla KK. Chitosan enhances the stability and targeting of immuno-nanovehicles to cerebro-vascular deposits of Alzheimer's disease amyloid protein. *Nanomedicine*. 2012;8(2):250–260.
80. Yusuf M, Khan M, Khan RA, Ahmed B. Preparation, characterization, in vivo and biochemical evaluation of brain targeted Piperine solid lipid nanoparticles in an experimentally induced Alzheimer's disease model. *J Drug Target*. 2013;21(3):300–311.
81. Sachdeva AK, Misra S, Kaur IP, Chopra K. Neuroprotective potential of sesamol and its loaded solid lipid nanoparticles in ICV-STZ-induced cognitive deficits: behavioral and biochemical evidence. *Eur J Pharmacol*. 2015;747:132–140.

82. Kakkar V, Kaur IP. Evaluating potential of curcumin loaded solid lipid nanoparticles in aluminium induced behavioural, biochemical and histopathological alterations in mice brain. *Food Chem Toxicol*. 2011;49(11):2906–2913.
83. Esposito E, Fantin M, Marti M, Drechsler M, Paccamiccio L, Mariani P, et al. Solid lipid nanoparticles as delivery systems for bromocriptine. *Pharm Res*. 2008;25(7):1521–1530.
84. Garg T, K Goyal A. Liposomes: targeted and controlled delivery system. *Drug Deliv Lett*. 2014;4(1):62–71.
85. Daraee H, Etemadi A, Kouhi M, Alimirzalu S, Akbarzadeh A. Application of liposomes in medicine and drug delivery. *Artif Cells Nanomed Biotechnol*. 2014;44:1–11.
86. Sharma P, Banerjee R, Narayan KP. Liposomes for controlled drug delivery: drugs of the future. *J Pharm Res*. 2014;8(5):637–641.
87. Kumar KS, Bhowmik D, Deb L. Recent trends in liposomes used as novel drug delivery system. *Pharma Innov*. 2012;1(1):29.
88. Akbarzadeh A, Rezaei-Sadabady R, Davaran S, Joo SW, Zarghami N, Hanifehpour Y, et al. Liposome: classification, preparation, and applications. *Nanoscale Res Lett*. 2013;8(1):102.
89. Laouini A, Jaafar-Maalej C, Limayem-Blouza I, Sfar S, Charcosset C, Fessi H. Preparation, characterization and applications of liposomes: state of the art. *J Colloid Sci Biotechnol*. 2012;1(2):147–168.
90. Tyagi S, Sharma PK, Malviya R. Advancement and patents on liposomal drug delivery. *Global J Pharmacol*. 2015;9(2):166–173.
91. Tikshdeep C, Sonia A, Bharat P, Abhishek C. Liposome drug delivery: a review. *Int J Pharm Chem Sci*. 2012;1(3):1103–1113.
92. Pattni BS, Chupin VV, Torchilin VP. New developments in liposomal drug delivery. *Chem Rev*. 2015;115(19):10938–10966.
93. Gentile E, Cilurzo F, Di Marzio L, Carafa M, Anna Ventura C, Wolfram J, et al. Liposomal chemotherapeutics. *Future Oncol*. 2013;9(12):1849–1859.
94. Dua J, Rana A, Bhandari A. Liposome: methods of preparation and applications. *Int J Pharm Stud Res*. 2012;3:14–20.
95. Sherry M, Charcosset C, Fessi H, Greige-Gerges H. Essential oils encapsulated in liposomes: a review. *J Liposome Res*. 2013;23(4):268–275.
96. Spuch C, Navarro C. Liposomes for targeted delivery of active agents against neurodegenerative diseases (Alzheimer's disease and Parkinson's disease). *J Drug Deliv*. 2011;2011:469679.
97. Di Stefano A, Carafa M, Sozio P, Pinnen F, Braghiroli D, Orlando G, et al. Evaluation of rat striatal L-dopa and DA concentration after intraperitoneal administration of L-dopa prodrugs in liposomal formulations. *J Control Release*. 2004;99(2):293–300.
98. Di Stefano A, Sozio P, Iannitelli A, Marianecci C, Santucci E, Carafa M. Maleic- and fumaric-diamides of (O,O-diacetyl)-L-Dopa-methylester as anti-Parkinson prodrugs in liposomal formulation. *J Drug Target*. 2006;14(9):652–661.
99. Arumugam K, Subramanian G, Mallayasamy S, Averineni R, Reddy M, Udupa N. A study of rivastigmine liposomes for delivery into the brain through intranasal route. *Acta Pharm*. 2008;58(3):287–297.
100. Mutlu NB, Değim Z, Yılmaz Ş, Eşsiz D, Nacar A. New perspective for the treatment of Alzheimer diseases: liposomal rivastigmine formulations. *Drug Dev Ind Pharm*. 2011;37(7):775–789.
101. Gobbi M, Re F, Canovi M, Beeg M, Gregori M, Sesana S, et al. Lipid-based nanoparticles with high binding affinity for amyloid-β 1-42 peptide. *Biomaterials*. 2010;31(25):6519–6529.
102. Canovi M, Markoutsa E, Lazar AN, Pampalakis G, Clemente C, Re F, et al. The binding affinity of anti-Aβ1-42 MAb-decorated nanoliposomes to Aβ1-42 peptides in vitro and to amyloid deposits in post-mortem tissue. *Biomaterials*. 2011;32(23):5489–5497.

103. Xu L, Zhang H, Wu Y. Dendrimer advances for the central nervous system delivery of therapeutics. *ACS Chem Neurosci.* 2013;5(1):2–13.

104. Tomalia DA. Birth of a new macromolecular architecture: dendrimers as quantized building blocks for nanoscale synthetic organic chemistry. *Aldrichim Acta.* 2004;37(2):39–57.

105. Abbasi E, Aval SF, Akbarzadeh A, Milani M, Nasrabadi HT, Joo SW, et al. Dendrimers: synthesis, applications, and properties. *Nanoscale Res Lett.* 2014;9(1):1–10.

106. Wu LP, Ficker M, Christensen JB, Trohopoulos PN, Moghimi SM. Dendrimers in medicine: therapeutic concepts and pharmaceutical challenges. *Bioconjugate Chem.* 2015;26(7): 1198–1211.

107. Klajnert B, Cangiotti M, Calici S, Majoral JP, Caminade AM, Cladera J, et al. EPR study of the interactions between dendrimers and peptides involved in Alzheimer's and prion diseases. *Macromol Biosci.* 2007;7(8):1065–1074.

108. Klajnert B, Cortijo-Arellano M, Bryszewska M, Cladera J. Influence of heparin and dendrimers on the aggregation of two amyloid peptides related to Alzheimer's and prion diseases. *Biochem Biophys Res Commun.* 2006;339(2):577–582.

109. Wasiak T, Ionov M, Nieznanski K, Nieznanska H, Klementieva O, Granell M, et al. Phosphorus dendrimers affect Alzheimer's (Aβ1-28) peptide and MAP-Tau protein aggregation. *Mol Pharm.* 2012;9(3):458–469.

110. Klementieva O, Benseny-Cases Nr, Gella A, Appelhans D, Voit B, Cladera J. Dense shell glycodendrimers as potential nontoxic anti-amyloidogenic agents in Alzheimer's disease. Amyloid–dendrimer aggregates morphology and cell toxicity. *Biomacromolecules.* 2011;12(11): 3903–3909.

111. Rekas A, Lo V, Gadd GE, Cappai R, Yun SI. PAMAM Dendrimers as Potential Agents against Fibrillation of α-synuclein, a Parkinson's disease-related protein. *Macromol Biosci.* 2009;9(3):230–238.

112. Milowska K, Malachowska M, Gabryelak T. PAMAM G4 dendrimers affect the aggregation of α-synuclein. *Int J Biol Macromol.* 2011;48(5):742–746.

113. Torchilin V. Targeted polymeric micelles for delivery of poorly soluble drugs. *Cell Mol Life Sci.* 2004;61(19–20):2549–2559.

114. Gothwal A, Khan I, Gupta U. Polymeric micelles: recent advancements in the delivery of anticancer drugs. *Pharm Res.* 2016;33:18–39.

115. Oerlemans C, Bult W, Bos M, Storm G, Nijsen JFW, Hennink WE. Polymeric micelles in anticancer therapy: targeting, imaging and triggered release. *Pharm Res.* 2010;27(12):2569–2589.

116. Xu W, Ling P, Zhang T. Polymeric micelles, a promising drug delivery system to enhance bioavailability of poorly water-soluble drugs. *J Drug Deliv.* 2013;2013:340315.

117. Kabanov AV, Chekhonin V, Alakhov VY, Batrakova E, Lebedev A, Melik-Nubarov N, et al. The neuroleptic activity of haloperidol increases after its solubilization in surfactant micelles: micelles as microcontainers for drug targeting. *FEBS Lett.* 1989;258(2):343–345.

118. Lu X, Ji C, Xu H, Li X, Ding H, Ye M, et al. Resveratrol-loaded polymeric micelles protect cells from Aβ-induced oxidative stress. *Int J Pharm.* 2009;375(1–2):89–96.

119. Hirlekar R, Yamagar M, Garse H, Vij M, Kadam V. Carbon nanotubes and its applications: a review. *Asian J Pharm Clin Res.* 2009;2(4):17–27.

120. Singh B, Baburao C, Pispati V, Pathipati H, Muthy N, Prassana S, et al. Carbon nanotubes. A novel drug delivery system. *Int J Res Pharm Chem.* 2012;2(2):523–532.

121. Saifuddin N, Raziah A, Junizah A. Carbon nanotubes: a review on structure and their interaction with proteins. *J Chem.* 2013;2013:676815.

122. Grobert N. Carbon nanotubes–becoming clean. *Mater Today.* 2007;10(1):28–35.

123. Eatemadi A, Daraee H, Karimkhanloo H, Kouhi M, Zarghami N, Akbarzadeh A, et al. Carbon nanotubes: properties, synthesis, purification, and medical applications. *Nanoscale Res Lett.* 2014;9(1):1–13.

124. Usui Y, Haniu H, Tsuruoka S, Saito N. Carbon nanotubes innovate on medical technology. *Med Chem.* 2012;2(1):1–6.

125. Wang JT-W, Al-Jamal KT. Functionalized carbon nanotubes: revolution in brain delivery. *Nanomedicine.* 2015;10(17):2639–2642.

126. Yang Z, Zhang Y, Yang Y, Sun L, Han D, Li H, et al. Pharmacological and toxicological target organelles and safe use of single-walled carbon nanotubes as drug carriers in treating Alzheimer disease. *Nanomedicine.* 2010;6(3):427–441.

127. Li H, Luo Y, Derreumaux P, Wei G. Carbon nanotube inhibits the formation of β-sheet-rich oligomers of the Alzheimer's amyloid-β (16-22) peptide. *Biophys J.* 2011;101(9):2267–2276.

128. Xue X, Wang L-R, Sato Y, Jiang Y, Berg M, Yang D-S, et al. Single-walled carbon nanotubes alleviate autophagic/lysosomal defects in primary glia from a mouse model of Alzheimer's disease. *Nano Lett.* 2014;14(9):5110–5117.

129. Vitale F, Summerson SR, Aazhang B, Kemere C, Pasquali M. Neural stimulation and recording with bidirectional, soft carbon nanotube fiber microelectrodes. *ACS Nano.* 2015;9(4):4465–4474.

130. Donaldson K, Stone V, Clouter A, Renwick L, MacNee W. Ultrafine particles. *Occup Environ Med.* 2001;58(3):211–216.

131. Morimoto Y, Kobayashi N, Shinohara N, Myojo T, Tanaka I, Nakanishi J. Hazard assessments of manufactured nanomaterials. *J Occup Health.* 2010;52(6):325–334.

132. Oberdörster G, Sharp Z, Atudorei V, Elder A, Gelein R, Kreyling W, et al. Translocation of inhaled ultrafine particles to the brain. *Inhalat Toxicol.* 2004;16(6–7):437–445.

133. Nel A, Xia T, Mädler L, Li N. Toxic potential of materials at the nanolevel. *Science.* 2006;311(5761):622–627.

134. Win-Shwe T-T, Fujimaki H. Nanoparticles neurotoxicity. *Int J Mol Sci.* 2011;12(9):6267–6280.

135. Pehlivan SB. Nanotechnology-based drug delivery systems for targeting, imaging and diagnosis of neurodegenerative diseases. *Pharm Res.* 2013;30(10):2499–2511.

136. Karmakar A, Zhang Q, Zhang Y. Neurotoxicity of nanoscale materials. *J Food Drug Anal.* 2014;22(1):147–160.

137. MatÉs JM, Pérez-Gómez C, De Castro IN. Antioxidant enzymes and human diseases. *Clin Biochem.* 1999;32(8):595–603.

138. Wang J, Sun P, Bao Y, Liu J, An L. Cytotoxicity of single-walled carbon nanotubes on PC12 cells. *Toxicol In Vitro.* 2011;25(1):242–250.

139. Pisanic TR, Blackwell JD, Shubayev VI, Fiñones RR, Jin S. Nanotoxicity of iron oxide nanoparticle internalization in growing neurons. *Biomaterials.* 2007;28(16):2572–2581.

140. Hussain SM, Javorina AK, Schrand AM, Duhart HM, Ali SF, Schlager JJ. The interaction of manganese nanoparticles with PC-12 cells induces dopamine depletion. *Toxicol Sci.* 2006;92(2):456–463.

141. Mitsushima D, Yamamoto S, Fukushima A, Funabashi T, Kobayashi T, Fujimaki H. Changes in neurotransmitter levels and proinflammatory cytokine mRNA expressions in the mice olfactory bulb following nanoparticle exposure. *Toxicol Appl Pharmacol.* 2008;226(2):192–198.

142. Zhang Q, Li M, Ji J, Gao F, Bai R, Chen C, et al. In vivo toxicity of nano-alumina on mice neurobehavioral profiles and the potential mechanisms. *Int J Immunopathol Pharmacol.* 2010;24(1 suppl):23S–29S.

143. Wu J, Wang C, Sun J, Xue Y. Neurotoxicity of silica nanoparticles: brain localization and dopaminergic neurons damage pathways. *ACS Nano.* 2011;5(6):4476–4489.

144. Hofmann-Amtenbrink M, Grainger DW, Hofmann H. Nanoparticles in medicine: current challenges facing inorganic nanoparticle toxicity assessments and standardizations. *Nanomedicine*. 2015;11(7):1689–1694.

145. Wiesing U, Clausen J. The clinical research of nanomedicine: a new ethical challenge?. *Nanoethics*. 2014;8(1):19–28.

146. Resnik DB, Tinkle SS. Ethical issues in clinical trials involving nanomedicine. *Contemp Clin Trials*. 2007;28(4):433–441.

147. Weissig V, Pettinger TK, Murdock N. Nanopharmaceuticals (part 1): products on the market. *Int J Nanomed*. 2014;9:4357.

148. Uchegbu IF, Siew A. Nanomedicines and nanodiagnostics come of age. *J Pharm Sci*. 2013;102(2):305–310.

149. Onoue S, Yamada S, Chan H-K. Nanodrugs: pharmacokinetics and safety. *Int J Nanomed*. 2014;9:1025–1037.

150. Mallapragada SK, Brenza TM, McMillan JM, Narasimhan B, Sakaguchi DS, Sharma AD, et al. Enabling nanomaterial, nanofabrication and cellular technologies for nanoneuromedicines. *Nanomedicine*. 2015;11(3):715–729.

151. Balakrishnan B, Nance E, Johnston MV, Kannan R, Kannan S. Nanomedicine in cerebral palsy. *Int J Nanomed*. 2013;8:4183–4195.

152. Maier-Hauff K, Ulrich F, Nestler D, Niehoff H, Wust P, Thiesen B, et al. Efficacy and safety of intratumoral thermotherapy using magnetic iron-oxide nanoparticles combined with external beam radiotherapy on patients with recurrent glioblastoma multiforme. *J Neurooncol*. 2011;103(2):317–324.

153. Beier CP, Schmid C, Gorlia T, Kleinletzenberger C, Beier D, Grauer O, et al. RNOP-09: pegylated liposomal doxorubicine and prolonged temozolomide in addition to radiotherapy in newly diagnosed glioblastoma-a phase II study. *BMC Cancer*. 2009;9(1):308.

154. Caraglia M, Addeo R, Costanzo R, Montella L, Faiola V, Marra M, et al. Phase II study of temozolomide plus pegylated liposomal doxorubicin in the treatment of brain metastases from solid tumours. *Cancer Chemother Pharmacol*. 2006;57(1):34–39.

155. Navajas A, Lassaletta Á, Morales A, López-Ibor B, Sábado C, Moscardó C, et al. Efficacy and safety of liposomal cytarabine in children with primary CNS tumours with leptomeningeal involvement. *Clin Transl Oncol*. 2012;14(4):280.

156. Benesch M, Siegler N, von Hoff K, Lassay L, Kropshofer G, Müller H, et al. Safety and toxicity of intrathecal liposomal cytarabine (Depocyte) in children and adolescents with recurrent or refractory brain tumors: a multi-institutional retrospective study. *Anti-Cancer Drugs*. 2009;20(9):794–799.

157. Lassaletta A, Lopez-Ibor B, Mateos E, Gonzalez-Vicent M, Perez-Martinez A, Sevilla J, et al. Intrathecal liposomal cytarabine in children under 4 years with malignant brain tumors. *J Neurooncol*. 2009;95(1):65–69.

158. Partap S, Murphy PA, Vogel H, Barnes PD, Edwards MS, Fisher PG. Liposomal cytarabine for central nervous system embryonal tumors in children and young adults. *J Neurooncol*. 2011;103(3):561–566.

159. Graham A, Walker R, Baird P, Hahn CN, Fazakerley JK. CNS gene therapy applications of the Semliki Forest virus 1 vector are limited by neurotoxicity. *Mol Ther*. 2006;13(3):631–635.

160. Yoshida J, Mizuno M, Fujii M, Kajita Y, Nakahara N, Hatano M, et al. Human gene therapy for malignant gliomas (glioblastoma multiforme and anaplastic astrocytoma) by in vivo transduction with human interferon β gene using cationic liposomes. *Human Gene Ther*. 2004;15(1):77–86.

161. Daloiso V, Spagnolo AG. Ethics research committees in reviewing nanotechnology clinical trials protocols. Responsibility in Nanotechnology Development. Dordrecht: Springer; 2014:[pp. 97–109].

162. Swierstra T, Rip A. Nano-ethics as NEST-ethics: patterns of moral argumentation about new and emerging science and technology. *Nanoethics*. 2007;1(1):3–20.

163. Wolf SM. Introduction: the challenge of nanomedicine human subjects research: protecting participants, workers, bystanders, and the environment. *J Law Med Ethics*. 2012;40(4): 712–715.

164. Fatehi L, Wolf SM, McCullough J, Hall R, Lawrenz F, Kahn JP, et al. Recommendations for nanomedicine human subjects research oversight: an evolutionary approach for an emerging field. *J Law Med Ethics*. 2012;40(4):716–750.

165. Resnik DB. Responsible conduct in nanomedicine research: environmental concerns beyond the common rule. *J Law Med Ethics*. 2012;40(4):848–855.

166. Bawa R. FDA and nanotech: baby steps lead to regulatory uncertainty. Bio-Nanotechnology: A Revolution in Food, Biomedical and Health Sciences. Chichester, West Sussex; Ames, IA: Wiley-Blackwell; 2013.

167. De Jong WH, Borm PJ. Drug delivery and nanoparticles: applications and hazards. *Int J Nanomed*. 2008;3(2):133.

168. Bawarski WE, Chidlowsky E, Bharali DJ, Mousa SA. Emerging nanopharmaceuticals. *Nanomedicine*. 2008;4(4):273–282.

169. Bawa R. Regulating nanomedicine-can the FDA handle it?. *Curr Drug Deliv*. 2011;8(3): 227–234.

4

PART

Brain-Targeted Experimental Models

CHAPTER 8

In Vitro CNS Models

Asli Kara, MSc*, Naile Ozturk, BSc, Imran Vural, PhD****
**Hitit University, Corum, Turkey; **Hacettepe University, Ankara, Turkey*

1 INTRODUCTION

In vitro systems have an important role in enabling the understanding of the properties of the blood–brain barrier (BBB) and developing new techniques.[1,2] For many years, researchers have tried to develop an in vitro culture model for the BBB due to many ethical reasons. In vitro models, compared to in vivo models, are preferred for their high throughput capacity, lower costs, reduced number of animals used in experiments, simplified experimental settings, and easier standardization of the experimental conditions.[3] A perfect and effective in vitro model should mimic all characteristics of an in vivo setting, for example, the cell model must have reproducible solute permeability, display a restrictive paracellular pathway, exhibit physiologically realistic architecture, have functional expression of transporters, and be easy to culture.[4,5]

Although there are other in vitro models in the literature, this chapter only focuses on explaining cell-based in vitro BBB models.

1.1 Development of In Vitro Cell Culture Models

The first approach to obtain an in vitro BBB model was performed by isolating brain microvessels to investigate the physiological and pathological situations of the BBB at cellular, subcellular, and molecular levels.[6] Thereafter a new generation for studying the cell-based models proceeded with brain capillary endothelial cells (BCECs) that were isolated as the first successful viable cells for cell culture.[7–9] A large number of species were used to isolate and cultivate the BCECs, including mouse, rat, cow, sheep, pig, monkey, and human.[10,11] Researches assume that isolated BCECs are more suitable as a model system to study the BBB. This is because BCECs are easier to culture and similar results can be obtained when compared with living BBB systems.[12]

The next stage of cell growth was carried out on specific filters.[13] Unfortunately, after passaging cultured BCECs, they tend to lose some specific features of the

CONTENTS

Nanotechnology Methods for Neurological Diseases and Brain Tumors. http://dx.doi.org/10.1016/B978-0-12-803796-6.00008-3

in vivo conditions, such as reduced activity of tight junctions (TJs), low transendothelial electrical resistance (TEER), and loss of specific transporters and enzymes.[14] To this end, researchers focused on developing an in vitro model mimicking in vivo conditions as much as possible. For this reason, the initial coculturing models were developed by culturing BCECs along with astrocytes to mimic the in vivo BBB features.[15] After these studies, new primary culture models, immortalized cell lines, and mono/coculture studies were developed, which are described in detail in the following sections.[16] The classification of the models was simplified to enable easier understanding of the developments. Generally, the in vitro systems are classified into two categories, namely the two- and three-dimensional (2D and 3D) models. 2D models are based on an endothelial monolayer that is used to establish the barrier functions directly. On the other hand, 3D models exhibit a dynamic structure, including cellular movement and organization, which allows cell–cell interaction, and therefore seems to be more promising in terms of the mimicking in vivo features. Nevertheless, both models have some disadvantages, which are important for determining the most suitable model before starting relevant studies.[17] Although significant progress has been reported over the last 10 years, the most suitable and specific in vitro BBB model has not been determined yet. Instead, researchers have focused on improving and optimizing selected models for specific reasons.[18,19]

2 CELLULAR STRUCTURE OF THE BBB

BCECs are the main actors in the generation of the BBB, but their interaction with basement membrane, pericytes, perivascular astrocytes, microglial cells, and neurons ensures effective capillary biology and barrier function.[20–22] Moreover, BCECs form a neurovascular unit with these cells, which is needed for the function and health of the central nervous system (CNS).[23]

BCECs have special properties that are important for the generation of the BBB. Unlike other endothelial cells of the body, BCECs have less endocytic vesicles, no fenestrations, and TJs possessing higher electrical resistance. Moreover, paracellular flux is restricted by TJs, and transcellular flux is limited by the low abundance of endocytic vesicles.[24] These specific characteristics provide a more restrictive permeability, particularly for hydrophilic molecules. Small molecules, such as O_2 and CO_2 gases and small lipophilic compounds, can diffuse through the membrane of BCECs.[25] Small hydrophilic molecules are transported by specific transport systems, which allow nutrients to pass through the BCEC cell membrane, efflux some molecules from the brain, and prevent toxic or harmful molecules from entering. The entry of large hydrophilic molecules, such as peptides and proteins, is restricted to receptor-mediated transcytosis.[26–28] BCECs contain a high number of mitochondria, which is another specific characteristic.[29,30] This specialty provides increased energy both for active transport of molecules and for enzymes that form a metabolic barrier by breaking down compounds.[23,31,32]

BCECs are surrounded with the basement membrane, which anchors the cells and provides integrity for brain capillary vessels. BCECs, pericytes, and astrocytes are responsible for the formation of the basement membrane. Structural proteins (collagen and elastin), specialized proteins (fibronectin and laminin), and proteoglycans are main components of the basement membrane. Extracellular matrix proteins, cell adhesion molecules (CAMs), and signaling proteins are also part of the basement membrane.[21] CAMs and matrix adhesion receptors are found on matrix proximate faces of BCECs, as well as in other brain cells mediating cell anchoring to the basement membrane.[33]

Furthermore, pericytes are embedded in the basement membrane. One layer of the basement membrane lies between the pericytes and abluminal surface of the BCECs, and another layer is located between pericytes and the luminal surface of the astrocyte endfeet.[23,33,34] Pericytes cover the brain capillaries, which is approximately 30% of the cerebral capillary surface, and that the ratio of pericyte to endothelial cell is higher in CNS (1:1 to 1:3) compared to other tissues (1:100 in human skeletal muscle tissue).[23,33,35,36] Recent studies have shown that pericytes have a role in the integrity and regulation of the BBB.[14,37–40] It is also demonstrated that pericytes regulate functional aspects of the BBB, such as the formation of TJs and vesicle trafficking in endothelial cells during BBB development.[14] Regulation of BBB-specific gene expression in endothelial cells and stimulation of polarization of astrocyte endfeet are proposed to be two ways by which pericytes function at the BBB.[38]

Astrocytes lay their endfeet to the outer surface of the brain capillary endothelium and form a lacework of fine lamellae.[25] They cover more than 99% of the BCECs,[41] and it is hypothesized that they induce BBB phenotype of brain endothelium.[25,33] Grafting and in vitro cell culture studies have been conducted to support this hypothesis, and emerging evidence from these studies confirm that astrocytes contribute to the BBB properties of brain endothelial cells. It was observed from grafting studies of avascular tissue of a 3-day-old quail brain to a chick embryo brain that neural tissue has an inductive influence on BBB formation.[42] In another study, nonnervous system endothelial cells were induced by cultured astrocyte implantation and they formed tightened vessels.[43] In vitro cell culture studies are usually in agreement with grafting experiments in relation to the role of astrocytes on BBB induction. Furthermore, primary or immortalized brain endothelial cells were grown on filter inserts as monolayers, and astrocytes (primary/immortalized) were grown on the bottom of the filter inserts or on the bottom of the wells without direct contact to endothelial cells to obtain cocultures. These cocultures showed that the tightness of TJs is induced by astrocytes and some enzymes, and that the expression of transport proteins, such as gamma-glutamyl transpeptidase (γ-GT) and P-glycoprotein (P-gp), is regulated by astrocytes.[1,25,44–49]

Microglia, immunocompetent cells located in the brain, are involved in immune responses of the CNS.[20] These cells are classified in two forms: resting and activated microglia. Resting microglia have small bodies and long, thin processes; activated microglia exhibit a phagocytic morphology, and have short processes.[23] Due to their perivascular location and interactions with endothelial cells, it is considered they might contribute to special BBB properties.[50]

Neurons are in close contact with BCECs, astrocytes, and pericytes. Additionally, BCECs and astrocytic processes are innervated by noradrenergic, serotonergic, cholinergic, and GABA-ergic neurons.[22,33] Moreover, it was found that neurons induce the expression of enzymes that are unique for BCECs, and thereby regulate the function of blood vessels in response to metabolic requirements.[51] Furthermore, in a coculture cell model it was found that neurons induce BCECs to synthesize and sort occludin at the cell periphery.[52] However, their effects on BBB properties are not clear yet.

3 TIGHT JUNCTIONS AND IN VITRO PARACELLULAR BARRIER CHARACTERIZATION

The BBB is the main stringent barrier that originates from endothelial cells within CNS microvessels and brain capillaries. The barrier is impermeable to hydrophilic molecules larger than 400 Da.[27,53,54] This barrier has some special features, such as providing the most stringent structure and having a significant role in regulating the transition of some materials, including ions, molecules, proteins, and cells.[54,55] The primary role of the BBB is to maintain homeostasis between the blood and brain by restricting cells; the barrier is also responsible for the microenvironment that maintains the blood composition and protects the neural tissue from toxic substances.[10,17] Furthermore, the BBB has three different obstacles that comprise the physical, transport, and metabolic barriers and has specific properties, such as uptake and efflux transport.[56,57]

The transition of molecules and cells across the BBB is achieved by two pathways known as the paracellular (junctional) and the transendothelial pathways. Transendothelial transition is mediated by transporter proteins, and ensures that the nutritional needs of the brain are met. The physical barrier (gate function) is the restrictive paracellular barrier, which comprises the interendothelial TJs and adherens junctions.[50,58] These interendothelial TJs and adherens junctions form a continuous network between the adjacent endothelial cells, which means that paracellular transition is regulated by TJs, which are created by a group of several specific TJ proteins.[19,58] Along with the latter, TJs mediate the passive diffusion of molecules (including electrolytes, other water-soluble compounds, and xenobiotic substances), and contribute to cell polarity called fence function, which prevents the free motion of some macromolecules (lipids and proteins) due to the separation of the apical and basolateral membranes.[59,60] The main function of TJs is to maintain high transendothelial resistance at the BBB.[61] However, TJs may bind to the two adjacent cells to form tricellular junctions, where the corners of three endothelial and epithelial cells meet.[10,17,19] TJs comprise of two molecular components: transmembrane and cytoplasmic proteins. Transmembrane proteins can be classified into three groups: occludin,[62] members of the claudin family,[63] and junctional adhesion molecules (JAMs).[64] Claudin and occludin proteins have an essential role in selective paracellular permeability, and the presence of these three proteins in cells ensures effective barrier functions.[65] Furthermore, the cytoplasmic proteins include zonula occludens (ZO-1, ZO-2, and ZO-3) and cigulin or junction-associated coiled coil protein (JACOP)/paracingulin.[66] Transmembrane (e.g., claudin and occludin) and cytoplasmic proteins (e.g., ZO-1) interact with each other, and then with actin cytoskeleton to form multiprotein complexes, such that this structure forms a seal.[61,67] These proteins play a crucial role in paracellular permeability; loss of one of these proteins or changes in their expression cause paracellular leakage, leading to alterations in BBB function.[27] In addition, these transmembrane proteins serve as a marker to determine the successful formation of BBB.[67] The transmembrane proteins of adherens junctions are cadherin proteins [mainly vascular endothelial (VE)-cadherin, also known as Cadherin-5] and nectins linked to afadin/AF6.[68] VE-cadherin is a marker of microvascular integrity. The adherens junctions provide the primary contact between endothelial cells and are necessary for the formation of TJs.[69] Adherens junctions are formed by the interactions of different cadherin molecules in the presence of a Ca^{2+} ions.[58,70,71] Consequently, tight and adherens junctions are essential structures, which are interconnected with endothelial cells by their protein complexes, and provide high electrical resistance and low paracellular permeability.

Determination of the integrity and characteristics of in vitro paracellular barriers is carried out by means of three parameters: transepithelial/TEER, permeability measurement by tracer hydrophilic molecules with known molecular weights, and immunostaining of known markers (occludin, claudin, and ZO-1).[67,72] Immunostaining of cell type–specific markers, such as ZO-1, Claudin-5, and occludin, gives qualitative results to help determine the barrier integrity of an endothelial and or epithelial monolayer.[73]

A simple method to measure the paracellular permeability of the barrier is achieved by tracer compounds, which have various molecular weights.[74]

Although radiolabeled sucrose (molecular weight: 342 Da) and mannitol (molecular weight: 182 Da) are commonly used tracer molecules, fluorescent markers, such as lucifer yellow (molecular weight: 443 Da), sodium fluorescein (molecular weight: 376 Da), inulin (molecular weight: 5 kDa), dextrans, and serum albumin (molecular weight: 67 kDa), can be used to measure less stringent layers. Biotin–albumin is also used for the quantification of serum protein permeability with serum albumin. Furthermore, for macromolecular permeability, fluorescence-labeled marker proteins, such as fluorescein isothiocyanate (FITC)-labeled substances (FITC–dextran and FITC–inulin), and enzymes, such as horseradish peroxidase (HRP) have also been used.[75]

Permeability is expressed by means of the endothelial permeability coefficient P_e (cm/s).[76] This method has some disadvantages: exhibits short half-life time; has short-term storage; is affected by few environmental changes (pH and temperature); shows interaction of fluorescent tracers with the fluorescent substrates, which affects the barrier integrity; and shows differences between in vivo and in vitro permeability results.[5,16,73,75] In studying the paracellular transport by tracers, the selective tracer compound should not interfere with influx or efflux transporters, endothelial receptors (as a ligand), and endothelial enzymes (as a substrate).[10]

TEER is a quantitative method that offers the greatest selectivity. It measures electrical resistance across the cell monolayer and determines barrier integrity and permeability.[10,19,77] Measurement of TEER is based on the culturing of a monolayer on semipermeable filters.[78] This system has membranes that have two compartments: the upper compartment is indicated as the apical side (blood side) and the lower compartment is indicated as the basolateral side (brain side). There are two systems to measure the TEER, which include the EVOM voltmeter (volt–ohm resistance meters) (World Precision Instruments, Millipore, etc.), voltage-measuring electrodes "chopstick," and impedance spectroscopy (CellZscope chamber–type electrodes).[79] There is an important difference between the two systems, while CellZscope (NanoAnalytics, Münster, Germany) provides continuous monitoring of the electrical resistances up to 24 wells, chopsticks allows only one measurement, which can be made only at certain time points.[18] There are two "chopstick" electrodes that measure the TEER, one of them is placed at the apical side and the other one is placed in the basolateral compartment (Fig. 8.1). The important point here is that the results expressed in ohm centimeter square ($\Omega \cdot cm^2$).[79,80] Extracellular Ca^{2+} ion concentrations affect TEER values, hence these values differ among different systems.[23] For example, in vivo BBB models show high TEER values, such as 1500–1800 $\Omega \cdot cm^2$, whereas for in vitro models, the range of 150–200 $\Omega \cdot cm^2$ is sufficient to study permeability or drug transportation.[5,16]

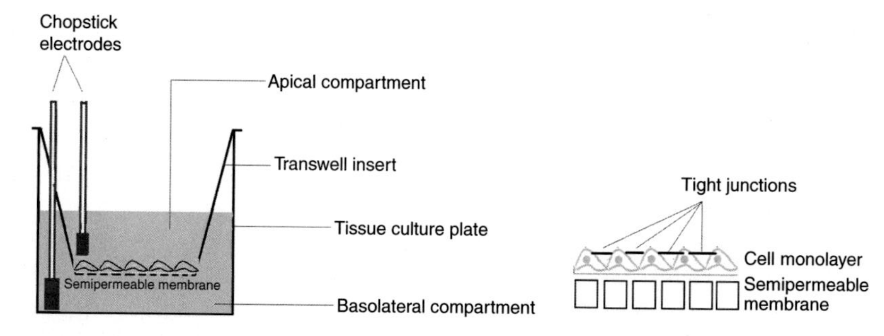

FIGURE 8.1 Transepithelial/transendothelial electrical resistance (TEER) measurement by chopstick electrode method.

The advantages of the TEER measurement method are that it does not interfere with other cell compounds, and it is easy to determine the culture day (with optimum tightness) on which the experiment should be carried out. Also, living cells can be monitored during growing and differentiating phases.[73,75]

4 CELL CULTURE MODELS

4.1 2D In Vitro BBB Models: Monoculture and Coculture

Basic and simple 2D models are static systems of monocultures based on monolayers of cerebral endothelial cells.[1,81] 2D models form the majority of the in vitro BBB models that are important for the pharmacology, transport, migration, and metabolic activity of the BBB.[82] The 2D system is based on semipermeable microporous membranes, such as Transwell polycarbonate or polyethylene terephthalate apparatus (e.g., Corning, Lowell, Massachusetts), which separate the apical and basolateral compartments by a vertical side-by-side diffusion system and offer better resolution when quantification of a minimum amount of compound is required.[28,41]

These Transwell systems are the most commonly and widely used cell-based models, which have different functions and allow the exchange of solutes by porous membranes. These models are obtained from different sources, including murine, rat, bovine, porcine, monkey, and human cerebral endothelial cells grown on the apical compartment of the membrane, which are subsequently dipped in their specific growth media.[50] Advantages provided by these models include simple and easy processing, which allows relative high throughput screenings for both drug permeability and binding affinity measurement. Furthermore these models also create high and pure cell populations with relative viability.[19] The disadvantage of these models, on the other hand, is the lack of a number of critical features that are important to form a complete and reliable BBB model in vitro. An absence of stimulating factors, which are derived from astrocytes, pericytes, and other parenchymal cells (e.g., neurons), causes acceleration of endothelial dedifferentiation after serial cell passages and therefore causes the loss of BBB characteristics, such as lower TEER.[83,84] However, transport mechanisms, cellular transmigration processes, and transendothelial resistance studies are suitable for this model.[85] These models generally show TEER values above 150 $\Omega \cdot cm^2$ and therefore may be useful for investigating the BBB permeability of potential drug candidates.[86]

4.1.1 Noncerebral-Based In Vitro Cell Culture Models

Cell culture models are based on two cell types namely cerebral endothelial cells and noncerebral epithelial cells.[87] Cerebral-originating cells are based on primary cultures of BCECs, immortalized cell lines alone, or coculture models. Noncerebral-originating cells are types of cell culture surrogates in BBB models, including epithelial cell lines. These models exhibit distinct morphological characteristics and have different cell contents (e.g., lipid composition of cell membrane) compared to cerebral endothelial cells, but may be used as BBB models.[3,88]

Difficulties about insufficient barrier properties of immortalized brain endothelial cell lines and the lack of barrier tightness to study BBB permeability have led to the development of epithelial cells of noncerebral origin.[50] Among these, Madin–Darby canine kidney (MDCK) cell line, human epithelial colorectal adenocarcinoma (Caco-2) cell line, and human urinary bladder carcinoma (ECV304) cell line are the most widely used epithelial cell lines.[50,89]

MDCK is a widely used cell line for TJ studies because of its similarity with endothelial cell TJs. This model has displayed decent paracellular permeability and easy growing features, and is therefore suitable for use in epithelial cell trafficking and polarity studies.[89,90] These cells can be easily transfected with multidrug resistance gene (MDR1) resulting in an overexpression of P-gp.[71,91] The latter has been used as a general permeability screen for passive and P-gp–mediated CNS penetrability of compounds.[92–94] Recently, the permeability of a stereoisomeric prodrug across MDCK–MDR1 cell membrane was investigated to determine its affinity to P-gp and peptide transporters. Results showed that the prodrug approach was an effective strategy to overcome P-gp–mediated efflux and to improve transcorneal permeability.[95]

The ECV304 cell line, an alternative model for BBB, expresses sufficient brain endothelial features and is tested in many permeability studies. Hurst and Fritz[96] studied cocultures of ECV304 with C6 glioma cells. They reported that in the presence of C6 coculture, the ECV304 model developed a low TEER value of about 100–180 $\Omega \cdot cm^2$, with a sucrose permeability coefficient as low as 5.3×10^{-6} cm/s.[96] Besides, the researchers also cocultured ECV304 with astrocytes to simulate an in vitro BBB model with acceptable results.[97] Generally, researchers determined similar results for TEER and permeability coefficient values.[98,99] Even in the presence of TJ modulators, these low values failed to generate a tight barrier, which limited the use of this in vitro model in drug permeability studies.[100]

Caco-2 cells are epithelial cells, which have been developed and extensively used primarily for the prediction of intestinal absorption and drug transport studies.[101,102] According to literature there are a limited number of reports investigating the use of Caco-2 as an in vitro model for the BBB. Generally, Caco-2 cells are being investigated in comparative studies of BBB models.[11,103] As intestine and brain vascular–derived cells exhibit completely different features, the use of Caco-2 cells for BBB modeling is limited.[104] Albeit low, these models also express P-gp and hence can identify P-gp substrates. Researchers also determined that P-gp levels of Caco-2 can be increased by vinblastine treatment, which improves the testing of ligands for efflux transporters.[105] Many comparison studies were performed with different cell lines. For example, when BCEC/astrocyte coculture and Caco-2 cells were compared, Caco-2 cells showed a poor response in in vitro BBB permeation studies ($r = 0.68$ and 0.86 for Caco-2 cells and BCEC/astrocytes, respectively).[106] Moreover, three different cell lines (MDCK–MDR1, Caco-2, and primary bovine endothelial cells) were studied with nine reference drugs to investigate drug permeability. Their results were correlated with an in vivo model. Results showed that all correlation coefficient values were high and close to each other: $r = 0.99$, 0.91, and 0.85 for primary bovine endothelial cells, Caco-2, and MDCK–MDR1, respectively.[103]

4.1.2 Cerebral-Based In Vitro Cell Culture Models

4.1.2.1 Primary Brain Endothelial Cell Culture

Drug delivery studies and consequent use of a large number of animals in experiments have led to the development of endothelial cell culture–based in vitro models from the early 1990s. The first in vitro BBB model was created by isolating microvessels from rat brain tissue.[107] This model has many advantages, such as being easily accessible, well established, and being able to maintain the many structural and functional properties of the BBB. Moreover, this model has been used for many applications, such as receptor, transporter, and signaling mechanisms and uptake or efflux studies.[5] Despite the use of brain microvessels for many applications, the models are not suitable for transport studies because isolation of microvessels is labor intensive and difficult. Additionally, the poor viability of microvessels limited the use of this model in drug screening research.[17,108] To this end, this system led to the development of a new era of cultured BBB cell models. Cell-based models have many advantages, such as simplicity, cost effectiveness, high flexibility, and easy control. Generally, these models are based on two main types of culture models: primary or immortalized cell cultures.[88,90,109] In 1978 the use of primary brain endothelial cells, isolated from rat BCECs, mimicked a phenotype that was the most similar to in vivo conditions and the best in vitro BBB models.[8] Many characteristics of these cells were genetically programmed, and they exhibited excellent similarity to BBB characteristics in early passages.[88] Primary brain endothelial cells can be isolated from different sources, including bovine, porcine, rat, or human.[110] Nevertheless, the use of primary cells is limited because of some disadvantages. First, the technique is difficult, expensive, and time consuming.[84] Second, cells lose their BBB properties due to the number of passages and the overgrowth of brain endothelial cells during long-term culture. Also, cells show rapid differentiation, lose their phenotype, and are easily contaminated with other cells.[111] Altogether, it is difficult to gain a large amount of cells in a homogenous cell culture. Besides, primary cells disturb the development of tight monolayers.[112]

Hence, due to the ease of use and low costs, new immortalized endothelial and epithelial cell culture methods have been developed from different sources.[10,113] However, immortalized cell lines create a leaky barrier, thus their use is limited in permeability studies.[114]

Primary bovine brain endothelial cells, which have been established from bovine brain gray matter, were the first cells incorporated in BBB models with measured permeability, and are widely used and accepted as vitro cell culture models.[115] This model is the most suitable model, comprising monolayers and TJs, for high-yield pharmacological screening.[5] These systems have been used for several applications, such as BBB transport and drug permeability studies.[116] Primary bovine endothelial cells were found to be the best characterized model by Cecchelli's group who developed a coculture system, and determined a comparatively high TEER value compared to a monoculture system, which exhibited a leaky paracellular barrier of 500–800 $\Omega \cdot cm^2$.[117]

The most widely accepted second model was developed a few years after primary bovine brain cells. This model, consisting of primary porcine brain cells, is a well-established model with high permeability.[118,119] Franke et al. first developed primary porcine brain cell culture with a high TEER value of 1800 $\Omega \cdot cm^2$ and low permeability of 0.2×10^{-6} to 1.8×10^{-6} cm/s.[120] Also, the researchers explored that the primary porcine brain cell–based BBB model exhibited a high TEER value when serum-free culture medium and hydrocortisone were used, without the need for coculture.[120,121] However, not all porcine-based BBB models exhibit high TEER values; a publication showed an important variation in TEER (75–800 $\Omega \cdot cm^2$) and P_e values (3.3×10^{-6} to 33×10^{-6} cm/s) for sucrose.[10] The model has been used as a screening method for the penetration of small drugs[104] and nanocarriers,[122] as well as for investigating the expression of multidrug-resistant and transport proteins,[118] and also for drug transport and targeting studies.[123]

Rat and mouse brain endothelial cell systems have a low yield compared to other species, including bovine and porcine cells (which have 200 million cells per model); for example, an amount of 1–2 million endothelial cells per rat brain can be isolated. Although both models exhibit the same barrier properties, their use for transendothelial permeability measurements were limited by their leaky integrity (TEER 150–200 $\Omega \cdot cm^2$) due to a small malfunctioning caused by contaminating pericytes.[118] Recently, a better model has been developed by coculturing rat cells with glial cells, resulting in higher TEER values of ~500 $\Omega \cdot cm^2$ or higher.[124] Similar results were obtained by using primary endothelial and glial cells, resulting in a TEER value of 778 $\Omega \cdot cm^2$.[125] Recently, coculturing rat primary endothelial cells with astrocytes resulted in acceptable results for BBB functions.[126] The advantages of rodent models include the availability of experimental animals, such as the use of transgenic and gene-targeted mouse models; the wide range of available antibodies; and the possibility to easily compare these models with in vivo results.[10] Additionally, studies with primary cultured rat and mouse systems comprise receptor modulation, drug uptake studies, and drug transport studies.[127,128]

The first isolation of microvessel endothelial cells faced problems, such as poor availability of human brain tissue. The use of human-origin primary cultures was limited because of ethical reasons and difficulties to procure human brain tissue by autopsy and biopsy procedures.[129] Other disadvantages of these human origin–based models are low yield of cells and low TEER values (~120–180 $\Omega \cdot cm^2$).[130] Even so, these models have been used in some diseases and other research areas, such as neuroinflammation,[131] multiple sclerosis,[132] epilepsy,[72] Alzheimer's disease,[133] HIV-related CNS disorders,[134] drug transport,[135] nanoparticle permeation,[136] and receptor-mediated mechanisms in permeability modulation.[137]

4.1.2.2 Immortalized Cell Lines

Due to the disadvantages of primary cell cultures, their use as a biological assay material was limited.[10,57,75] For this reason, various new immortalized brain endothelial cell lines have been developed last 20 years. These cell lines are easily established, need small amount of tissue, provide reproducible results, and are sufficiently tight.[115,138,139] However, due to a fast isolation procedure, faster growth rates, and no contamination risk by fibroblasts, these models are preferred for studying brain tumors and neurodegenerative diseases.[140] Nevertheless, these models have a major disadvantage, when cells are grown as a cell monolayer on a porous membrane, they do not form complete TJs and therefore they have weak barrier.[16] The first-generation immortalized cell lines were developed by transfection with the polyoma virus T antigen, adenovirus EIA gene, or SV40 large T antigen. Recently, immortalized cell lines have been established by transfection with SV40 large T antigen gene (tsA58 T antigen gene).[138,141,142] Several immortalized cell models have been studied in the literature; however, Table 8.1 shows the most used cell lines currently.

Table 8.1 TEER Values of the Most Commonly Used Noncerebral, Primary Cultures, and Immortalized Cell Lines of the Blood–Brain Barrier (BBB)

	Sources or Cell Lines	TEER Values (Ω·cm²)	References
Noncerebral-based cultures	MDCK	~1500	[143]
	ECV304	250–350	[144]
	Caco-2	1400–2400	[145]
Primary cultures	Bovine	100–200	[146]
	Porcine	70–120	[147]
	Rat	100–150	[148]
	Mouse	~200	[149]
	Human	100–180	[150]
Immortalized cell lines	Bovine: SV-BEC	~40	[16]
	Porcine: PBMEC/C1-2	250–300	[16]
	Rat: RBE4	10–150	[16]
	Mouse: TM-BBB	~100–120	[151]
	Human: hCMEC/D3	30–50	[152]

Bovine and porcine models have very limited use in experimental studies because of the availability of good primary cultured endothelial cells. The most commonly used bovine endothelial cell line is tBBEC-117[49] and the porcine endothelial cell line is PBMEC/C1-2.[153]

Most developed models have been derived from rats.[57] There are several immortalized rat endothelial cells, including RBE4, RBEC1, TR-BBB13, and GP8/GPNT cell lines. Disadvantages of these cell lines are low TEER values and a rather leaky monolayer formation. According to published literature, TEER values were not available and permeability studies of drugs were not adequate, showing a range between 11×10^{-6} and 214×10^{-6} cm/s for sucrose permeability through RBE4 cells.[57,154] Therefore, immortalized rat cell lines are specially used for several studies, such as endothelial cell uptake assay via P-gp,[155] carrier-mediated transport,[156] and signaling.[155]

Besides, several immortalized mouse endothelial cell lines have been developed as rat cell lines. For permeability studies, MB114 was the first used brain endothelial cell line.[157,158] Then, mouse brain endothelial cells were widely used for BBB studies, including signaling,[159] transport mechanism,[160] and permeability research.[161] Among these, the best characterized cell lines are bEnd3,[162] bEnd5,[163] and TM-BBB,[142] which are especially used for permeability assays.

Although there is a good model covering the immortalized human model entirely, hCMEC/D3 cell line is the most widely used and best characterized model for infectiological studies with human pathogens and drug delivery studies. The advantages of this cell line are that it is a source that can indefinitely proliferate and preserve its differentiating properties after repetitive passages.[84] hCMEC/D3 cell line was introduced for expression of normal endothelial markers.[164,165] hCMEC/D3 monolayers exhibit low TEER values under static culture cultures, around 30–50 $\Omega \cdot cm^2$.[57,165] The study on the effect of glucocorticoids on TJ proteins of the human BBB yielded a TEER value of 300 $\Omega \cdot cm^2$, when compared to hCMEC/D3 cell line alone (69 ± 19.3 $\Omega \cdot cm^2$).[166] Coculture of hCMEC/D3 with astrocytes increased TEER values from 30 to 60 $\Omega \cdot cm^2$, while hCMEC/D3 showed a lower TEER value of 39 $\Omega \cdot cm^2$.[167] Recently, the effects of smoking and hyperglycemic conditions on in vitro hCMEC/D3 were studied. Researchers determined that cytoplasmic proteins, such as ZO-1, are down-regulated under hyperglycemic conditions and/or cigarette smoke extracts over an exposure time of 24 h. The results showed that when endothelial cells are exposed to these two factors, it leads to the release of angiogenic and inflammatory factors.[168]

4.1.3 BBB Models of Cocultures

Brain endothelial cells are the main components of the BBB, but other cells of the neurovascular unit have important roles in the maintenance and induction of the BBB. Therefore, cocultures of BECs with other neurovascular unit cells, such as astrocytes, pericytes, and neurons, are used to obtain different BBB models.

In Transwell models using different cell types, contact and noncontact coculture models can be established (Fig. 8.2).

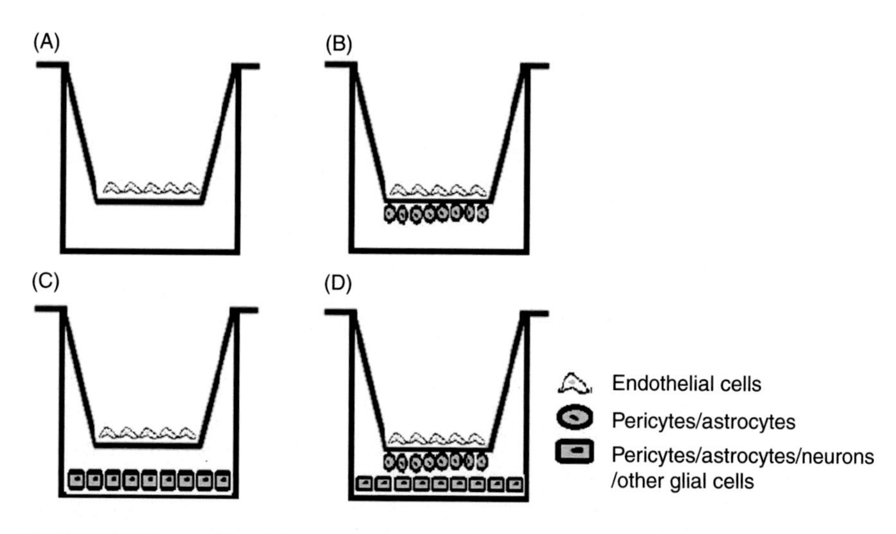

FIGURE 8.2 Transwell models of the BBB.
(A) Monoculture of endothelial cells. (B) Contact coculture of endothelial cells and pericytes/astrocytes. (C) Noncontact coculture of endothelial cells and pericytes/astrocytes. (D) Triple coculture of endothelial cells and pericytes/astrocytes/neurons/other glial cells.

4.1.3.1 Coculture of Endothelial Cells With Astrocytes

The role of astrocytes in inducing BBB properties has been researched widely,[169] and consequently coculture studies of astrocytes and endothelial cells have been conducted to achieve a model that has more specific BBB properties than the monoculture of BCECs. In these studies, usually primary endothelial cells and astrocytes[170] are used; however, immortalized cells can be used as well.[44] Moreover, cells can derive from the same origin[171] or different origins.[46] For example, a coculture model of human brain microvascular endothelial cells and human astrocytes increased TEER values.[172] In another study, Gaillard et al.[173] established a coculture of bovine primary BCECs with rat astrocytes, and they obtained higher TEER values (134–386 $\Omega \cdot cm^2$) compared to monoculture of BCECs (92 $\Omega \cdot cm^2$). Also, it was found that TEER values were higher when astrocytes were cultured on the bottom of the filter compared to when cultured at the bottom of the well.[173] Other studies also indicate that direct contact between endothelial cells and astrocytes is necessary to achieve good TEER values.[45,78,167,172] In another coculture study, the effect of the basement membrane on the BBB properties was investigated using collagen I and IV mixture and Matrigel as the basement membrane. Coculture of bEnd3 cell line (mouse endothelial cells) and rat astrocytes with collagen I and IV mixture as the basement membrane showed the lowest permeability to water and other solutions, but there was no significant difference between the coculture and the bEnd3 monoculture, which were cultured without collagen I and IV mixture coating.[174] Another approach to obtain BBB phenotype on endothelial cells, which are induced by astrocytes, is the use of astrocyte-conditioned media. It is well known that astrocytes are able to secrete a range of chemical agents [transforming growth factor-β (TGFβ), glial-derived neurotrophic factor (GDNF), basic fibroblast growth factor (bFGF), and angiopoetin 1], which can induce BBB properties.[48,169] Therefore, astrocyte-derived soluble factors secreted into the culture medium are used to create an improved BBB model.[175,176]

4.1.3.2 Coculture of Endothelial Cells With Pericytes

Pericytes have a critical effect on BBB integrity and regulation; disruption or lack of pericyte–endothelial interactions may lead to an increase in BBB permeability.[14,38] Due to the latter, the inclusion of pericytes in BBB cell culture models is a sensible approach. In a pericyte–endothelial coculture model, it was found that pericytes increase TEER values of monolayers, especially when in contact with endothelial cells.[177] Nakagawa and coworkers[177] developed a coculture model of rat pericytes with rat BCECs, and it was found that pericytes increased TEER values of BCEC monolayers by 400%. Endothelial cells in contact with pericytes showed higher TEER values than those not in contact pericytes. In this study, astrocyte–endothelial cell coculture and triple coculture (astrocyte–pericyte–endothelial cell) were also established. The increase in TEER values was higher in pericyte-incorporated models. Furthermore, triple coculture models with pericytes–endothelial contact achieved the highest TEER values.[40] Contrary results were obtained for the barrier integrity–increasing properties of pericytes in a coculture study of primary porcine BCECs with primary porcine brain capillary pericytes. In this study, vascular endothelial growth factor (VEGF) and matrix metalloproteinases (MMPs) secreted by pericytes were suggested to be barrier integrity–decreasing proteins. The contrary findings about the impact of the pericytes on barrier integrity were explained by the differences between species, preparations, the ratio between endothelial cells and pericytes, and the culture conditions.[178] In a recent study, it was observed that porcine pericytes, when cultured with porcine BCECs in contact culture, increase claudin-5 expression levels, and that rat pericytes increase P-gp and occludin expression levels when cultured in a noncontact coculture with porcine BCECs. However, there was no significant difference between couture of porcine BCECs–porcine/rat astrocytes, and triple culture of porcine BCECs–porcine/rat astrocytes–porcine/rat pericytes (contact) regarding gene expression, TEER, and permeability.[179]

4.1.3.3 Coculture of Endothelial Cells With Neurons and Microglia

Coculture of BCECs with neurons is another approach that is used to study the induction of BBB characteristics. Activity of a BBB-related enzyme, γ-GT, was found to increase when BCECs were cultured with plasma membranes isolated from primary neurons,[51] and the expression of occludin was induced on rat brain endothelial cells when they were cultured with neurons.[52] When neurons were cultured with BCECs in the presence of astrocytes, it was found that the neuron-induced synthesis and peripheral localization of occludin is precocious compared to when BCECs are cultured with neurons only.[180] In the light of this observation, a three cell–type BBB model (BCECs, neurons, and astrocytes) was constructed to investigate additive effects of neurons and astrocytes on paracellular permeability. This study revealed that when BCECs were cocultured with both astrocytes and neurons, the paracellular flux of sucrose was reduced to one-third as compared to that of BCECs cultured alone.[181] Furthermore, a triple coculture of BCECs, astrocytes, and neurons was found to have the highest TEER value, γ-GT enzyme activity, P-gp, and ZO-1 levels and the lowest permeability for HRP compared to cocultures of astrocytes and BCECs or neurons and BCECs.[182]

Permeability regulated by glial cells was investigated using BCECs and conditioned media containing soluble factors derived from glial cells (astrocytes and microglia) under basal culture conditions. These factors induced TJ formation and reduced the permeability of bovine serum albumin by 70% when compared with BCECs, which were grown in regular culture media.[183] However, it was found that activated microglia increased the BBB vascular permeability by soluble factors, such as tumor necrosis factor-α (TNF-α),[183,184] or by producing reactive oxygen species through NADPH oxidase.[185] Also, the P-gp function of rat BCECs was decreased when cultured with lipopolysaccharide-activated microglia.[186]

4.1.3.4 Coculture of Stem Cells/Stem Cell–Derived Cells

Stem cells or stem cell–derived cells have been used in various studies for BBB modeling as an alternative to primary cells.[187] BBB properties were induced when BCECs were cocultured with differentiating embryonic neural progenitor cells (NPC),[188] or NPC-derived astrocytes and neurons.[189] Also, codifferentiated neural cells with endothelial cells, derived from human pluripotent stem cells, were found to acquire BBB properties. The coculture of these endothelial cells with astrocytes provided a maximum TEER value of $1450 \pm 140\ \Omega \cdot cm^2$.[190] Coculture of endothelial cells, differentiated from human hematopoietic stem cells, and pericytes resulted in a reproducible model that maintained BBB properties for at least 20 days.[191] Human pluripotent stem cell–derived BCECs treated with retinoic acid and cocultured with primary human brain pericytes, human astrocytes, and neurons derived from human NPCs generated a human BBB model yielding a TEER value of $\sim 5000\ \Omega \cdot cm^2$.[192]

4.2 3D In Vitro BBB Models

3D spheroid models can be used to mimic the in vivo organization of cells and their microenvironment. Thus, discrepancies between in vivo BBB and in vitro BBB models can be reduced. A 3D model of BBB was obtained by culturing collagen–cell suspension droplets (either rat BCECs alone or rat BCECs cocultured with astrocytes and pericytes) on Petri dishes. This model displayed characteristics of BBB and was used to investigate the effect of astrocytes and pericytes on BBB development.[37] In another study, Matrigel and 3D hanging-drop spheroid models were used to investigate associations between BCECs, astrocytes, and pericytes, with the advantage of free and spontaneous cell interaction.[193] Therefore 3D models are convenient and exciting models to study BBB function and regulation.

4.3 Dynamic Models of the BBB

Endothelial barrier properties of blood vessels are affected by fluid-induced shear stress (approximately 4 dyn/cm^2).[194] This phenomenon led to the development of dynamic systems instead of static models because of the changes in the physiological forces. Different dynamic models have been developed recently.[41,84]

In the cone plate apparatus method, a cone and plate viscometer were used to generate a shear stress in vitro at the bottom of the plate.[195] In this model, when endothelial cells are grown as a monolayer on the bottom of the plate, the shear stress was generated by a rotating cone. Due to the lack of astrocytes, pericytes, and other cells, this model does not show similarities with the in vivo model of the BBB, hence its use in BBB research was limited in terms of reliability.[41,84]

Dynamic in vitro models create a tunable shear stress by a continuous flow of culture medium on endothelial luminal surfaces, and they use artificial capillary-like apparatus made up of 3D hollow fibers (made of thermoplastic polymers, such as polypropylene, polysulfone, etc.), which mimics capillaries.[28] Also, this system allows the coculturing of endothelial cells (animal or human, both primary culture and immortalized cell line) with astrocytes (or glial cells). The aim of the development of an in vitro dynamic model is to mimic physiological stress in blood vessels and to closely resemble the in vitro BBB both functionally and anatomically.[194] A comparative study showed that the dynamic model exhibited a 10-fold higher TEER value when compared with the Transwell coculture model, and 5- to 10-fold lower permeability values in response to sucrose and phenytoin, respectively.[196] Results show that this model may be advantageous in BBB studies due to its tighter barrier functions. Despite the advantages, its use in BBB research on a large scale has been limited. Direct visualization is difficult, and there is need for high cell numbers ($>1 \times 10^6$) and technical equipment. Further, in comparison to other models, for dynamic models the steady-state time is longer (generally 9–12 days) than coculture models (generally 3–4 days).[194]

In this system, the hollow fiber apparatus is represented by a bundle of porous (pore size of 200 nm) pronectin/fibronectin-coated polypropylene tubes that provide the connection between cocultured cells. Thus they allow the culture of cerebral endothelial cells in a lumen of fibers. Astrocytes are located on the outer surface (perivascular/parenchymal side of the brain microcapillaries) of the same fibers.[87,197] This fiber system is suspended in a sealed chamber. The artificial capillaries are continuously in touch with a medium source through a flow path, which consists of a gas-permeable silicon tubing system for the exchange of O_2 and CO_2.[198] A variable speed pumping mechanism generates the intraluminal pulsatile flow, which maintains the shear stress and intraluminal pressure according to the diameter of hollow fibers and the viscosity of the medium (generally between 5 and 23 dynes/cm^2).[199] The dynamic in vitro BBB model was studied for its multiple functional properties and physiological responses.[200–203] The best characterized dynamic model, based on coculture of bovine aortic endothelial cells and glial cells, has been developed by the Janigro's group. When using bovine aortic endothelial cells in coculture with astrocytes, they obtained TEER values greater than 500 $\Omega \cdot cm^2$. Also, flow has played an essential role in endothelial cell differentiation by decreasing cell division after 20 days in culture.[204] Then, a model was developed to humanize the dynamic in vitro BBB model by the coculturing of normal human microvascular endothelial cells (HBMECs) with human brain astrocytes (HAs). The results showed TEER values greater than 1000 $\Omega \cdot cm^2$, with a sucrose permeability below 2×10^{-7} cm/s.[205] Cucullo et al.[201] obtained similar high TEER values (~1200 $\Omega \cdot cm^2$ with or without astrocytes) in a dynamic model while studying an immortalized cell line, a novel human brain endothelial cell line (HCMEC/D3cocultured with astrocytes, which closely mimicked the BBB in vivo. When compared with the dynamic systems, static Transwell systems of the same cells exhibited 15- to 20-fold lower TEER values (ranging from 60 to 80 $\Omega \cdot cm^2$). This study has attracted attention because of the effect of the shear stress on the TEER value.[201] Siddharthan et al.[176] developed a flow-based in vitro BBB model using HBMEC and human fetal astrocytes (HFA). Similar results were obtained for the TEER values (1400–1500 $\Omega \cdot cm^2$) by other researchers as well.[176] HFAs significantly reduced the permeability of an in vitro BBB model cocultured with primary human brain endothelial cells. Subsequently, Cucullo et al. developed a further dynamic BBB model to obtain a suitable model for transendothelial trafficking of immune cells, such as THP-1 monocytes. They used hollow fibers having microholes of 2–4 μm pore size. The TEER values and permeability of sucrose were unaffected by the change in the diameter of the pores. According to the latter, the new model may explain that BBB phenotype is not significantly affected by the properties of the supporting material structure.[194] More recently, a new model was developed to understand the pathogenesis of neurological diseases and pharmacokinetics of drugs. The system is based on a novel artificial vascular system, which mimics venous segments in vivo. For this, instead of coculture with astrocytes, which were used in the classical dynamic in vitro BBB model, smooth muscles cells were cocultured with endothelial cells as a venule module. This module provides a hemodynamic microenvironment, with a shear stress causing lower pressure than that found in capillaries.[206]

4.4 Microfluidic BBB Models

Static Transwell models are not suitable for mimicking dynamic blood flow. Therefore, microfluidic device–based BBB models have been developed to solve this problem.[207–210] One of the developed microfluidic BBB models is called the μBBB. This model comprises four PDMS substrates, two glass layers, and a porous polycarbonate membrane sandwiched between the PDMS layers present at the center. Cells are cultured on the porous membrane that is at the intersection of the two perpendicularly crossing channels, which produce a dynamic flow. Multiple electrodes are embedded to monitor the TEER across the barrier. The cell lines b.End3 (endothelial) and C8D1A (astrocyte) were cocultured on both sides of the porous membrane through flowing cell suspensions. TEER values of μBBB cocultures were over 250 $\Omega \cdot cm^2$ compared to static Transwell cultures (25 $\Omega \cdot cm^2$) of the same cells.[207]

4.5 Disease Models

In vitro cell culture models of CNS diseases are useful for investigating the mechanism of the diseases. An in vitro model to examine mechanisms of Parkinson's disease (PD) pathology was developed. Human neuroblastoma cells were exposed to a low concentration of rotenone, which inhibits chronic systemic complex I and induces features of PD.[211] To investigate the effect of inflammatory/immune mechanisms in dopaminergic cell injury relevant to PD, an in vitro model of nigral injury was developed. In this model, injury of the dopaminergic cell line (MES 23.5 cells) and dopaminergic neurons was induced by lipopolysaccharide-activated microglia.[212]

5 COMPARISON OF IN VITRO CELL CULTURE MODELS

In vitro models have greatly contributed to BBB research because of the many advantages they offer. Although there are a lot of in vitro models, there is no perfect model. Each single cell model has unique features and disadvantages (Table 8.2). For example, when epithelial cell models were involved in the efflux transport of drug candidates, brain endothelial cell culture models more suitably mimicked the properties of BBB.[88] An ideal in vitro BBB model should mimic all in vivo characteristics.[28] Theoretically it is possible to create a complete BBB model; however, in practice this has not been achieved yet. This may be due to the extracellular environment, which lacks normal brain microenvironment, and because simple environments do not provide cells with enough blood and brain contents.[83] The cell-based models have the potential for transcellular and paracellular drug diffusion, metabolism, and transportation. Additionally, the models are useful for investigating physiology and permeation mechanisms.[83,84]

6 CONCLUSIONS AND FUTURE OUTLOOK

Cell culture–based BBB models are used to study both BBB permeability and brain physiology and pathology. Although some functional cell-based in vitro models of BBB have been established, their practical use is limited. The primary cultures of BCECs are inconvenient for high-throughput assays because they are not suitable for subculturing and storage for future use. Besides immortalized cell lines do not provide the desired tight monolayers. Surrogate BBB models, such as MDCK–MDR1 and Caco-2 cell lines can be used for preliminary permeability screening assays, but they lack some properties of BCECs.

Considering that transport proteins of the BBB differ among species, using cells of human origin is also important to eliminate the effect of species-specific differences in the BBB model. On the other hand, human pluripotent stem cells

Table 8.2 Advantages and Disadvantages of Different Cell Models Incorporated in In Vitro BBB Systems

Cell Models	Advantages	Disadvantages	References
Noncerebral-based cell models	• Easy to transfect • High permeable • Easy to standardize • Use in drug transport studies	• BBB modeling is limited because of the differences between the endothelial cells	[3,213]
Isolated microvessels	• High available • Easy access • Maintain structural and cellular characteristic of BBB (cell differentiation) • Many methods to purify brain capillary are present	• Difficult to isolate because of presence of neuronal and/or other contaminants • Labor intensive • Generally have poor viability • Not suitable for directional transport • Drug permeability screening studies and cellular migration	[17,214]
Monoculture Transwell system	• Simple, common, and widely used model • Suitable to study BBB–immune cell interactions • Used at a large scale (e.g., biochemical studies and physiological investigations) • Easy to culture (in Petri dishes) • Suitable for transmigration processes • Relative high throughput screening • Low costs • Potential for using pure cells • More viable than capillaries	• Loss of BBB characteristics with serial passages (polarization and barrier function) • Relatively low TEER • Adequate paracellular barrier • Absence of stimulating factors (e.g., pericytes and astrocytes) • Missing endothelial differentiation	[19,215,216]
Coculture 2D Transwell systems	• Expression of more stringent TJs • Higher TEER • Lower permeability • Association with neurovascular units to better mimic the anatomic structure of the in vivo BBB model • Suitable for permeability studies, cell–cell interaction	• Lack of shear stress • Do not exhibit BBB phenotype	[83,217,41]

Dynamic in vitro models	• Exposed to shear stress • Better mimic the in vivo endothelial phenotype • Resemble in vivo BBB in terms of anatomic and physiological properties • Induce expression of significant factors (transporters, ion channels, and efflux proteins) • High differentiation degree of cells • High TEER • Low permeability	• Difficult to visualize endothelial morphology directly • High cell number • Require technical support • Longer steady-state time • Difficult to control capillaries and cell morphology • High-throughput pharmacological studies are not suitable • High cost	[176,194,218]
Microfluidic models	• 3D cell arrangement • Low thickness membrane or pillars • Suitable for transmigration/trafficking studies • Application of shear stress • Closely mimic the in vivo BBB models • Short time to reach steady-state TEER • Computer controlled parameters	• Limited publication • Expensive system • Difficult to set up • Require high cell number • Need to be to establish • Visualization difficulties • Advanced imaging techniques • Lower TEER	[17,19,41]

can be the answer for a fully and true human BBB cell culture model. The stem cells provide an easier way for obtaining cells of human origin, and it is expected in the future that a validated fully human BBB model can be obtained for BBB studies.

Different approaches have been utilized to mimic the in vivo interactions of cells from the neurovascular unit, as well as its dynamic flow. The most used system is the Transwell model, but other systems, such as dynamic, 3D, and microfluidic models, have been developed as well. The microfluidic models are suitable for coculture of different cells, as they provide dynamic conditions in vitro. The development of high-throughput microfluidic BBB models can be promising for permeability screening studies. However, 3D spheroid BBB models seem to be in the developing phase, especially for studying BBB physiology and pathology.

Developing cell culture models of the BBB that reflect the in vivo conditions are necessary both for the research of brain physiology and pathology, as well as for predicting BBB permeability. Therefore, it is important to develop simple, reproducible, and validated BBB cell culture models for a better understanding of the CNS and BBB drug permeability.

Abbreviations

2D Two dimensional
3D Three dimensional
BBB Blood–brain barrier
BCEC Brain capillary endothelial cell
bFGF Basic fibroblast growth factor
Caco-2 Human epithelial adenocarcinoma
CAM Cell adhesion molecule
CNS Central nervous system
FITC Fluorescein isothiocyanate
GDNF Glial-derived neurotrophic factor
γ-GT Gamma-glutamyl transpeptidase
HAs Human brain astrocytes
HBMECs Human microvascular endothelial cells
HFA Human fetal astrocytes
HRP Horseradish peroxidase
JACOP Junction-associated coiled coil protein
JAMs Junctional adhesion molecules
MDCK Madin–Darby canine kidney
MDR1 Multidrug resistant gene
MMPs Matrix metalloproteinases
NPC Neural progenitor cell
PD Parkinson's disease
P-gp P-glycoprotein

TEER	Transendothelial electrical resistance
TGFβ	Transforming growth factor-β
TJ	Tight junction
VE	Vascular endothelial
VEGF	Vascular endothelial growth factor
ZO	Zonula occludens

References

1. Rubin LL, Hall DE, Porter S, Barbu K, Cannon C, Horner HC, Janatpour M, Liaw CW, Manning K, Morales J. A cell culture model of the blood-brain barrier. *J. Cell Biol.* 1991;115(6):1725-1735.
2. Pardridge W. *Brain Drug Targeting: The Future of Brain Drug Development*. Cambridge: Cambridge University Press; 2001.
3. Lundquist S, Renftel M. The use of in vitro cell culture models for mechanistic studies and as permeability screens for the blood–brain barrier in the pharmaceutical industry—background and current status in the drug discovery process. *Vascul Pharmacol.* 2002;38(6):355–364.
4. Rubin LL, Staddon J. The cell biology of the blood-brain barrier. *Annu Rev Neurosci.* 1999;22:11–28.
5. Gumbleton M, Audus KL. Progress and limitations in the use of in vitro cell cultures to serve as a permeatbility screen for the blood-brain barrier. *J Pharm Sci.* 2001;90(11):1681–1698.
6. Joó F. The blood-brain barrier in vitro: ten years of research on microvessels isolated from the brain. *Neurochem Int.* 1985;7(1):1–25.
7. Brendel K, Meezan E, Carlson EC. Isolated brain microvessels: a purified, metabolically active preparation from bovine cerebral cortex. *Science.* 1974;185:953–955.
8. Panula P, Joó F, Rechardt L. Evidence for the presence of viable endothelial cells in cultures derived from dissociated rat brain. *Experientia.* 1978;34(1):95–97.
9. Joó F. The cerebral microvessels in culture, an update. *J Neurochem.* 1992;58(1):1–17.
10. Deli MA, Ábrahám CS, Kataoka Y, Niwa M. Permeability studies on in vitro blood-brain barrier models: physiology, pathology, and pharmacology. *Cell Mol Neurobiol.* 2005;25(1): 59–127.
11. Garberg P, Ball M, Borg N, Cecchelli R, Fenart L, Hurst RD, Lindmark T, Mabondzo A, Nilsson JE, Raub TJ, Stanimirovic D, Terasaki T, Öberg JO, Österberg T. In vitro models for the blood-brain barrier. *Toxicol In Vitro.* 2005;19(3):299–334.
12. Begley DJ, Bradbury MWB, Kreuter J. *The Blood-Brain Barrier and Drug Delivery to the CNS.* New York, NY: Marcel Dekker; 2000.
13. Grasset E, Pinto M, Dussaulx E, Zweibaum A, Desjeux JF. Epithelial properties of human colonic carcinoma cell line Caco-2: electrical parameters. *Am J Physiol.* 1984;247(3 pt 1):C260–C267.
14. Daneman R, Zhou L, Kebede A, Barres B. Pericytes are required for blood-brain barrier integrity during embryogenesis. *Nature.* 2010;468(7323):562–566.
15. DeBault LE, Cancilla PA. gamma-Glutamyl transpeptidase in isolated brain endothelial cells: induction by glial cells in vitro. *Science.* 1980;207(4431):653–655.
16. Reichel A, Begley DJ, Abbott NJ. An overview of in vitro techniques for blood-brain barrier studies. *Blood Brain Barrier Res Protoc.* 2003;:307–324.
17. Ruck T, Bittner S, Meuth S. Blood-brain barrier modeling: challenges and perspectives. *Neural Regen Res.* 2015;10(6):889.

18. Wilhelm I, Fazakas C, Krizbai IA. In vitro models of the blood-brain barrier. *Acta Neurobiol Exp (Wars)*. 2011;71(1):133–138.
19. Wilhelm I, Krizbai IA. In vitro models of the blood-brain barrier for the study of drug delivery to the brain. *Mol Pharm*. 2014;11(7):1949–1963.
20. Abbott NJ. Physiology of the blood–brain barrier and its consequences for drug transport to the brain. International Congress Series; vol.1277. Amsterdam: Elsevier; 2005:3–18.
21. Carvey PM, Hendey B, Monahan AJ. The blood-brain barrier in neurodegenerative disease: a rhetorical perspective. *J Neurochem*. 2009;111(2):291–314.
22. Weiss N, Miller F, Cazaubon S, Couraud PO. The blood-brain barrier in brain homeostasis and neurological diseases. *Biochim Biophys Acta Biomembr*. 2009;1788(4):842–858.
23. Cardoso FL, Brites D, Brito MA. Looking at the blood-brain barrier: molecular anatomy and possible investigation approaches. *Brain Res Rev*. 2010;64(2):328–363.
24. Tajes M, Ramos-Fernández E, Weng-Jiang X, Bosch-Morató M, Guivernau B, Eraso-Pichot A, Salvador B, Fernàndez-Busquets X, Roquer J, Muñoz FJ. The blood-brain barrier: structure, function and therapeutic approaches to cross it. *Mol Membr Biol*. 2014;31(5):152–167.
25. Abbott NJ. Astrocyte–endothelial interactions and blood–brain barrier permeability. *J Anat*. 2002;200(5):523–534.
26. Abbott NJ, Friedman A. Overview and introduction: the blood-brain barrier in health and disease. *Epilepsia*. 2012;53(s6):1–6.
27. Abbott NJ, Patabendige KAA, Dolman DEM, Yusof SR, Begley DJ. Structure and function of the blood-brain barrier. *Neurobiol Dis*. 2010;37(1):13–25.
28. Abbott NJ. Blood-brain barrier structure and function and the challenges for CNS drug delivery. *J Inherit Metab Dis*. 2013;36(3):437–449.
29. Oldendorf WH, Cornford ME, Brown WJ. The large apparent work capability of the blood-brain barrier: a study of the mitochondrial content of capillary endothelial cells in brain and other tissues of the rat. *Ann Neurol*. 1977;1(5):409–417.
30. Cornford EM, Hyman S. Blood–brain barrier permeability to small and large molecules. *Adv Drug Deliv Rev*. 1999;36(2–3):145–163.
31. de Boer AG, Gaillard PJ. Blood–brain barrier dysfunction and recovery. *J. Neural Transm*. 2006;113(4):445–462.
32. Hawkins BT, Davis TP. The blood-brain barrier/neurovascular unit in health and disease. *Pharmacol Rev*. 2005;57(2):173–185.
33. Sá-Pereira I, Brites D, Brito MA. Neurovascular unit: a focus on pericytes. *Mol Neurobiol*. 2012;45(2):327–347.
34. Wong AD, Ye M, Levy AF, Rothstein JD, Bergles DE, Searson PC. The blood-brain barrier: an engineering perspective. *Front Neuroeng*. 2013;6:512–523.
35. Armulik A, Abramsson A, Betsholtz C. Endothelial/pericyte interactions. *Circ Res*. 2005;97(6):7.
36. Armulik A, Genové G, Betsholtz C. Pericytes: developmental, physiological, and pathological perspectives, problems, and promises. *Dev Cell*. 2011;21(2):193–215.
37. Al Ahmad A, Taboada CB, Gassmann M, Ogunshola OO. Astrocytes and pericytes differentially modulate blood-brain barrier characteristics during development and hypoxic insult. *J Cereb Blood Flow Metab*. 2011;31(2):693–705.
38. Armulik A, Genové G, Mäe M, Nisancioglu MH, Wallgard E, Niaudet C, He L, Norlin J, Lindblom P, Strittmatter K, Johansson BR, Betsholtz C. Pericytes regulate the blood-brain barrier. *Nature*. 2010;468(7323):557–561.
39. Bell RD, Winkler EA, Sagare AP, Singh I, LaRue B, Deane R, Zlokovic BV. Pericytes control key neurovascular functions and neuronal phenotype in the adult brain and during brain aging. *Neuron*. 2010;68(3):409–427.

40. Nakagawa S, Deli Ma, Nakao S, Honda M, Hayashi K, Nakaoke R, Kataoka Y, Niwa M. Pericytes from brain microvessels strengthen the barrier integrity in primary cultures of rat brain endothelial cells. *Cell Mol Neurobiol*. 2007;27(6):687–694.

41. He Y, Yao Y, Tsirka SE, Cao Y. Cell-culture models of the blood-brain barrier. *Stroke*. 2014;45(8):2514–2526.

42. a Stewart P, Wiley MJ. Developing nervous tissue induces formation of blood-brain barrier characteristics in invading endothelial cells: a study using quail—chick transplantation chimeras. *Dev Biol*. 1981;84(1):183–192.

43. Janzer RC, Raff MC. Astrocytes induce blood–brain barrier properties in endothelial cells. *Nature*. 1987;325(6101):253–257.

44. Cantrill CA, Skinner RA, Rothwell NJ, Penny JI. An immortalised astrocyte cell line maintains the in vivo phenotype of a primary porcine in vitro blood-brain barrier model. *Brain Res*. 2012;1479:17–30.

45. Demeuse P, Kerkhofs A, Struys-Ponsar C, Knoops B, Remacle C, Van Den Bosch De Aguilar P. Compartmentalized coculture of rat brain endothelial cells and astrocytes: a syngenic model to study the blood-brain barrier. *J Neurosci Methods*. 2002;121(1):21–31.

46. Gaillard PJ, van der Sandt IC, Voorwinden LH, Vu D, Nielsen JL, de Boer AG, Breimer DD. Astrocytes increase the functional expression of P-glycoprotein in an in vitro model of the blood-brain barrier. *Pharm Res*. 2000;17(10):1198–1205.

47. Hamm S, Dehouck B, Kraus J, Wolburg-Buchholz K, Wolburg H, Risau W, Cecchelli R, Engelhardt B, Dehouck MP. Astrocyte mediated modulation of blood-brain barrier permeability does not correlate with a loss of tight junction proteins from the cellular contacts. *Cell Tissue Res*. 2004;315(2):157–166.

48. Haseloff RF, Blasig IE, Bauer HC, Bauer H. In search of the astrocytic factor(s) modulating blood-brain barrier functions in brain capillary endothelial cells in vitro. *Cell Mol Neurobiol*. 2005;25(1):25–39.

49. Sobue K, Yamamoto N, Yoneda K, Hodgson ME, Yamashiro K, Tsuruoka N, Tsuda T, Katsuya H, Miura Y, Asai K, Kato T. Induction of blood-brain barrier properties in immortalized bovine brain endothelial cells by astrocytic factors. *Neurosci Res*. 1999;35(2):155–164.

50. Bicker J, Alves G, Fortuna A, Falcão A. Blood-brain barrier models and their relevance for a successful development of CNS drug delivery systems: a review. *Eur J Pharm Biopharm*. 2014;87(3):409–432.

51. Tontsch U, Bauer HC. Glial cells and neurons induce blood-brain barrier related enzymes in cultured cerebral endothelial cells. *Brain Res*. 1991;539(2):247–253.

52. Savettieri G, Di Liegro I, Catania C, Licata L, Pitarresi GL, D'Agostino S, Schiera G, De Caro V, Giandalia G, Giannola LI, Cestelli A. Neurons and ECM regulate occludin localization in brain endothelial cells. *Neuroreport*. 2000;45(8):2514–2526.

53. Misra A, Ganesh S, Shahiwala A, Shah SP. Drug delivery to the central nervous system: a review. *J Pharm Pharm Sci*. 2003;6(2):252–273.

54. Engelhardt B, Sorokin L. The blood-brain and the blood-cerebrospinal fluid barriers: function and dysfunction. *Semin Immunopathol*. 2009;31(4):497–511.

55. Haseloff RF, Dithmer S, Winkler L, Wolburg H, Blasig IE. Transmembrane proteins of the tight junctions at the blood–brain barrier: structural and functional aspects. *Semin Cell Dev Biol*. 2014;38:16–25.

56. Abbott NJ. Prediction of blood-brain barrier permeation in drug discovery from in vivo, in vitro and in silico models. *Drug Discov Today Technol*. 2004;1(4):407–416.

57. Deli MA. Blood-brain barrier models. *Handb Neurochem Mol Neurobiol*. 2007;:29–55.

58. Chen Y, Liu L. Modern methods for delivery of drugs across the blood-brain barrier. *Adv Drug Deliv Rev*. 2012;64(7):640–665.

59. Butt AM, Jones HC, Abbott NJ. Electrical resistance across the blood-brain barrier in anaesthetized rats: a developmental study. *J Physiol*. 1990;429:47–62.
60. Tietz S, Engelhardt B. Brain barriers: crosstalk between complex tight junctions and adherens junctions. *J Cell Biol*. 2015;209(4):493–506.
61. Vorbrodt AW, Dobrogowska DH. Molecular anatomy of interendothelial junctions in human blood-brain barrier microvessels. *Folia Histochem Cytobiol*. 2004;42(2):67–75.
62. Furuse M. Occludin: a novel integral membrane protein localizing at tight junctions. *J Cell Biol*. 1993;123(6):1777–1788.
63. Furuse M. Claudin-1 and -2: novel integral membrane proteins localizing at tight junctions with no sequence similarity to occludin. *J Cell Biol*. 1998;141(7):1539–1550.
64. Martin-Padura I. Junctional adhesion molecule, a novel member of the immunoglobulin superfamily that distributes at intercellular junctions and modulates monocyte transmigration. *J Cell Biol*. 1998;142(1):117–127.
65. Huber D, Balda MS, Matter K. Occludin modulates transepithelial migration of neutrophils. *J Biol Chem*. 2000;275(8):5773–5778.
66. Haskins J, Gu L, Wittchen ES, Hibbard J, Stevenson BR. ZO-3 a novel member of the MAGUK protein family found at the tight junction, interacts with ZO-1 and occludin. *J Cell Biol*. 1998;141(1):199–208.
67. Wolff A, Antfolk M, Brodin B, Tenje M. In vitro blood-brain barrier models-an overview of established models and new microfluidic approaches. *J Pharm Sci*. 2015;104(9):2727–2746.
68. Breier G, Breviario F, Caveda L, Berthier R, Schnürch H, Gotsch U, Vestweber D, Risau W, Dejana E. Molecular cloning and expression of murine vascular endothelial-cadherin in early stage development of cardiovascular system. *Blood*. 1996;87(2):630–641.
69. Wolburg H, Lippoldt A. Tight junctions of the blood-brain barrier: development, composition and regulation. *Vascul Pharmacol*. 2002;38(6):323–337.
70. Brown RC, Davis TP. Calcium modulation of adherens and tight junction function: a potential mechanism for blood-brain barrier disruption after stroke. *Stroke*. 2002;33(6):1706–1711.
71. Gartzke D, Fricker G. Establishment of optimized MDCK cell lines for reliable efflux transport studies. *J Pharm Sci*. 2014;103(4):1298–1304.
72. Bernas MJ, Cardoso FL, Daley SK, Weinand ME, Campos AR, Ferreira AJG, Hoying JB, Witte MH, Brites D, Persidsky Y, Ramirez SH, Brito MA. Establishment of primary cultures of human brain microvascular endothelial cells to provide an in vitro cellular model of the blood-brain barrier. *Nat Protoc*. 2010;5(7):1265–1272.
73. Srinivasan B, Kolli AR, Esch MB, Abaci HE, Shuler ML, Hickman JJ. TEER measurement techniques for in vitro barrier model systems. *J Lab Autom*. 2015;20(2):107–126.
74. Avdeef A. *Absorption and Drug Development*. Hoboken, NJ: John Wiley and Sons, Inc.; 2003.
75. Hammarlund-Udenaes M, de Lange ECM, Thorne RG. *Drug Delivery to the Brain Physiological Concepts, Methodologies and Approaches*. New York, NY: Springer; 2014.
76. Audus KL, Borchardt RT. Characteristics of the large neutral amino acid transport system of bovine brain microvessel endothelial cell monolayers. *J Neurochem*. 1986;47(2):484–488.
77. Zucco F, Batto G, Bises J, Chambaz A, Chiusolo R, Consalvo H, Cross G, Dal Negro I, de Angelis G, Fabre F, Guillou S, Hoffman L, Laplanche E, Morel M, Pincon-Raymond P, Prieto L, Turco G, Ranaldi M, Rousset Y, Sambuy ML, Scarino F, Torreilles A, Stammati. An interlaboratory study to evaluate the effects of medium composition on the differentiation and barrier function of Caco-2 cell lines. *Altern Lab Anim*. 2005;33(6):603–618.
78. Malina KCK, Cooper I, Teichberg VI. Closing the gap between the in-vivo and in-vitro blood-brain barrier tightness. *Brain Res*. 2009;1284:12–21.

79. Benson K, Cramer S, Galla H-J. Impedance-based cell monitoring: barrier properties and beyond. *Fluids Barriers CNS*. 2013;10(1):5.

80. Hartmann C, Zozulya A, Wegener J, Galla H-J. The impact of glia-derived extracellular matrices on the barrier function of cerebral endothelial cells: an in vitro study. *Exp Cell Res*. 2007;313:1318–1325.

81. Dehouck MP, Méresse S, Delorme P, Fruchart JC, Cecchelli R. An easier, reproducible, and mass-production method to study the blood-brain barrier in vitro. *J Neurochem*. 1990;54(5):1798–1801.

82. Furie MB. Cultured endothelial cell monolayers that restrict the transendothelial passage of macromolecules and electrical current. *J Cell Biol*. 1984;98(3):1033–1041.

83. Ogunshola OO. In vitro modeling of the blood-brain barrier: simplicity versus complexity. *Curr Pharm Des*. 2011;17:2755–2761.

84. Naik P, Cucullo L. In vitro blood-brain barrier models: current and perspective technologies. *J Pharm Sci*. 2012;101(4):1337–1354.

85. Bittner S, Ruck T, Schuhmann MK, Herrmann AM, ou Maati HM, Bobak N, Göbel K, Langhauser F, Stegner D, Ehling P, Borsotto M, Pape H-C, Nieswandt B, Kleinschnitz C, Heurteaux C, Galla H-J, Budde T, Wiendl H, Meuth SG. Endothelial TWIK-related potassium channel-1 (TREK1) regulates immune-cell trafficking into the CNS. *Nat Med*. 2013;19(9):1161–1165.

86. Gaillard PJ, De Boer AG. Relationship between permeability status of the blood-brain barrier and in vitro permeability coefficient of a drug. *Eur J Pharm Sci*. 2000;12(2):95–102.

87. Stanness KA, Guatteo E, Janigro D. A dynamic model of the blood-brain barrier in vitro. *Neurotoxicology*. 1996;17(2):481–496.

88. Hellinger É, Veszelka S, Tóth AE, Walter F, Kittel Á, Bakk ML, Tihanyi K, Háda V, Nakagawa S, Dinh Ha Duy T, Niwa M, Deli Ma, Vastag M. Comparison of brain capillary endothelial cell-based and epithelial (MDCK-MDR1, Caco-2, and VB-Caco-2) cell-based surrogate blood-brain barrier penetration models. *Eur J Pharm Biopharm*. 2012;82(2):340–351.

89. Fletcher NF, Callanan JJ. *Cell Culture Models of the Blood-Brain Barrier: New Research*. Hauppauge, NY: Nova Science Publishers; 2012:[Chapter 8, vol. 43, no. 1].

90. Veszelka S, Kittel Á, Deli MA. Tools of modelling blood–brain barrier penetrability. Solubility, Delivery and ADME Problems of Drugs and Drug Candidates. Washington: Bentham Science; 2011:[pp. 166–188].

91. Pastan I, Gottesman MM, Ueda K, Lovelace E, V Rutherford A, Willingham MC. A retrovirus carrying an MDR1 cDNA confers multidrug resistance and polarized expression of P-glycoprotein in MDCK cells. *Proc Natl Acad Sci USA*. 1988;85(12):4486–4490.

92. Carrara S, Reali V, Misiano P, Dondio G, Bigogno C. Evaluation of in vitro brain penetration: optimized PAMPA and MDCKII-MDR1 assay comparison. *Int J Pharm*. 2007;345(1–2):125–133.

93. Summerfield SG, Read K, Begley DJ, Obradovic T, Hidalgo IJ, Coggon S, Lewis AV, Porter RA, Jeffrey P. Central nervous system drug disposition: the relationship between in situ brain permeability and brain free fraction. *J Pharmacol Exp Ther*. 2007;322(1):205–213.

94. Saaby L, Tfelt-Hansen P, Brodin B. The putative P-gp inhibitor telmisartan does not affect the transcellular permeability and cellular uptake of the calcium channel antagonist verapamil in the P-glycoprotein expressing cell line MDCK II MDR1. *Pharmacol Res Perspect*. 2015;3(4):e00151.

95. Sheng Y, Yang X, Wang Z, Mitra AK. Stereoisomeric prodrugs to improve corneal absorption of prednisolone: synthesis and in vitro evaluation. *AAPS PharmSciTech*. 2015;17(3):718–726.

96. Hurst RD, Fritz IB. Properties of an immortalised vascular endothelial/glioma cell co-culture model of the blood-brain barrier. *J Cell Physiol*. 1996;167(1):81–88.

97. Wang G-Y, Wang N, Liao H-N. Effects of muscone on the expression of P-gp, MMP-9 on blood–brain barrier model in vitro. *Cell Mol Neurobiol*. 2015;35(8):1105–1115.

98. V Ramsohoye P, Fritz IB. Preliminary characterization of glial-secreted factors responsible for the induction of high electrical resistances across endothelial monolayers in a blood-brain barrier model. *Neurochem Res*. 1998;23(12):1545–1551.

99. Wang Q, Luo W, Zhang W, Liu M, Song H, Chen J. Involvement of DMT1+IRE in the transport of lead in an in vitro BBB model. *Toxicol In Vitro*. 2011;25(4):991–998.

100. Barar J, Gumbleton M, Asadi M, Omidi Y. Barrier functionality and transport machineries of human ECV304 cells. *Med Sci Monit*. 2010;16(1):BR52–BR60.

101. Artursson P, Palm K, Luthman K. Caco-2 monolayers in experimental and theoretical predictions of drug transport. *Adv Drug Deliv Rev*. 2001;46(1):27–43.

102. Hubatsch I, Ragnarsson EGE, Artursson P. Determination of drug permeability and prediction of drug absorption in Caco-2 monolayers. *Nat Protoc*. 2007;2(9):2111–2119.

103. Hakkarainen JJ, Jalkanen AJ, Kääriäinen TM, Keski-Rahkonen P, Venäläinen T, Hokkanen J, Mönkkönen J, Suhonen M, Forsberg MM. Comparison of in vitro cell models in predicting in vivo brain entry of drugs. *Int J Pharm*. 2010;402(1–2):27–36.

104. Lohmann C, Hüwel S, Galla H-J. Predicting blood-brain barrier permeability of drugs: evaluation of different in vitro assays. *J Drug Target*. 2002;10(4):263–276.

105. Hellinger E, Bakk ML, Pócza P, Tihanyi K, Vastag M. Drug penetration model of vinblastine-treated Caco-2 cultures. *Eur J Pharm Sci*. 2010;41(1):96–106.

106. Lundquist S, Renftel M, Brillault J, Fenart L, Cecchelli R, Dehouck M-P. Prediction of drug transport through the blood-brain barrier in vivo: a comparison between two in vitro cell models. *Pharm Res*. 2002;19(7):976–981.

107. Joó F, Karnushina I. A procedure for the isolation of capillaries from rat brain. *Cytobios*. 1973;8(29):41–48.

108. Bickel U. How to measure drug transport across the blood-brain barrier. *NeuroRx*. 2005;2(1):15–26.

109. Hakkarainen JJ. *In Vitro Cell Models in Predicting Blood-Brain Barrier Permeability of Drugs*. Finland: University of Eastern Finland; 2013 [Doctoral dissertation].

110. Stanimirovic DB, Bani-Yaghoub M, Perkins M, Haqqani AS. Blood–brain barrier models: in vitro to in vivo translation in preclinical development of CNS-targeting biotherapeutics. *Exp Opin Drug Discov*. 2015;10(2):141–155.

111. Abbott NJ, Hughes CC, a Revest P, Greenwood J. Development and characterisation of a rat brain capillary endothelial culture: towards an in vitro blood-brain barrier. *J Cell Sci*. 1992;103:23–37.

112. Parkinson FE, Hacking C. Pericyte abundance affects sucrose permeability in cultures of rat brain microvascular endothelial cells. *Brain Res*. 2005;1049(1):8–14.

113. Navone SE, Marfia G, Invernici G, Cristini S, Nava S, Balbi S, Sangiorgi S, Ciusani E, Bosutti A, Alessandri G, Slevin M, a Parati E. Isolation and expansion of human and mouse brain microvascular endothelial cells. *Nat Protoc*. 2013;8(9):1680–1693.

114. Urich E, Lazic SE, Molnos J, Wells I, Freskgård P-O. Transcriptional profiling of human brain endothelial cells reveals key properties crucial for predictive in vitro blood-brain barrier models. *PLoS One*. 2012;7(5):e38149.

115. Bowman PD, Ennis SR, Rarey KE, Betz AL, Goldstein GW. Brain microvessel endothelial cells in tissue culture: a model for study of blood-brain barrier permeability. *Ann Neurol*. 1983;14(4):396–402.

116. Wallace BK, Foroutan S, O'Donnell ME. Ischemia-induced stimulation of Na-K-Cl cotransport in cerebral microvascular endothelial cells involves AMP kinase. *AJP Cell Physiol*. 2011;301(2):C316–C326.

117. Cecchelli R, Dehouck B, Descamps L, Fenart L, Buée-Scherrer V, Duhem C, Lundquist S, Rentfel M, Torpier G, Dehouck MP. In vitro model for evaluating drug transport across the blood-brain barrier. *Adv Drug Deliv Rev*. 1999;36(2–3):165–178.

118. Patabendige A, Skinner Ra, Abbott NJ. Establishment of a simplified in vitro porcine blood–brain barrier model with high transendothelial electrical resistance. *Brain Res*. 2013;1521: 1–15.

119. Patabendige A, Abbott NJ. Primary porcine brain microvessel endothelial cell isolation and culture. *Curr Protoc Neurosci*. 2014;:3–27.

120. Franke H, Galla HJ, Beuckmann CT. An improved low-permeability in vitro-model of the blood-brain barrier: transport studies on retinoids, sucrose, haloperidol, caffeine and mannitol. *Brain Res*. 1999;818(1):65–71.

121. Franke H, Galla HJ, Beuckmann CT. Primary cultures of brain microvessel endothelial cells: a valid and flexible model to study drug transport through the blood-brain barrier in vitro. *Brain Res Protoc*. 2000;5(3):248–256.

122. Qiao R, Jia Q, Hüwel S, Xia R, Liu T, Gao F, Galla H-J, Gao M. Receptor-mediated delivery of magnetic nanoparticles across the blood–brain barrier. *ACS Nano*. 2012;6(4):3304–3310.

123. Freese C, Reinhardt S, Hefner G, Unger RE, Kirkpatrick CJ, Endres K. A novel blood-brain barrier co-culture system for drug targeting of Alzheimer's disease: establishment by using acitretin as a model drug. *PLoS One*. 2014;9(3):e91003.

124. Watson PMD, Paterson JC, Thom G, Ginman U, Lundquist S, Webster CI. Modelling the endothelial blood-CNS barriers: a method for the production of robust in vitro models of the rat blood-brain barrier and blood-spinal cord barrier. *BMC Neurosci*. 2013;14(1):59.

125. Coisne C, Dehouck L, Faveeuw C, Delplace Y, Miller F, Landry C, Morissette C, Fenart L, Cecchelli R, Tremblay P, Dehouck B. Mouse syngenic in vitro blood-brain barrier model: a new tool to examine inflammatory events in cerebral endothelium. *Lab Invest*. 2005;85(6): 734–746.

126. Burkhart A, Thomsen LB, Thomsen MS, Lichota J, Fazakas C, Krizbai I, Moos T. Transfection of brain capillary endothelial cells in primary culture with defined blood–brain barrier properties. *Fluids Barriers CNS*. 2015;12(1):19.

127. Sipos I, Dömötör E, Abbott NJ, Adam-Vizi V. The pharmacology of nucleotide receptors on primary rat brain endothelial cells grown on a biological extracellular matrix: effects on intracellular calcium concentration. *Br J Pharmacol*. 2000;131(6):1195–1203.

128. Pardridge WM. Drug transport across the blood–brain barrier. *J Cereb Blood Flow Metab*. 2012;32(11):1959–1972.

129. Dorovini-Zis K, Prameya R, Bowman PD. Culture and characterization of microvascular endothelial cells derived from human brain. *Lab Invest*. 1991;64(3):425–436.

130. Mukhtar M, Pomerantz RJ. Development of an in vitro blood-brain barrier model to study molecular neuropathogenesis and neurovirologic disorders induced by human immunodeficiency virus type 1 infection. *J Hum Virol*. 2000;3(6):324–334.

131. Parikh NU, Aalinkeel R, Reynolds JL, Nair BB, Sykes DE, Mammen MJ, Schwartz SA, Mahajan SD. Galectin-1 suppresses methamphetamine induced neuroinflammation in human brain microvascular endothelial cells: neuroprotective role in maintaining blood brain barrier integrity. *Brain Res*. 2015;1624:175–187.

132. Larochelle C, Cayrol R, Kebir H, Alvarez JI, Lécuyer M-A, Ifergan I, Viel É, Bourbonnière L, Beauseigle D, Terouz S, Hachehouche L, Gendron S, Poirier J, Jobin C, Duquette P, Flanagan K, Yednock T, Arbour N, Prat A. Melanoma cell adhesion molecule identifies encephalitogenic T lymphocytes and promotes their recruitment to the central nervous system. *Brain*. 2012;135(pt 10):2906–2924.

133. Giri R, Selvaraj S, Miller CA, Hofman F, Yan SD, Stern D, V Zlokovic B, Kalra VK. Effect of endothelial cell polarity on beta-amyloid-induced migration of monocytes across normal and AD endothelium. *Am J Physiol Cell Physiol.* 2002;283(3):C895–C904.

134. Ramirez SH, Heilman D, Morsey B, Potula R, Haorah J, Persidsky Y. Activation of peroxisome proliferator-activated receptor (PPAR) Suppresses Rho GTPases in human brain microvascular endothelial cells and inhibits adhesion and transendothelial migration of HIV-1 infected monocytes. *J Immunol.* 2008;180(3):1854–1865.

135. Riganti C, Salaroglio IC, Pinzòn-Daza ML, Caldera V, Campia I, Kopecka J, Mellai M, Annovazzi L, Couraud P-O, Bosia A, Ghigo D, Schiffer D. Temozolomide down-regulates P-glycoprotein in human blood–brain barrier cells by disrupting Wnt3 signaling. *Cell Mol Life Sci.* 2014;71(3):499–516.

136. Gil ES, Wu L, Xu L, Lowe TL. β-Cyclodextrin-poly(β-amino ester) nanoparticles for sustained drug delivery across the blood-brain barrier. *Biomacromolecules.* 2012;13(11):3533–3541.

137. Sade H, Baumgartner C, Hugenmatter A, Moessner E, Freskgård P-O, Niewoehner J. A Human blood-brain barrier transcytosis assay reveals antibody transcytosis influenced by pH-dependent receptor binding. *PLoS One.* 2014;9(4):e96340.

138. Durieu-Trautmann O, Foignant-Chaverot N, Perdomo J, Gounon P, Strosberg a##D, Couraud PO. Immortalization of brain capillary endothelial cells with maintenance of structural characteristics of the blood-brain barrier endothelium. *Vitr Cell Dev Biol.* 1991;27A(10): 771–778.

139. Eigenmann DE, Xue G, Kim KS, V Moses A, Hamburger M, Oufir M. Comparative study of four immortalized human brain capillary endothelial cell lines, hCMEC/D3, hBMEC, TY10, and BB19, and optimization of culture conditions, for an in vitro blood-brain barrier model for drug permeability studies. *Fluids Barriers CNS.* 2013;10(1):33.

140. Manda KR, Banerjee A, Banks WA, Ercal N. Highly active antiretroviral therapy drug combination induces oxidative stress and mitochondrial dysfunction in immortalized human blood-brain barrier endothelial cells. *Free Radic Biol Med.* 2011;50(7):801–810.

141. a Muruganandam LM, Herx R, Monette JP, Durkin DB. Stanimirovic, Development of immortalized human cerebromicrovascular endothelial cell line as an in vitro model of the human blood-brain barrier. *FASEB J.* 1997;11(13):1187–1197.

142. Terasaki T, Hosoya K. Conditionally immortalized cell lines as a new in vitro model for the study of barrier functions. *Biol Pharm Bull.* 2001;24(2):111–118.

143. Cho MJ, Thompson DP, Cramer CT, Vidmar TJ, Scieszka JF. The Madin Darby canine kidney (MDCK) epithelial cell monolayer as a model cellular transport barrier. *Pharm Res.* 1989;6(1):71–77.

144. Suda K, Rothen-Rutishauser B, Günthert M, Wunderli-Allenspach H. Phenotypic characterization of human umbilical vein endothelial (ECV304) and urinary carcinoma (T24) cells: endothelial versus epithelial features. *In Vitro Cell Dev Biol Anim.* 2001;37(8): 505–514.

145. Hilgendorf C, Spahn-Langguth H, Regårdh CG, Lipka E, Amidon GL, Langguth P. Caco-2 versus Caco-2/HT29-MTX co-cultured cell lines: permeabilities via diffusion, inside-and outside-directed carrier-mediated transport. *J Pharm Sci.* 2000;89(1):63–75.

146. Abbruscato TJ, Davis TP. Combination of hypoxia/aglycemia compromises in vitro blood-brain barrier integrity. *J Pharmacol Exp Ther.* 1999;289(2):668–675.

147. Fischer S, Wobben M, Kleinstück J, Renz D, Schaper W. Effect of astroglial cells on hypoxia-induced permeability in PBMEC cells. *Am J Physiol Cell Physiol.* 2000;279(4):C935–C944.

148. De Vries HE, Blom-Roosemalen MCM, Van Oosten M, De Boer AG, Van Berkel TJC, Breimer DD, Kuiper J. The influence of cytokines on the integrity of the blood-brain barrier in vitro. *J Neuroimmunol.* 1996;64(1):37–43.

149. Weidenfeller C, Schrot S, Zozulya A, Galla H-J. Murine brain capillary endothelial cells exhibit improved barrier properties under the influence of hydrocortisone. *Brain Res.* 2005;1053(1–2):162–174.

150. Daniels BP, Cruz-Orengo L, Pasieka TJ, Couraud P-O, Romero IA, Weksler B, Cooper JA, Doering TL, Klein RS. Immortalized human cerebral microvascular endothelial cells maintain the properties of primary cells in an in vitro model of immune migration across the blood brain barrier. *J Neurosci Methods.* 2013;212(1):173–179.

151. Hosoya K, Tetsuka K, Nagase K, Tomi M, Saeki S, Ohtsuki S, Creation NI, Yanai N, Obinata M, Kikuchi A, Okano T, Takanaga H, Corporation T, Biopharmacy M, Growth C, Brain M, Endothelial C, Line C. Conditionally Immortalized brain capillary endothelial cell lines established from a transgenic mouse harboring temperature-sensitive simian virus 40 large T-antigen gene. *AAPS J.* 2000;2(3):69–79.

152. Markoutsa E, Pampalakis G, Niarakis A, Romero IA, Weksler B, Couraud P-O, Antimisiaris SG. Uptake and permeability studies of BBB-targeting immunoliposomes using the hCMEC/D3 cell line. *Eur J Pharm Biopharm.* 2011;77(2):265–274.

153. Neuhaus W, Stessl M, Strizsik E, Bennani-Baiti B, Wirth M, Toegel S, Modha M, Winkler J, Gabor F, Viernstein H, Noe CR. Blood–brain barrier cell line PBMEC/C1-2 possesses functionally active P-glycoprotein. *Neurosci Lett.* 2010;469(2):224–228.

154. Rist RJ, Romero IA, Chan MWK, Couraud PO, Roux F, Abbott NJ. F-actin cytoskeleton and sucrose permeability of immortalised rat brain microvascular endothelial cell monolayers: Effects of cyclic AMP and astrocytic factors. *Brain Res.* 1997;768(1–2):10–18.

155. Vilas-Boas V, Silva R, Guedes-de-Pinho P, Carvalho F, Bastos ML, Remião F. RBE4 cells are highly resistant to paraquat-induced cytotoxicity: studies on uptake and efflux mechanisms. *J Appl Toxicol.* 2014;34(9):1023–1030.

156. Tega Y, Akanuma S, Kubo Y, Terasaki T, Hosoya K. Blood-to-brain influx transport of nicotine at the rat blood-brain barrier: involvement of a pyrilamine-sensitive organic cation transport process. *Neurochem Int.* 2013;62(2):173–181.

157. Hart CP, Van Dyk MN, Moore LF, Shasby DM SA. No differential opening of the brain endothelial barrier following neutralization of the endothelial luminal anionic charge in vitro. *J Neuropathol Exp Neurol.* 1987;46:141–153.

158. a Wijsman J, Shivers RR. Immortalized mouse brain endothelial cells are ultrastructurally similar to endothelial cells and respond to astrocyte-conditioned medium. *Vitr Cell Dev Biol Anim.* 1998;34(10):777–784.

159. Lee H-T, Chang Y-C, Tu Y-F, Huang C-C. CREB activation mediates VEGF-A's protection of neurons and cerebral vascular endothelial cells. *J Neurochem.* 2010;113(1):79–91.

160. Wen CJ, Zhang LW, Al-Suwayeh SA, Yen TC, Fang JY. Theranostic liposomes loaded with quantum dots and apomorphine for brain targeting and bioimaging. *Int J Nanomedicine.* 2012;7:1599–1611.

161. Yang T, Roder KE, Abbruscato TJ. Evaluation of bEnd5 cell line as an in vitro model for the blood–brain barrier under normal and hypoxic/aglycemic conditions. *J Pharm Sci.* 2007;96(12):3196–3213.

162. Lindsaywilliams R. Embryonic lethalities and endothelial tumors in chimeric mice expressing polyoma virus middle T oncogene. *Cell.* 1988;52(1):121–131.

163. Wagner E, Risau W. Oncogenes in the study of endothelial cell growth and differentiation. *Semin Cancer Biol.* 1994;5(2):137–145.

164. Weksler BB, Subileau E, Perrière N, Charneau P, Holloway K, Leveque M, Tricoire-Leignel H, Nicotra S, Bourdoulous P, Turowski DK, Male F, Roux J, Greenwood I, a Romero PO, Couraud. Blood-brain barrier-specific properties of a human adult brain endothelial cell line. *FASEB J.* 2005;19(13):1872–1874.

165. Weksler B, a Romero I, Couraud P-O. The hCMEC/D3 cell line as a model of the human blood brain barrier. *Fluids Barriers CNS*. 2013;10(1):16.

166. Förster C, Burek M, Romero IA, Weksler B, Couraud P-O, Drenckhahn D. Differential effects of hydrocortisone and TNF (on tight junction proteins in an in vitro model of the human blood-brain barrier). *J Physiol*. 2008;586(7):1937–1949.

167. Hatherell K, Couraud PO, Romero IA, Weksler B, Pilkington GJ. Development of a three-dimensional, all-human in vitro model of the blood-brain barrier using mono-, co-, and tri-cultivation Transwell models. *J Neurosci Methods*. 2011;199(2):223–229.

168. Prasad S, Sajja RK, Park JH, Naik P, Kaisar MA, Cucullo L. Impact of cigarette smoke extract and hyperglycemic conditions on blood–brain barrier endothelial cells. *Fluids Barriers CNS*. 2015;12(1):18.

169. Abbott NJ, Rönnbäck L, Hansson E. Astrocyte-endothelial interactions at the blood-brain barrier. *Nat Rev Neurosci*. 2006;7(1):41–53.

170. Megard I, Garrigues A, Orlowski S, Jorajuria S, Clayette P, Ezan E, Mabondzo A. A co-culture-based model of human blood-brain barrier: application to active transport of indinavir and in vivo-in vitro correlation. *Brain Res*. 2002;927(2):153–167.

171. Jeliazkova-Mecheva VV, Bobilya DJ. A porcine astrocyte/endothelial cell co-culture model of the blood-brain barrier. *Brain Res Protoc*. 2003;12(2):91–98.

172. Kuo Y-C, Lu C-H. Effect of human astrocytes on the characteristics of human brain-microvascular endothelial cells in the blood-brain barrier. *Colloids Surf B*. 2011;86(1):225–231.

173. Gaillard PJ, Voorwinden LH, Nielsen JL, Ivanov A, Atsumi R, Engman H, Ringbom C, de Boer AG, Breimer DD. Establishment and functional characterization of an in vitro model of the blood–brain barrier, comprising a co-culture of brain capillary endothelial cells and astrocytes. *Eur J Pharm Sci*. 2001;12(3):215–222.

174. Li G, Simon MJ, Cancel LM, Shi ZD, Ji X, Tarbell JM, Morrison B, Fu BM. Permeability of endothelial and astrocyte cocultures: in vitro blood-brain barrier models for drug delivery studies. *Ann Biomed Eng*. 2010;38(8):2499–2511.

175. Fitsanakis VA, Piccola G, Aschner JL, Aschner M. Manganese transport by rat brain endothelial (RBE4) cell-based Transwell model in the presence of astrocyte conditioned media. *J Neurosci Res*. 2005;81(2):235–243.

176. Siddharthan V, V Kim Y, Liu S, Kim KS. Human astrocytes/astrocyte-conditioned medium and shear stress enhance the barrier properties of human brain microvascular endothelial cells. *Brain Res*. 2007;1147:39–50.

177. Hayashi K, Nakao S, Nakaoke R, Nakagawa S, Kitagawa N, Niwa M. Effects of hypoxia on endothelial/pericytic co-culture model of the blood-brain barrier. *Regul Pept*. 2004;123(1):77–83.

178. Thanabalasundaram G, Pieper C, Lischper M, Galla H-J. Regulation of the blood–brain barrier integrity by pericytes via matrix metalloproteinases mediated activation of vascular endothelial growth factor in vitro. *Brain Res*. 2010;1347:1–10.

179. Thomsen LB, Burkhart A, Moos T. A triple culture model of the blood-brain barrier using porcine brain endothelial cells, astrocytes and pericytes. *PLoS One*. 2015;10(8):e0134765.

180. Schiera G, Bono E, Raffa MP, Gallo A, Pitarresi GL, Di Liegro I, Savettieri G, Pia M, Laura G, Di I. Synergistic effects of neurons and astrocytes on the differentiation of brain capillary endothelial cells in culture. *J Cell Mol Med*. 2003;7(2):165–170.

181. Schiera G, Sala S, Gallo A, Raffa MP, Pitarresi GL, Savettieri G, Di Liegro I. Permeability properties of a three-cell type in vitro model of blood-brain barrier. *J Cell Mol Med*. 2005;9(2):373–379.

182. Xue Q, Liu Y, Qi H, Ma Q, Xu L, Chen W, Chen G, Xu X. A novel brain neurovascular unit model with neurons, astrocytes and microvascular endothelial cells of rat. *Int J Biol Sci*. 2013;9(2):174–189.

183. Prat A, Biernacki K, Wosik K, Antel JP. Glial cell influence on the human blood-brain barrier. *Glia*. 2001;36(2):145–155.

184. Nishioku T, Matsumoto J, Dohgu S, Sumi N, Miyao K, Takata F, Shuto H, Yamauchi A, Kataoka Y. Tumor necrosis factor-α mediates the blood–brain barrier dysfunction induced by activated microglia in mouse brain microvascular endothelial cells. *J Pharmacol Sci*. 2010;112(2):251–254.

185. Sumi N, Nishioku T, Takata F, Matsumoto J, Watanabe T, Shuto H, Yamauchi A, Dohgu S, Kataoka Y. Lipopolysaccharide-activated microglia induce dysfunction of the blood-brain barrier in rat microvascular endothelial cells co-cultured with microglia. *Cell Mol Neurobiol*. 2010;30(2):247–253.

186. Matsumoto J, Dohgu S, Takata F, Nishioku T, Sumi N, Machida T, Takahashi H, Yamauchi A, Kataoka Y. Lipopolysaccharide-activated microglia lower P-glycoprotein function in brain microvascular endothelial cells. *Neurosci Lett*. 2012;524(1):45–48.

187. Lippmann ES, Al-Ahmad A, Palecek SP, Shusta E. Modeling the blood-brain barrier using stem cell sources. *Fluids Barriers CNS*. 2013;10(1):2.

188. Weidenfeller C, Svendsen CN, Shusta EV. Differentiating embryonic neural progenitor cells induce blood-brain barrier properties. *J Neurochem*. 2007;101(2):555–565.

189. Lippmann ES, Weidenfeller C, Svendsen CN, Shusta EV. Blood-brain barrier modeling with co-cultured neural progenitor cell-derived astrocytes and neurons. *J Neurochem*. 2011;119(3):507–520.

190. Lippmann ES, Azarin SM, Kay JE, Nessler RA, Wilson HK, Al-Ahmad A, Palecek SP, V Shusta E. Derivation of blood-brain barrier endothelial cells from human pluripotent stem cells. *Nat Biotechnol*. 2012;30(8):783–791.

191. Cecchelli R, Aday S, Sevin E, Almeida C, Culot M, Dehouck L, Coisne C, Engelhardt B, Dehouck MP, Ferreira L. A stable and reproducible human blood-brain barrier model derived from hematopoietic stem cells. *PLoS One*. 2014;9(6):e99733.

192. Lippmann ES, Al-Ahmad A, Azarin SM, Palecek SP, Shusta EV. A retinoic acid-enhanced, multicellular human blood-brain barrier model derived from stem cell sources. *Sci Rep*. 2014;4:4160.

193. Urich E, Patsch C, Aigner S, Graf M, Iacone R, Freskgård P-O. Multicellular self-assembled spheroidal model of the blood brain barrier. *Sci Rep*. 2013;3:1500.

194. Cucullo L, Marchi N, Hossain M, Janigro D. A dynamic in vitro BBB model for the study of immune cell trafficking into the central nervous system. *J Cereb Blood Flow Metab*. 2011;31(2):767–777.

195. Bussolari SR. Apparatus for subjecting living cells to fluid shear stress. *Rev Sci Instrum*. 1982;53(12):1851.

196. Santaguida S, Janigro D, Hossain M, Oby E, Rapp E, Cucullo L. Side by side comparison between dynamic versus static models of blood-brain barrier in vitro: a permeability study. *Brain Res*. 2006;1109(1):1–13.

197. Janigro D, Leaman SM, a Stanness K. Dynamic modeling of the blood-brain barrier: a novel tool for studies of drug delivery to the brain. *Pharm Sci Technol Today*. 1999;2(1): 7–12.

198. Krizanac-Bengez L, Mayberg MR, Cunningham E, Hossain M, Ponnampalam S, Parkinson FE, Janigro D. Loss of shear stress induces leukocyte-mediated cytokine release and blood-brain barrier failure in dynamic in vitro blood-brain barrier model. *J Cell Physiol*. 2006;206(1):68–77.

199. Koutsiaris AG, V Tachmitzi S, Batis N, Kotoula MG, Karabatsas CH, Tsironi E, Chatzoulis DZ. Volume flow and wall shear stress quantification in the human conjunctival capillaries and post-capillary venules in vivo. *Biorheology*. 2007;44(5–6):375–386.

200. McAllister MS, Krizanac-Bengez L, Macchia F, Naftalin RJ, Pedley KC, Mayberg MR, Marroni M, Leaman S, Stanness KA, Janigro D. Mechanisms of glucose transport at the blood-brain barrier: an in vitro study. *Brain Res*. 2001;904(1):20–30.

201. Cucullo L, Couraud P-O, Weksler B, Romero I-A, Hossain M, Rapp E, Janigro D. Immortalized human brain endothelial cells and flow-based vascular modeling: a marriage of convenience for rational neurovascular studies. *J Cereb Blood Flow Metab*. 2008;28(2): 312–328.

202. Devraj K, Klinger ME, Myers RL, Mokashi A, Richard A, Simpson IA. Glut-1 glucose transporters in the blood-brain barrier: differential phosphorylation. *J Neurosci Res*. 2013;89(12).

203. Molino Y, Jabès F, Lacassagne E, Gaudin N, Khrestchatisky M. Setting-up an in vitro model of rat blood-brain barrier (BBB): a focus on BBB impermeability and receptor-mediated transport. *JoVE*. 2014;28(88):e51278.

204. Cucullo L, McAllister MS, Kight K, Krizanac-Bengez L, Marroni M, Mayberg MR, Stanness Ka, Janigro D. A new dynamic in vitro model for the multidimensional study of astrocyte-endothelial cell interactions at the blood-brain barrier. *Brain Res*. 2002;951(2):243–254.

205. Cucullo L, Hossain M, Rapp E, Manders T, Marchi N, Janigro D. Development of a humanized in vitro blood-brain barrier model to screen for brain penetration of antiepileptic drugs. *Epilepsia*. 2007;48(3):505–516.

206. Cucullo L, Hossain M, Tierney W, Janigro D. A new dynamic in vitro modular capillaries-venules modular system: cerebrovascular physiology in a box. *BMC Neurosci*. 2013;14(1):18.

207. Booth R, Kim H. Characterization of a microfluidic in vitro model of the blood-brain barrier (μBBB). *Lab Chip*. 2012;12(10):1784–1792.

208. Yeon JH, Na D, Choi K, Ryu SW, Choi C, Park JK. Reliable permeability assay system in a microfluidic device mimicking cerebral vasculatures. *Biomed Microdevices*. 2012;14(6): 1141–1148.

209. Prabhakarpandian B, Shen M-C, Nichols JB, Mills IR, Sidoryk-Wegrzynowicz M, Aschner M, Pant K. SyM-BBB: a microfluidic blood brain barrier model. *Lab Chip*. 2013;13(6): 1093–1101.

210. Griep LM, Wolbers F, de Wagenaar B, ter Braak PM, Weksler BB, Romero IA, Couraud PO, Vermes I, van der Meer AD, van den Berg A. BBB ON CHIP: microfluidic platform to mechanically and biochemically modulate blood-brain barrier function. *Biomed Microdevices*. 2013;15:145–150.

211. Sherer TB, Betarbet R, Stout AK, Lund S, Baptista M, Panov A, Cookson MR, Greenamyre JT. An in vitro model of Parkinson' s disease: linking mitochondrial impairment to altered α-synuclein metabolism and oxidative damage. *Science*. 2002;22(16):7006–7015.

212. Le W, Rowe D, Xie W, Ortiz I, He Y, Appel SH. Microglial activation and dopaminergic cell injury: an in vitro model relevant to Parkinson's disease. *J Neurosci*. 2001;21(12): 8447–8455.

213. Di L, Kerns EH, Bezar IF, Petusky SL, Huang Y. Comparison of blood-brain barrier permeability assays: in situ brain perfusion, MDR1-MDCKII and PAMPA-BBB. *J Pharm Sci*. 2009;98(6):1980–1991.

214. Nicolazzo JA, Charman SA, Charman WN. Methods to assess drug permeability across the blood-brain barrier. *J Pharm Pharmacol*. 2006;58(3):281–293.

215. Smith M, Omidi Y, Gumbleton M. Primary porcine brain microvascular endothelial cells: biochemical and functional characterisation as a model for drug transport and targeting. *J Drug Target*. 2007;15(4):253–268.

216. Patabendige A, Skinner Ra, Abbott NJ. Establishment of a simplified in vitro porcine blood-brain barrier model with high transendothelial electrical resistance. *Brain Res*. 2013;1521: 1–15.

217. Lacombe O, Videau O, Chevillon D, Guyot A-C, Contreras C, Blondel S, Nicolas L, Ghettas A, Bénech H, Thevenot E, Pruvost A, Bolze S, Krzaczkowski L, Prévost C, Mabondzo A. In vitro primary human and animal cell-based blood-brain barrier models as a screening tool in drug discovery. *Mol Pharm*. 2011;8(3):651–663.

218. Neuhaus W, Lauer R, Oelzant S, Fringeli UP, Ecker GF, Noe CR. A novel flow based hollow-fiber blood–brain barrier in vitro model with immortalised cell line PBMEC/C1-2. *J Biotechnol*. 2006;125(1):127–141.

CHAPTER 9

In Vivo/In Situ Animal Models

Muhammed Abdur Rauf, PhD
Yeditepe University, Istanbul, Turkey

1 INTRODUCTION

Animal models of human diseases are relevant mainly for two reasons. First, many biological processes are evolutionarily conserved between animals and humans. Second, there are ethical concerns and difficulties in using humans for experimental purposes. Therefore, various animal models for neurological diseases and brain tumors have been developed for the evaluation of potential therapeutic agents both in the of free molecular form and as delivery systems. From an ethical perspective, the use of small animals involves lower cost and greater acceptability compared to larger animals.

A model is something simple made by scientists to understand something complicated; a good model reduces the complexity of the modeled system, while preserving its essential features.[1] The objectives of animal models are to create human diseases in animals to understand the disease processes and find appropriate treatment approaches. Three main validity domains[2] are usually used to assess the strength of the model in reproducing a given human neuropsychiatric condition: (1) construct validity, that is how the model was developed in terms of fidelity toward the given natural causes of the disease in humans; (2) face validity, which describes a model's reliability in terms of a symptom array and biological changes pertinent to the disease being studied; and (3) predictive validity, how well the model responds to the medication used to treat the given disorder in comparison to how human patients respond, and therein how valid the model is for testing of new compounds and biological targets for future research.[3]

In the context of biological sciences, in vitro (Latin; in the glass) models are cells or biological molecules outside their normal biological environment used usually in test tubes, flasks, Petri dishes, etc. On the other hand, in vivo (Latin; within the living) models are whole, living organisms usually including animals, humans, and plants. Ex vivo (Latin; out of the living) models

CONTENTS

Nanotechnology Methods for Neurological Diseases and Brain Tumors. http://dx.doi.org/10.1016/B978-0-12-803796-6.00009-5

refer to tissues isolated from an organism in an external environment with the minimum alteration of natural conditions. In situ (Latin; on site or in position) models are those used locally in the same place where the phenomenon is taking place without isolating it from other systems or altering the original conditions of the test. Thus, "animal models" covers both in vivo and in situ models in this discussion. In this chapter, animal models will be described for brain tumors, as well as neurological diseases, including Alzheimer's disease (AD), Parkinson's disease (PD), Huntington's disease (HT), depression, anxiety, schizophrenia, pain, and stroke.

2 ANIMAL MODELS OF ALZHEIMER'S DISEASE

The pathophysiology of AD[4] typically includes extracellular β-amyloid (Aβ) deposits, intracellular neurofibrillary tangles, senile plaques, and loss of neurons. Impaired cholinergic transmission and degeneration of cholinergic neurons in AD are imitated by pharmacological, lesion, and transgenic animal models.[5]

Scopolamine-induced amnesia is a pharmacological model. Scopolamine is a centrally active anticholinergic agent. Scopolamine influences delta, theta, alpha, and beta activity in electroencephalograms and produces similar memory deficits seen in the elderly, but it cannot induce the full range of deficits seen in patients with AD.[6]

Lesion models are created either by surgical cholinergic denervation (e.g., transection of the fimbria/fornix leading to cholinergic denervation of the hippocampus), by destroying discrete cholinergic nuclei by electrolytic method (e.g., the nucleus basalis magnocellularis or the septal nuclei), or by microinjection of a neurotoxin, such as quinolinic acid, ibotenic acid, or the cholinotoxin.[7]

Both pharmacological and lesion models do not approximate the complex molecular and cellular pathology of AD. As a consequence, several lines of transgenic models have been created by mutation of the genes coding for the amyloid precursor protein (APP), presenilin-1 (PS1), and presenilin-2 (PS2). Mutations of these genes change the processing of Aβ by APP and result in increased levels of the toxic Aβ42 form.[8] However, different lines of transgenic mouse differ in amyloid pathology, memory deficits, or degeneration of neurons. The most widely used mouse models of AD involve transgenic expression of mutated human APP (hAPP). The crossing of PS1 line (produces a mutated form of PS1) with a hAPP line produced the PSAPP line, which develops Aβ plaques earlier, has increased Aβ42 levels, and exhibits cognitive deficits.[9] However, they all lack the neurofibrillary tau pathology of AD. Lines containing mutant tau protein, a component of neurofibrillary tangles and Aβ, have been crossed with lines containing mutant hAPP and/or PS1 to produce lines that exhibit both the plaque and neurofibrillary tangle pathology, for example,

3xTg line, which develops plaque pathology before tangle pathology, just as observed in humans.[10]

3 ANIMAL MODELS OF PARKINSON'S DISEASE

In PD, progressive degeneration of dopaminergic neurons primarily in the substantia nigra pars compacta causes resting-state tremor, postural instability, slowness of movement, and freezing. 6-Hydroxydopamine (6-OHDA) or 1-methyl-4-phenyl-1,2,3,6-tetrahydropyridine (MPTP) induced killing of neurons is the basis of toxin-based models.[11] Though both toxins kill dopaminergic neurons by producing toxic reactive oxygen species, 6-OHDA does not produce the same cellular pathology as that observed in PD. MPTP is preferentially taken up into dopaminergic neurons, where it blocks the mitochondrial electron transport chain, decreases cellular ATP levels, and forms toxic reactive oxygen species. MPTP-treated monkeys show a remarkable similarity to PD. However, they do not lose monoamine neurons in the locus coeruleus or form Lewy bodies, which are characteristic features of PD.[11]

Several autosomal dominant genes, for example, alpha-synuclein, leucine rich–repeat kinase 2 (LRRK2), and autosomal recessive genes (e.g., parkin, DJ-1, and PINK1) have been linked to PD.[12] Alpha-synuclein is a major structural component of Lewy bodies. Of the several alpha-synuclein transgenic lines, only transgenic mice with the prion promoter show the full range of human's alpha-synuclein pathology. Knockout mice for the recessive genes are applicable only in the early, preneurodegenerative changes that occur in PD.[12]

4 ANIMAL MODELS OF HUNTINGTON'S DISEASE

Motor, cognitive, behavioral, and psychological impairments in HD are caused by mutation within the cytidylic acid, adenylic acid, and guanylic acid (poly-CAG) repeat region of the huntingtin gene. The transgenic mouse models include mice carrying human huntingtin gene with poly-CAG mutations, knockin mice with additional CAG repeats inserted into the existing CAG expansion in the murine gene, and mice that express the full-length human huntingtin gene along with the murine form.[13] In earlier studies of HD, before the discovery of the huntingtin mutation in 1993, toxin-induced models were most often used to study mitochondrial impairment and excitotoxicity-induced cell death.[14]

5 ANIMAL MODELS OF DEPRESSION

Acute or chronic exposure of rat or mouse to a stressor is used to elicit one or more symptoms of depression. Rodent forced swim test or tail suspension test are two popular models where acute stress is applied. In both of the tests, following an initial period of struggling, the animal eventually displays a floating or immobile posture, which is interpreted as an expression of behavioral despair. These models are able to detect a broad range of antidepressants.[15]

Chronic stress models include the chronic mild stress model, psychosocial stress models, or the olfactory bulbectomy model. In the chronic stress model, physical stressors, such as flashes of lights, restraint, or swim stress, are applied over a time period between 1 and 7 weeks. Whereas in the psychosocial model two or more rodents are allowed to interact socially and physically, such that one achieves dominant status and the others remain subordinate. In forced subordination strategy, reliably aggressive rodents (usually larger) are employed to consistently subordinate others. Following multiple defeat encounters, rodents display depression symptoms, many of which are reversed by chronic, but not acute, antidepressant administration.[16] In the olfactory bulbectomy model, animal olfactory bulbs are surgically removed, which results in the development of locomotor hyperactivity that can be reversed by chronic treatment with antidepressants. However, a bulbectomized rat model was reported that required only short-term (days), not chronic, antidepressant treatment.[17]

6 ANIMAL MODELS OF ANXIETY

Animal models of anxiety can be classified into three major categories: (1) ethological test models, (2) conflict test models, and (3) cognitive-based test models.[5,18] Ethological models combine a naturally motivated behavior, for example, exploration of the environment with a naturally aversive stimulus, such as a brightly lit space to create anxiety. Anxiolytics increase the animals' acceptance of aversive stimulus. Examples include the elevated plus maze, elevated zero maze, light–dark box, open field, ultrasonic vocalizations, and defensive burying.[5] Conflict tests involve a motivated behavior, for example, eating or drinking, with an aversive stimulus, such as a mild shock. In the Vogel punished drinking assay, anxiolytics increase water consumption. Other tests in this category include the Geller–Seifter test, four-plate test, novelty-suppressed feeding, and novelty-induced hypophagia.[5] Cognitive-based tests use classical conditioning to train an animal to associate a neutral stimulus, such as a tone, with an unpleasant stimulus, such as a mild shock, to change its natural behavior. An example is the fear conditioning test, where anxiolytics reduce the levels of freezing behavior or enhanced startle response to the conditioned stimulus.[5]

7 ANIMAL MODELS OF SCHIZOPHRENIA

A combination of positive symptoms (e.g., hallucinations and delusions), negative symptoms (e.g., emotional withdrawal and anhedonia), and cognitive dysfunction (e.g., impaired working memory and attention) are seen in schizophrenic patients. Although its specific cause is unknown, schizophrenia has a biologic basis, as evidenced by alterations in brain structure (e.g., enlarged cerebral ventricles, thinning of the cortex, and decreased size of the anterior hippocampus and other brain regions) and by changes in neurotransmitters, especially altered activity of dopamine and glutamate.[19]

Animal models of schizophrenia are classified into four categories: (1) genetic models, (2) pharmacological models, (3) lesion and developmental models, and (4) maternal immune activation models.[3] Several candidate genes have been associated with increased risk of schizophrenia, for example, disrupted in schizophrenia 1 (DISC-1), neuregulin-1 (NRG1), and Reelin. Transgenic mice with constitutive knockout, inducible knockout, and reduced expression or mutations of these genes have been created and each exhibit unique and complex phenotypes that include characteristics associated with schizophrenia.[20] *N*-Methyl-D-aspartate (NMDA) receptor antagonists (e.g., phencyclidine, ketamine, and dizocilpine) lead to schizophrenia symptoms in humans or worsen symptoms of schizophrenic patients.[21] Therefore, pharmacological models of schizophrenia are produced by injecting NMDA receptor antagonists into adult mice or rats. Though these models lack the neurodevelopmental component of schizophrenia, they are widely used because of their simplicity in production and the possibility to quickly test new pharmacological interventions. An excitotoxic (ibotenic acid or tetrodotoxin) lesion in the rat ventral hippocampus on postnatal day 7 shows symptoms observed in schizophrenia, as rats reach late puberty.[22] In maternal immune activation models, lipopolysaccharide or PolyI:C (a double-stranded synthetic RNA recognized by toll-like receptor 3 as a viral component) shows great reliability, as it shows a natural postpubertal outbreak of symptoms as observed in humans, suggestive of a neurodevelopmental course.[23]

8 ANIMAL MODELS OF PAIN

Models that are commonly used to evaluate different sensory modalities of pain include the von Frey and Randal–Sellito tests for mechanical sensation; tail flick, hot plate, and Hargreaves' tests for heat sensation; and the acetone, cold plate, and cold water tests for cold sensation.[5,24] Acute pain models include subcutaneous injection of formalin into the plantar tissue of the paw, intraperitoneal administration of an irritant (such as acetone), and surgical incisions in multiple parts of the animal body. In general, these models are sensitive to analgesics of different classes.[25] There are chronic models of arthritis, cancer, and neuropathy-related pain as well. The two most commonly used models of neuropathic pain are the "chronic constriction injury" and "spinal nerve ligation" models. Cancer pain models involve injection of sarcoma or carcinoma cells into various bones, for example, the femur, calcaneus, and humeral bones. The most common model of diabetic neuropathy–induced pain involves administration of streptozotocin that selectively destroys pancreatic beta cells and produces hyperalgesia within 2–3 weeks.[26–28]

9 ANIMAL MODELS OF STROKE

Like many other diseases, stroke in humans is a heterogeneous disorder with a complex pathophysiology. Therefore, mimicking all aspects of human stroke in one animal model is difficult. The most commonly used models of stroke involve occlusion of the middle cerebral artery (MCA), as the MCA and its branches are the cerebral vessels that are most often affected in human ischemic stroke, accounting for approximately 70% of infarcts.[29] Intraluminal suture model involves occluding the common carotid artery, introducing a suture directly into the internal carotid artery, and advancing the suture until it interrupts the blood supply to the MCA. Laser doppler flowmetry can be a useful tool for ensuring complete MCA occlusion.[30] The craniectomy model includes direct surgical MCA occlusion requiring a craniectomy and sectioning of the dura mater to expose the MCA.

As most large infarcts result from thromboembolism, embolic models have been developed. The thromboembolic clot model is based on the application of spontaneously formed clots or thrombin-induced clots from autologous blood.[31] This model is more convenient for investigating thrombolytic agents and pathophysiological processes after thrombolysis. The microsphere model applies spheres, which have a diameter of 20–50 μm, of different materials, such as dextran, superparamagnetic iron oxide, titanium dioxide, and ceramic. The procedure involves inserting microspheres into the MCA using a microcatheter, which are flushed passively into the cerebral circulation by the blood flow. The microsphere-induced stroke model results in multifocal and heterogeneous infarcts.[32]

10 ANIMAL MODELS OF BRAIN TUMOR

Gliomas are the most common primary brain tumors in adults, whereas in children medulloblastomas are the most common type.[33] Ideally, a good brain tumor model should: (1) display the same genetic lesions, anatomical location, histopathological features, and developmental time frame as the human tumor; (2) recapitulate intertumoral and intratumoral heterogeneity; and (3) be predictive of the patients' response to treatment.[34]

Gliomas can be induced in rats by injecting the alkylating agents *N*-methylnitrosourea or *N*-ethyl-*N*-nitrosourea. Such carcinogen-induced rat tumors have not been characterized at the molecular level and do not become invasive like the human tumors in many rats.[35] Xenograft models are generated by the transplantation of biopsies or cultured cells derived from human brain tumors into immunodeficient mice. During evaluation of antitumor interventions in this model, the following must be taken into consideration: lack of immune system of the host, changed genotype and phenotype from parent human tumors (particularly if cultured in serum), and transformation into noninvasive tumors.[36] In genetically engineered mouse models of brain tumors, gene alterations in human tumors affecting specific oncogenes or tumor suppressor genes are identified and introduced in the germline (knockout, knockin, and transgenic models) to allow for de novo tumor formation.[37–38] Although these models may better mimic tumorigenesis, deletion of a tumor suppressor gene in the entire organism is likely to lead to a wide spectrum of diseases, including cancers, precluding correct analysis of its implication in brain tumorigenesis.[34]

11 CONCLUSIONS

Animal models have significantly advanced the understanding and treatment of CNS disorders. Vertebrates, especially mammals, provide essential models for many specific human disease processes. Rats, mice, guinea pigs, and hamsters have been favored because of their small size, ease of handling, short life span, and high reproductive rate. A range of techniques are used to create animal models including usage of drug or chemical substances, application of physical or psychological stress, occlusion of specific artery, surgical denervation, transplantation of biopsies or cultured cells derived from human tumors, and genetic modification. Animal models do not entirely simulate human diseases with each model having its own advantages and disadvantages. Still, they can be used to address certain aspects of human diseases. Obviously, there is a necessity to fine-tune existing models or develop new ones.

Abbreviations

Aβ	β-Amyloid
AD	Alzheimer's disease
APP	Amyloid precursor protein
CAG	Cytidylic acid, adenylic acid, and guanylic acid
DISC-1	Disrupted in schizophrenia 1
hAPP	Human amyloid precursor protein
HD	Huntington's disease
LRRK2	Leucine rich–repeat kinase 2
MCA	Middle cerebral artery
MPTP	1-Methyl-4-phenyl-1,2,3,6-tetrahydropyridine
NMDA	*N*-Methyl-D-aspartate
NRG1	Neuregulin-1
6-OHDA	6-Hydroxydopamine
PD	Parkinson's disease
PS1	Presenilin-1
PS2	Presenilin-2

References

1. Segev I. Single neuron models: oversimple, complex, and reduced. *Trends Neurosci.* 1992;15:414–421.
2. Willner P. Validation criteria for animal models of human mental disorders: learned helplessness as a paradigm case. *Prog Neuropsychopharmacol Biol Psychiatry.* 1986;10(6):677–690.
3. Mattei D, Schweibold R, Wolf SA. Brain in flames—animal models of psychosis: utility and limitations. *Neuropsychiatr Dis Treat.* 2015;11:1313–1329.
4. Huang J. Alzheimer Disease, Merck Manual Professional Version. Available from: http://www.merckmanuals.com/professional/neurologic-disorders/delirium-and-dementia/alzheimer-disease
5. McGonigle P. Animal models of CNS disorders. *Biochem Pharmacol.* 2014;87:140–149.
6. Ebert U, Kirch W. Scopolamine models of dementia: electroencephalogram findings and cognitive performance. *Eur J Clin Invest.* 1998;28:944–949.
7. Toledana A, Alvarez MI. Lesion-induced vertebrate models of Alzheimer dementia. In: De Deyn PP, Van Dam D, eds. *Animal Models of Dementia*. New York, NY: Springer; 2010:295–345.
8. Hall AM, Roberson ED. Mouse models of Alzheimer's disease. *Brain Res Bull.* 2012;88:3–12.
9. Holcomb L, Gordon MN, McGowan E, Yu X, Benkovic S, Jantzen P, et al. Accelerated Alzheimer-type phenotype in transgenic mice carrying both mutant amyloid precursor protein and presenilin 1 transgenes. *Nat Med.* 1998;4:97–100.
10. Oddo S, Caccamo A, Shepherd JD, Murphy MP, Golde TE, Kayed R, et al. Triple transgenic model of Alzheimer's disease with plaques and tangles: intracellular Abeta and synaptic dysfunction. *Neuron.* 2003;39:409–421.
11. Dauer W, Przedborski S. Parkinson's disease: mechanisms and models. *Neuron.* 2003;39:889–909.
12. Chesselet MF, Fleming S, Mortazavi F, Meurers B. Strengths and limitations of genetic mouse models of Parkinson's disease. *Parkinsonism Relat Disord.* 2008;14:S84–S87.
13. Ferrante RJ. Mouse models of Huntington's disease and methodological considerations for therapeutic trials. *Biochim Biophys Acta.* 2009;1792:506–520.
14. Ramaswamy S, McBride JL, Kordower JH. Animal models of Huntington's disease. *ILAR J.* 2007;48(4):356–373.
15. Krishnan V, Nestler EJ. Animal models of depression: molecular perspectives. *Curr Top Behav Neurosci.* 2011;7:121–147.
16. Becker C, Zeau B, Rivat C, Blugeot A, Hamon M, Benoliel JJ. Repeated social defeat-induced depression-like behavioral and biological alterations in rats: involvement of cholecystokinin. *Mol Psych.* 2008;13(12):1079–1092.
17. Kelly JP, Wrynn AS, Leonard BE. The olfactory bulbectomized rat as a model of depression: an update. *Pharmacol Ther.* 1997;74:299–316.
18. Cryan JF, Sweeney FF. The age of anxiety: role of animal models of anxiolytic action in drug discovery. *Br J Pharmacol.* 2011;164:1129–1161.
19. Schulz SC. Schizophrenia, Merck Manual Professional Edition. Available from: http://www.merckmanuals.com/professional/psychiatric-disorders/schizophrenia-and-related-disorders/schizophrenia
20. Jaaro-Peled H, Ayhan Y, Pletnikov M, Sawa A. Review of pathological hallmarks of schizophrenia: comparison of genetic models with patients and nongenetic models. *Schizophr Bull.* 2010;36:301–313.

21. Anticevic A, Corlett PR, Cole MW, Savic A, Gancsos M, Tang Y, et al. *N*-Methyl-D-aspartate receptor antagonist effects on prefrontal cortical connectivity better model early than chronic schizophrenia. *Biol Psych*. 2015;77(6):569–580.

22. Lipska BK, Halim ND, Segal PN, Weinberger DR. Effects of reversible inactivation of the neonatal ventral hippocampus on behavior in the adult rat. *J Neurosci*. 2002;22(7):2835–2842.

23. Meyer U. Prenatal poly(I:C) exposure and other developmental immune activation models in rodent systems. *Biol Psych*. 2014;75(4):307–315.

24. Sandkuhler J. Models and mechanisms of hyperalgesia and allodynia. *Physiol Rev*. 2009;89: 707–758.

25. Mogil JS. Animal models of pain: progress and challenges. *Nature Rev Neurosci*. 2009;10: 283–294.

26. Berge O-G. Predictive validity of behavioral animal models for chronic pain. *Br J Pharm*. 2011;164:1195–1206.

27. Jaggi AS, Jain V, Singh N. Animal models of neuropathic pain. *Fund Clin Pharm*. 2011;25:1–28.

28. Courteix C, Eschalier A, Lavarenne J. Streptozocin-induced diabetic rats: behavioural evidence for a model of chronic pain. *Pain*. 1993;53:81–88.

29. Felix Fluri, Michael K, Schuhmann, Christoph, Kleinschnitz. Animal models of ischemic stroke and their application in clinical research. *Drug Des Dev Ther*. 2015;9:3445–3454.

30. Schmid-Elsaesser R, Zausinger S, Hungerhuber E, Baethmann A, Reulen HJ. A critical reevaluation of the intraluminal thread model of focal cerebral ischemia: evidence of inadvertent premature reperfusion and subarachnoid hemorrhage in rats by laser-doppler flowmetry. *Stroke*. 1998;29(10):2162–2170.

31. Overgaard K, Sereghy T, Boysen G, Pedersen H, Høyer S, Diemer NH. A rat model of reproducible cerebral infarction using thrombotic blood clot emboli. *J Cereb Blood Flow Metab*. 1992;12(3):484–490.

32. Gerriets T, Li F, Silva MD, Meng X, Brevard M, Sotak CH, Fisher M. The macrosphere model: evaluation of a new stroke model for permanent middle cerebral artery occlusion in rats. *J Neurosci Methods*. 2003;122(2):201–211.

33. Dolecek TA, Propp JM, Stroup NE, Kruchko C. CBTRUS statistical report: primary brain and central nervous system tumors diagnosed in the United States in 2005-2009. *Neuro Oncol*. 2012;14(suppl 5):v1–v49.

34. Simeonova I, Huillard E. In vivo models of brain tumors: roles of genetically engineered mouse models in understanding tumor biology and use in preclinical studies. *Cell Mol Life Sci*. 2014;71(20):4007–4026.

35. Barth RF, Kaur B. Rat brain tumor models in experimental neuro-oncology: the C6, 9L, T9, RG2, F98, BT4C, RT-2 and CNS-1 gliomas. *J Neurooncol*. 2009;94:299–312.

36. Huszthy PC, Daphu I, Niclou SP, Stieber D, Nigro JM, et al. In vivo models of primary brain tumors: pitfalls and perspectives. *Neuro Oncol*. 2012;14:979–993.

37. van Miltenburg MH, Jonkers J. Using genetically engineered mouse models to validate candidate cancer genes and test new therapeutic approaches. *Curr Opin Genet Dev*. 2012;22(1):21–27.

38. Cheon DJ, Orsulic S. Mouse models of cancer. *Annu Rev Pathol*. 2011;6:95–119.

CHAPTER 10

Microdialysis and Brain Perfusion

Margareta Hammarlund-Udenaes, PhD
Uppsala University, Uppsala, Sweden

1 INTRODUCTION

The distribution and elimination processes taking place when a nanocarrier (NC) is administered into the blood stream is described in Fig. 10.1. The NC follows the blood flow and will be distributed throughout the whole body. The release of drug to the free, pharmacologically active moiety can take place in all parts of the body. The aim of using a NC for brain targeting is to increase its specific delivery to the brain. Therefore, a successful NC delivery can be said to be present when the NC increases the drug release and/or uptake to the site of action in the brain in high enough quantities to accomplish a pharmacological action. This is a process where the rate of release from the NC competes with the rate of elimination of the released drug itself.

For the purpose of this chapter, no distinction is made between different types of NCs, but the techniques are discussed in relation to released versus NC-bound drug. Small molecular drugs and peptides are mainly discussed, even if there are examples where larger molecules have been studied with the techniques presented.

The quantitative analysis of the pharmacokinetic processes taking place after NC delivery of a drug is much more complicated than when the drug itself is administered. As illustrated in Fig. 10.1, both the NC and the released drug are present, while the effect will only be exerted by the released, nonprotein-bound (unbound) drug. The task from a methodological perspective to understand all processes is to separate the two entities. As the NC content in blood is generally far higher than the released drug, this is challenging and not possible with traditional sampling methods. The contents of the NC and released drug in the brain capillaries will significantly influence the analysis, although blood volume is only 2%–3% of the brain weight. The drug present in the brain parenchyma is even more difficult to measure if we would like to separate the NC contents from the released drug.

CONTENTS

Nanotechnology Methods for Neurological Diseases and Brain Tumors. http://dx.doi.org/10.1016/B978-0-12-803796-6.00010-1

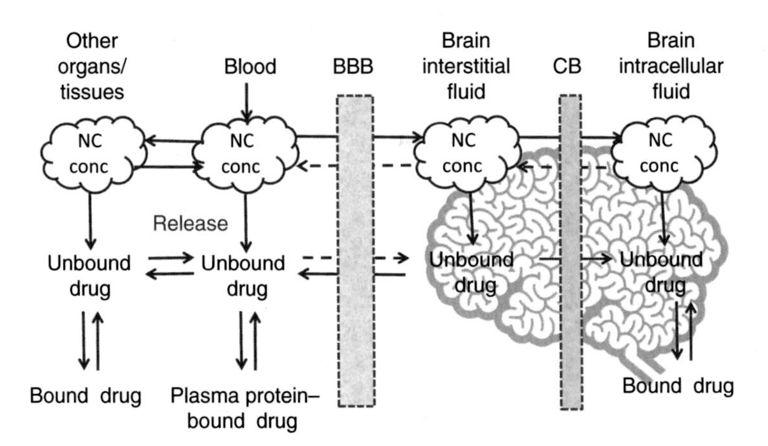

FIGURE 10.1 Partitioning of nanocarrier (NC) and released drug in the body.
BBB, Blood–brain barrier; *CB*, cellular barrier.

The site of NC release of drug in relation to the blood–brain barrier (BBB) may differ depending on the type of NC that is administered. The most common proof of increased delivery of a drug to the brain with a NC has been made by measuring the effect of the drug after nanoparticle administration, for example, the increased life span of mice after administering a cytotoxic drug in a nanoparticle or increased resistance to pain stimuli.[1,2] The transport of the complete nanoparticle across the BBB into the brain parenchyma has been shown with transmission electron microscopy[3] or confocal laser scanning fluorescence microscopy.[4] A combined study with pharmacokinetics and effect measurements of doxorubicin was performed by Gaillard et al.[5] Alternatively, the released drug from liposomal delivery has been measured in the brain with microdialysis[6,7] or open flow microperfusion.[8]

Quantitative microdialysis is able to measure the released drug, but not the drug included in the NC. It specifically measures the unbound moiety in the brain interstitial fluid and in blood, as well as in the interstitial fluid of peripheral tissues, depending on where the probes are inserted. It can be considered an advantage if NC delivery can be studied where only the released drug is measured. Total drug concentrations (NC content and unbound and bound drug) can be measured with plasma samples and whole-tissue samples. There are very few studies utilizing microdialysis measurements that have directly measured the released drug from an NC inside the brain.[6,7] Open flow microperfusion measures all contents in the interstitial fluid, as it does not have a membrane (see further). This will include the complete amount of the drug that is unbound and that is present in NCs, if these NCs are free in the interstitial fluid. It will not measure intracellular drug or drug bound to tissue components. The technique has been used to measure doxorubicin delivery advantage to the brain with liposomes.[8] In situ brain perfusion is a fast technique where delivery is measured within a few minutes. This technique therefore shows more of a direction of increased delivery in relation to a control than a clinically relevant quantitative estimation of the possible improvement.

2 TECHNIQUES IN RELATION TO MEASURED ENTITIES

The two aspects of delivery to the brain that can be measured are the extent of delivery and the rate of delivery. These are two distinct features are complementary and are not exchangeable when trying to understand brain drug delivery.

The extent of delivery is the ratio of drug concentration in the brain versus in the plasma, that is, how much of the drug has reached the brain and is present at steady state. The extent of delivery can be measured with total brain versus total plasma concentrations (K_p) or with unbound brain interstitial fluid (ISF) versus unbound plasma concentrations ($K_{p,uu}$).[9,10] The unbound ratio $K_{p,uu}$ specifically describes BBB transport, while the K_p ratio also includes binding of the drug in the brain and to plasma proteins.[11] If a drug is mainly passively transported, the $K_{p,uu}$ is close to unity. If there are active efflux transporters hindering the drug from entering the brain, the $K_{p,uu}$ ratio is <1, and if there are active uptake transporters dominating the transport, the ratio is >1. The lower the $K_{p,uu}$, the more efficient is the efflux. Loperamide, for example, has a $K_{p,uu} < 0.01$ in rats, meaning that less than 1% is present as an active drug in the brain ISF in relation to what is unbound in blood.[12] For a further reading about the details of these measures, please refer to Ref. [11].

The total brain-to-plasma ratio K_p is also called logBB or B/P ratio. It is easily measured by sampling whole brain (after perfusion of blood) and plasma. As mentioned earlier, it includes both the binding in the brain and in the plasma, apart from the extent of BBB transport, and is therefore a less "clean" measure of delivery, where different drugs cannot be compared directly regarding their BBB transport due to the presence of differences in binding in brain and/or plasma.

The rate of delivery can be measured by several techniques, of which the in situ brain perfusion is presented in this chapter. The permeability of a drug or a NC across the BBB can be related to the half-life in the plasma of the NC/drug, as enough plasma circulation time is needed if the BBB permeability is low.

3 MICRODIALYSIS

Microdialysis measurements can be made either at steady state during an intravenous infusion of a drug or by the administration of a drug orally or intravenously as a short infusion, and subsequent measurement of the area under the curve (AUC) by repeated sampling for at least three half-lives. The method mainly measures the extent of drug delivery. The rate of delivery can be calculated if modeling is performed.[13,14] The unbound ratio $K_{p,uu}$ can be directly measured by implanting a probe in the brain and another one in a vein, sampling the dialysate at regular intervals for several hours. Alternatively regular plasma samples can be taken and values can be recalculated based on the plasma protein binding to obtain the unbound ratio. The venous microdialysis probe is necessary to separate the NC from the free drug in plasma, which is one of the important aspects involved in understanding the quantitative influence of NC delivery across the BBB.

Due to the problem of lipophilic compounds sticking to tubings and membranes, other techniques have been developed, combining K_p estimations with in vitro brain slices or brain homogenates and plasma protein–binding measurements. This technique is called the combinatory mapping approach (CMA) and can also be further used to understand intracellular drug distribution.[15] However, it is not possible to use to separate out encapsulated and released drugs with this technique.

3.1 Aspects of Microdialysis in Relation to NC Studies

Microdialysis can be performed over several hours and even days. It can thereby give detailed information of drug transport across the BBB for slow processes over time. The cutoff of the membranes is either 20 or 100 kDa, limiting the possibility of sampling NCs. The advantage, as mentioned earlier, is that microdialysis measures the released and unbound drug at the site of action, thereby making this technique unique for NC studies. It can thereby be very valuable when evaluating the in vivo quantitative success of an NC system, especially where there is no clear pharmacodynamic readout.

3.2 Methodological Aspects

The crucial element in microdialysis is to obtain quantitative measurements of tissue concentrations. Two aspects are important in this regard: (1) checking that the drug of interest does not stick to the tubings or membranes used and (2) measuring in vivo recovery to calibrate the measured concentration to the factual unbound extracellular concentrations present in the tissue. General flow rate through the microdialysis probe is in the range of 0.3–2 μL/min. There is always a trade-off between the flow rate and the collection intervals of the dialysate in relation to the detection limit of the analytical method.

3.2.1 *In Vitro Preparations*

Many drugs show significant sticking to tubings and probes, making quantitative measurements with microdialysis impossible. An in vitro check of tubing and probe adsorption is therefore essential before starting in vivo studies, saving time and resources. The test is first performed with the tubing itself, and thereafter with the probe present. A possible design is to have a 1-m long tubing, infusing a drug solution through the tubing, and sampling at the end of the tubing every 10 min for a total of 1 h with a flow rate of 1 μL/min. If there is a drop in concentration of the first samples compared with the inlet concentration, adsorption to the tubing is occuring. Adding 0.25 or 0.5% albumin to the perfusate or changing to another tubing material may improve the situation. If the drug of interest has a high plasma protein binding, the addition of albumin to the perfusate may yield erroneous results (unpublished observations). A study of loss and gain across the probe membrane should thereafter be performed in vitro.[16,17] Here, the drug is added either to the perfusate or to the buffer in the Eppendorf tube where the probe is placed. The outgoing dialysate is analyzed, for example every 10 min for 1 h, after which there is a washout with no drug present in any of the two buffers for 1 h, and thereafter the drug-containing buffer is added to the other buffer for 1 h. In vivo studies can be initiated if there is minimal or no sticking of components to the tubing and if loss and gain are equal.

3.2.2 In Vivo Preparations

The surgery and implantation of microdialysis probes should be made at least a day before the study. Even if a guide cannula is implanted the day before and the probe is placed on the same day as the experiment, there is too much leakage for the measurements to be considered reliable. This significantly influences the results after NC administration and would give false positive information.

In vivo calibration is essential for the estimation of the in vivo relevant recovery across the microdialysis probe. In vitro recovery cannot be used to calibrate the concentrations, as the surrounding buffer fluid is different from the tissue being studied in vivo, giving rise to different recovery values. In vivo recovery can be estimated when the drug of interest is given as a retrodialysis infusion (in the perfusate) before the drug is administered, and calculating the loss across the membrane.[18] An even better approach is when the drug of interest is used in a deuterated form with a difference in molecular weight of 3 or more, as a retrodialysis calibrator throughout the study, and is analyzed together with the drug by LC–MS/MS.[19] The advantage with having the calibrator present during the whole study is that possible deviations in recovery across the membrane, although very uncommon, can be observed and accounted for. The advantage with using a deuterated form of the drug of interest is that it is as close as possible to the drug being studied. Other compounds used as calibrators, although rather similar in structure, may behave quite differently at the BBB, thereby resulting in erroneous recoveries. The reason for differences between similar compounds is that they can be prone to be transported in different ways, resulting in different movements of the compounds in the tissue and thereby different recoveries.

3.3 Microdialysis Study Design of Nanodelivery to the Brain

Several aspects need to be taken into consideration when microdialysis is used to study the influence of a NC on the delivery of its cargo to the brain. Either the drug can be administered to a separate group than the NC group or a crossover study design can be used. A crossover design requires that the drug should be eliminated from the body before the NC is administered (or the opposite). As the NC can be expected to have a much longer half-life than the drug itself, it is more probable that the drug will be administered before the NC. Together with the microdialysis technique this may pose difficulties, as it is advisable to perform the two experiments on sequential days. The crossover experimental design is exemplified in two publications, where the opioid peptide (D-Ala2, *N*-MePhe4, gly-ol)-enkephalin (DAMGO) was studied.[6,7] Due to the very short, 9-min half-life of DAMGO, it was possible to administer the two formulations on the same day (Fig. 10.2).[6] However, DAMGO had to be administered before the liposomal formulation due to the long half-life of the latter (6.9 h). In Fig. 10.2 it can be seen that the difference between total liposomal and free drug was very large in the plasma (130-fold). It can also

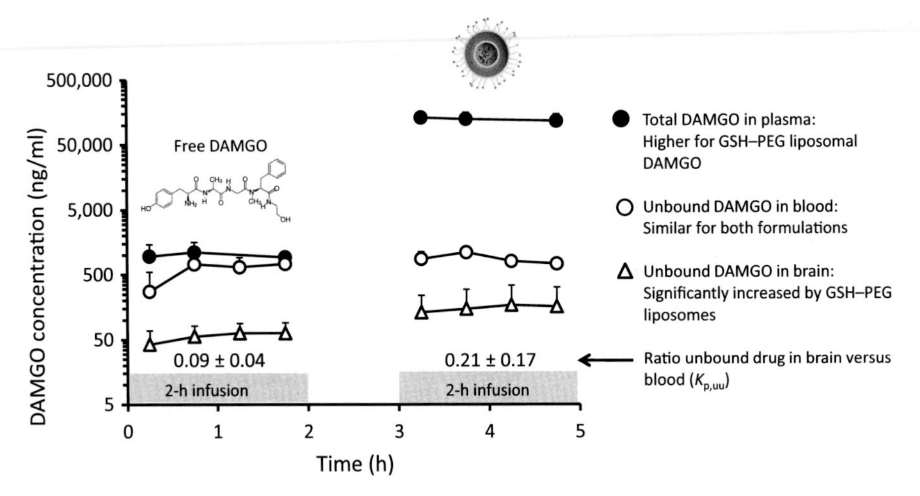

FIGURE 10.2 Brain and blood microdialysis results showing the change in brain penetration of the opioid peptide (D-Ala2, *N*-MePhe4, gly-ol)-enkephalin (DAMGO) administered alone (left) and encapsulated in a gluthatione-tagged PEGylated (GSH–PEG) liposome (right).

$K_{p,uu}$ is the ratio of unbound DAMGO in the brain to that in the blood. The graph represents individual data in rats. *Data reproduced from Lindqvist A, Rip J, Gaillard PJ, Bjorkman S, Hammarlund-Udenaes M. Enhanced brain delivery of the opioid peptide DAMGO in glutathione pegylated liposomes: a microdialysis study.* Mol Pharm. *2012;10(5):1533–1541.*[6]

be seen that the unbound drug concentration was similar, independent of the entity administered, which was an aim with the study design. A constant infusion of both the drug itself and the NC can be used to clearly observe possible differences in BBB transport and brain delivery. The $K_{p,uu}$ values indicate that DAMGO is kept out of the brain quite efficiently.

4 OPEN FLOW MICROPERFUSION

The open flow microperfusion technique has been established as a development from the push–pull technique for CNS applications.[8,20,21] It is still a new technique, which is of reduced general use in the scientific community, but with promising results regarding the measurement of NC delivery.[8] It also measures the extent of transport of a drug or NC to the brain. The method is similar to microdialysis, as a tubing is implanted in a specific location in the brain, however, without the use of a membrane, but with an inlet and outlet tubing with a flow of CNS buffer. The interstitial fluid is therefore sampled directly and no recovery estimations are needed. The animals should recover for 15 days between the surgery and experimental procedure, with a dummy probe inserted. It has been shown that the BBB needs that much time to recover.[20] During the procedure, there is a constant flow of buffer into the parenchymal tissue and a constant withdrawal of fluid from another tube tip in the close vicinity of the incoming tube tip. It is still uncertain what is measured, as there is no membrane present. Everything that is dissolved in the interstitial fluid is sampled, whereas what is intracellularly located or bound to tissue components is not. NCs would be sampled given that they are present in the interstitial fluid.

It is argued that an advantage of this method is that components do not stick to the membrane (as there are none).[20–22] However, there are not any scientific publications yet showing if any component sticks to the tubing on the way out from the device, which is still a problem and could be the case. It is therefore advised that the sticking of components to the used tubings is tested in vitro prior to in vivo studies, also when using the open flow microperfusion technique.

Open flow microperfusion may be a better in vivo technique than microdialysis for quantifying the presence of NC in the brain tissue, given that the tissue concentration of the NC is high enough for detection.[21] It is also a technique that can sample larger molecules in general, as there is no membrane hindering transport.

In a comparative study of doxorubicin, Birngruber et al. showed a fivefold higher uptake of gluthatione-tagged PEGylated liposomes than the commercially available product Caelyx across the BBB by the used of the open flow microperfusion technique (Fig. 10.3).

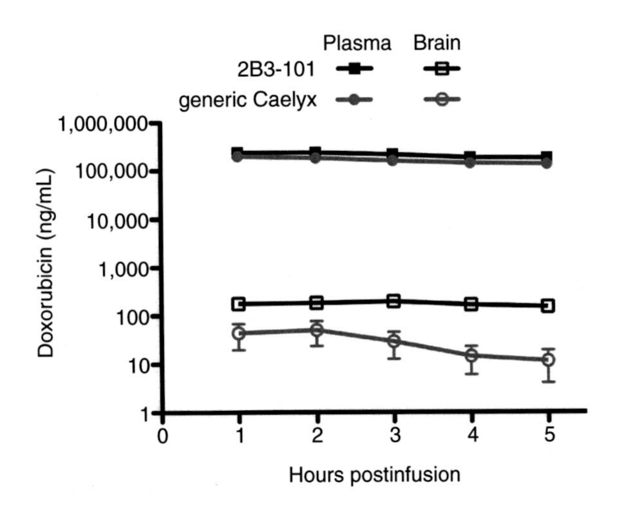

FIGURE 10.3 Open flow microperfusion measurements of concentrations of doxorubicin in plasma (total concentrations, including drug encapsulated in liposomes) and in brain after intravenous administration of doxorubicin in glutathione-tagged PEGylated liposomes (2B3-101, $n = 4$) or commercially available PEGylated liposomes (Caelyx/Doxil, $n = 8$), showing 4.8-fold higher brain concentrations with the use of glutathione-tagged liposomes in rats.

Values are represented as mean ± SEM. *From Birngruber T, Raml R, Gladdines W, Gatschelhofer C, Gander E, Ghosh A, Kroath T, Gaillard PJ, Pieber TR, Sinner F. Enhanced doxorubicin delivery to the brain administered through glutathione PEGylated liposomal doxorubicin (2B3-101) as Compared with generic Caelyx/Doxil—a cerebral open flow microperfusion pilot study.* J Pharm Sci. *2014;103(7):1945–1948.*[8]

5 IN SITU BRAIN PERFUSION

Several studies have used in situ brain perfusion to study NCs, both nanoparticles and liposomes.[23–30] In situ brain perfusion measures the rate of delivery to the brain by sampling whole-brain tissue at different time points, shortly after the start of a carotid infusion with a buffer solution containing the drug (and NC) and a radioactive non-BB–permeable marker, such as sucrose or inulin.[31,32] The nonpermeable marker is used to compensate for vascular volume in the sample analysis. Koziara et al. compared paclitaxel uptake with nanoparticles both with and without postexperimental washout for 30 s and after a 45-s perfusion.[30] Due to the rapid procedure, preferably less than 2 min, in situ brain perfusion does not quantitatively measure the transport of NCs to the brain parenchyma, but can indicate differences in the rate of uptake during the time of the study, which also indicates that there will be differences in the uptake during steady state.

6 DISCUSSION

There is still very sparse in vivo information on the possible quantitative improvement in delivery of drugs to the brain using NCs. When studying brain drug delivery with means other than those discussed in this chapter, the most common way of presenting the data is as a percentage of administered dose. The use of this measure makes it difficult to understand the pharmacokinetic processes of the NC distribution and release, as it is a static measure at one (or a few) time point(s).

Microdialysis and open flow microperfusion have shown to be good instruments for the study of quantitative aspects of NC delivery[6–8]; although, more studies have measured only pharmacodynamic consequences of the delivery, showing improved effects.[1,2] It is likely that liposomes and nanoparticles will behave differently at the BBB, having different stabilities for passing through membranes. In any instance the drug has to be released from the NC to exert its action within the brain. The success of a NC formulation will depend on the release rate of the drug from the NC, on competing transport rates of the drug to and from the brain, and on its elimination rate from the body. The NC formulation, in general, prolongs the half-life of the drug in plasma, thereby also prolonging the exposure at the BBB and improving the chance for BBB penetration for low-permeability compounds.

From a methodological perspective it is important to have a pharmacokinetic view of the processes taking place, by also measuring the free drug present in plasma, as is done in the microdialysis examples.[6,7] It is less clear whether the open flow microperfusion study can separate the two entities, as it samples interstitial fluid directly. Based on the knowledge of doxorubicin BBB transport using a comparative study design, it is likely that it is possible to measure a true difference between the formulations.[8]

In conclusion, it will be important to collect more quantitative in vivo information regarding the degree of improvement in delivery of a drug to the brain by the use of an NC, something that can be done with the methods discussed in this chapter.

Abbreviations

AUC Area under the plasma concentration time curve
BBB Blood–brain barrier
B/P ratio Ratio of total brain to total plasma concentrations (similar to K_p)
CMA Combinatory mapping approach (to study BBB transport and intracellular distribution of drugs)
DAMGO (D-Ala2, *N*-MePhe4, gly-ol)-enkephalin, a synthetic opioid peptide
ISF Interstitial fluid (in the brain)
K_p Ratio of total brain to total plasma concentrations
$K_{p,uu}$ Ratio of unbound brain interstitial fluid concentrations to unbound plasma concentrations
LC–MS/MS Liquid chromatography mass spectrometry
logBB Ratio of total brain to total plasma concentrations (similar to K_p)
NC Nanocarrier
PEG Polyethylene glycol

References

1. Ulbrich K, Knobloch T, Kreuter J. Targeting the insulin receptor: nanoparticles for drug delivery across the blood-brain barrier (BBB). *J Drug Target*. 2011;19(2):125–132.
2. Petri B, Bootz A, Khalansky A, Hekmatara T, Muller R, Uhl R, et al. Chemotherapy of brain tumour using doxorubicin bound to surfactant-coated poly(butyl cyanoacrylate) nanoparticles: revisiting the role of surfactants. *J Control Release*. 2007;117(1):51–58.
3. Zensi A, Begley D, Pontikis C, Legros C, Mihoreanu L, Wagner S, et al. Albumin nanoparticles targeted with Apo E enter the CNS by transcytosis and are delivered to neurons. *J Control Release*. 2009;137(1):78–86.
4. Helm F, Fricker G. Liposomal conjugates for drug delivery to the central nervous system. *Pharmaceutics*. 2015;7(2):27–42.
5. Gaillard PJ, Appeldoorn CC, Dorland R, van Kregten J, Manca F, Vugts DJ, et al. Pharmacokinetics, brain delivery, and efficacy in brain tumor-bearing mice of glutathione PEGylated liposomal doxorubicin (2B3-101). *PLoS One*. 2014;9(1):e82331.
6. Lindqvist A, Rip J, Gaillard PJ, Bjorkman S, Hammarlund-Udenaes M. Enhanced brain delivery of the opioid peptide DAMGO in glutathione pegylated liposomes: a microdialysis study. *Mol Pharm*. 2013;10(5):1533–1541.
7. Lindqvist A, Rip J, van Kregten J, Gaillard PJ, Hammarlund-Udenaes M. In vivo functional evaluation of increased brain delivery of the opioid peptide DAMGO by glutathione-PEGylated liposomes. *Pharm Res*. 2016;33(1):177–185.
8. Birngruber T, Raml R, Gladdines W, Gatschelhofer C, Gander E, Ghosh A, et al. Enhanced doxorubicin delivery to the brain administered through glutathione PEGylated liposomal doxorubicin (2B3-101) as compared with generic Caelyx/Doxil—a cerebral open flow microperfusion pilot study. *J Pharm Sci*. 2014;103(7):1945–1948.
9. Gupta A, Chatelain P, Massingham R, Jonsson EN, Hammarlund-Udenaes M. Brain distribution of cetirizine enantiomers: comparison of three different tissue-to-plasma partition coefficients: K(p), K(p,u), and K(p,uu). *Drug Metab Dispos*. 2006;34(2):318–323.
10. Hammarlund-Udenaes M, Friden M, Syvanen S, Gupta A. On the rate and extent of drug delivery to the brain. *Pharm Res*. 2008;25(8):1737–1750.
11. Hammarlund-Udenaes M. Pharmacokinetic concepts in brain drug delivery. In: Hammarlund-Udenaes M, de Lange ECM, Thorne RG, eds. *Drug Delivery to the Brain Physiological Concepts, Methodologies and Approaches. AAPS Advances in the Pharmaceutical Sciences Series*. New York, Heidelberg, Dordrecht, London: Springer; 2014:127–161.
12. Friden M, Ducrozet F, Middleton B, Antonsson M, Bredberg U, Hammarlund-Udenaes M. Development of a high-throughput brain slice method for studying drug distribution in the central nervous system. *Drug Metab Dispos*. 2009;37(6):1226–1233.
13. Tunblad K, Hammarlund-Udenaes M, Jonsson EN. An integrated model for the analysis of pharmacokinetic data from microdialysis experiments. *Pharm Res*. 2004;21(9):1698–1707.
14. Bostrom E, Simonsson US, Hammarlund-Udenaes M. In vivo blood-brain barrier transport of oxycodone in the rat: indications for active influx and implications for pharmacokinetics/pharmacodynamics. *Drug Metab Dispos*. 2006;34(9):1624–1631.
15. Loryan I, Sinha V, Mackie C, Van Peer A, Drinkenburg W, Vermeulen A, et al. Mechanistic understanding of brain drug disposition to optimize the selection of potential neurotherapeutics in drug discovery. *Pharm Res*. 2014;31(8):2203–2219.
16. Chaurasia CS, Muller M, Bashaw ED, Benfeldt E, Bolinder J, Bullock R, et al. AAPS-FDA workshop white paper: microdialysis principles, application and regulatory perspectives. *Pharm Res*. 2007;24(5):1014–1025.

17. Hammarlund-Udenaes M. Microdialysis in CNS PKPD research: unraveling unbound concentrations. In: Müller M, ed. *Microdialysis in Drug Development. AAPS Advances in the Pharmaceutical Sciences Series 4*. New York, NY: Springer; 2013:83–102.
18. Bouw MR, Hammarlund-Udenaes M. Methodological aspects of the use of a calibrator in in vivo microdialysis-further development of the retrodialysis method. *Pharm Res*. 1998;15(11):1673–1679.
19. Bengtsson J, Bostrom E, Hammarlund-Udenaes M. The use of a deuterated calibrator for in vivo recovery estimations in microdialysis studies. *J Pharm Sci*. 2008;97(8):3433–3441.
20. Birngruber T, Ghosh A, Hochmeister S, Asslaber M, Kroath T, Pieber TR, et al. Long-term implanted cOFM probe causes minimal tissue reaction in the brain. *PLoS One*. 2014;9(3):e90221.
21. Birngruber T, Ghosh A, Perez-Yarza V, Kroath T, Ratzer M, Pieber TR, et al. Cerebral open flow microperfusion: a new in vivo technique for continuous measurement of substance transport across the intact blood-brain barrier. *Clin Exp Pharmacol Physiol*. 2013;40(12):864–871.
22. Ghosh A, Birngruber T, Sattler W, Kroath T, Ratzer M, Sinner F, et al. Assessment of blood-brain barrier function and the neuroinflammatory response in the rat brain by using cerebral open flow microperfusion (cOFM). *PLoS One*. 2014;9(5):e98143.
23. Alyautdin RN, Tezikov EB, Ramge P, Kharkevich DA, Begley DJ, Kreuter J. Significant entry of tubocurarine into the brain of rats by adsorption to polysorbate 80-coated polybutylcyanoacrylate nanoparticles: an in situ brain perfusion study. *J Microencapsul*. 1998;15(1):67–74.
24. Lockman PR, Koziara JM, Mumper RJ, Allen DD. Nanoparticle surface charges alter blood-brain barrier integrity and permeability. *J Drug Target*. 2004;12(9–10):635–641.
25. Lockman PR, Mumper RJ, Khan MA, Allen DD. Nanoparticle technology for drug delivery across the blood-brain barrier. *Drug Dev Ind Pharm*. 2002;28(1):1–13.
26. Gosk S, Vermehren C, Storm G, Moos T. Targeting anti-transferrin receptor antibody (OX26) and OX26-conjugated liposomes to brain capillary endothelial cells using in situ perfusion. *J Cereb Blood Flow Metab*. 2004;24(11):1193–1204.
27. van Rooy I, Hennink WE, Storm G, Schiffelers RM, Mastrobattista E. Attaching the phage display-selected GLA peptide to liposomes: factors influencing target binding. *Eur J Pharm Sci*. 2012;45(3):330–335.
28. van Rooy I, Mastrobattista E, Storm G, Hennink WE, Schiffelers RM. Comparison of five different targeting ligands to enhance accumulation of liposomes into the brain. *J Control Release*. 2011;150(1):30–36.
29. Ko YT. Nanoparticle-mediated delivery of oligonucleotides to the blood-brain barrier: in vitro and in situ brain perfusion studies on the uptake mechanisms. *J Drug Target*. 2013;21(9):866–873.
30. Koziara JM, Lockman PR, Allen DD, Mumper RJ. Paclitaxel nanoparticles for the potential treatment of brain tumors. *J Control Release*. 2004;99(2):259–269.
31. Smith QR, Allen DD. In situ brain perfusion technique. *Methods Mol Med*. 2003;89:209–218.
32. Takasato Y, Rapoport SI, Smith QR. An in situ brain perfusion technique to study cerebrovascular transport in the rat. *Am J Physiol*. 1984;247(3 pt 2):H484–H493.

CHAPTER 11

Neuroimaging: Techniques and General Applications

Elif Bulut, MD, Ayca Akgoz, MD, Kader Karlı Oguz, MD
Hacettepe University, Ankara, Turkey

CONTENTS

1 INTRODUCTION

Diverse techniques with improved performance exist in neuroimaging and enable a very detailed morphological and functional evaluation of the brain. These techniques vary from noninvasive, such as ultrasound, to high-resolution imaging of very close histopathologic specimens, such as magnetic resonance imaging (MRI), especially in high field systems.

2 ULTRASONOGRAPHY

Ultrasonography (US) is a radiation-free, easy accessible, and relatively inexpensive imaging modality with high temporal resolution. In neuroradiology, US has a primary role in fetal imaging; most of the brain and spinal cord malformations can be detected with prenatal US.[1] Although MRI remains a mainstay of neuroimaging, cranial US is usually the first-line imaging technique during the neonatal period.[2–4] Neonatal neurosonography, advanced with modern technology, has proven to be diagnostically accurate when compared to MRI. Furthermore, it does not require sedation and can be easily performed at the bedside. These are the main advantages of cranial US over MRI in newborns in whom many of the neurologic conditions need urgent imaging.

Non- or poorly ossified parts, such as the fontanels, the poorly ossified parts of the temporal bone, or the cartilaginous posterior arches of the vertebral spine, are the acoustic windows used for US imaging.[2,4] Microcurved or phased array/sector probes (frequency ranging from 5 to 10 MHz) and linear transducers (frequency ranging from 5/8 to 18 MHz) for a detailed assessment of the brain surface are available.[3,4] Standard pediatric cranial US starts with basic grayscale imaging in the coronal and sagittal planes via anterior fontanel. The ventricular system, choroid plexus, and particularly midline anatomic structures

Nanotechnology Methods for Neurological Diseases and Brain Tumors. http://dx.doi.org/10.1016/B978-0-12-803796-6.00011-3

(corpus callosum, basal ganglia, and thalamus) can be visualized by this approach. Additional approaches, such as transtemporal, transoccipital, and via mastoid fontanel, are essential for a detailed evaluation of the posterior fossa.[2] Doppler sonography (DS) can be applied for screening vascular structures. Patency of the vascular tree is assessed using color DS and flow characteristics, such as peak systolic velocity, end diastolic velocity, and resistive index, are measured by focused spectral analysis in relevant vessels.[2,3] Power DS can be added to the procedure to search for perfusion impairment.[2,4] Modern methods, such as harmonic imaging and three-dimensional (3D) US, further expand the imaging potential.[4]

One of the major clinical applications of cranial US in the neonatal period is brain hemorrhage. This condition is commonly seen in premature babies and starts in the subependymal germinal matrix (GM), which surrounds the lateral ventricle.[3–5] It is seen as a hyperechoic focus usually bulging upward into the floor of the lateral ventricle from the caudothalamic groove. Detection of hemorrhage and its extension enables staging GM hemorrhage and prognosis in the patients. Other indications of cranial US in neonates and infants include evaluation for possible or suspected hypoxic ischemic brain injury, vascular abnormalities, congenital malformations, signs and/or symptoms of CNS disorders (such as seizures), follow-up or surveillance of previously documented abnormalities (including prenatal abnormalities), and screening before surgical procedures.[3–5]

US can also be used to assess intracerebral vessels and brain parenchyma in adults.[6–8] However, imaging in adults is limited due to dependency of US on adequate acoustic bone window. Transcranial doppler (TCD) involves the use of low-frequency US waves (≤2 MHz) to measure cerebral blood velocity in the major intracranial arteries through transtemporal, transorbital, or suboccipital windows.[7] The current applications of TCD include vasospasm in sickle cell disease, subarachnoid hemorrhage, and intracranial arterial stenosis or occlusion. TCD is also used in brain stem death, increased intracranial pressure, intraoperative monitoring, cerebral microembolism, and autoregulatory testing.[7] Since the early 1990s, improvements in transducer technologies and signal processing have enabled better sonographic visualization of brain parenchyma. Nowadays, US is increasingly used to assess brainstem and subcortical brain structures for the diagnosis of movement disorders. The most widely recognized sonographic finding of movement disorders is increased echogenicity of the substantia nigra in the midbrain.[8] This finding has a highly predictive value for the diagnosis of Parkinson's disease (PD). Other sonographic findings, such as hypoechogenicity of the brainstem raphe and hyperechogenicity of the lentiform nucleus, might help the differential diagnosis of PD and other movement disorders.

3 COMPUTED TOMOGRAPHY

Computed tomography (CT), which uses X-rays to produce cross-sectional images, is extensively utilized in neuroimaging. Indications for its use should be selected carefully to minimize radiation exposure, particularly in children. CT in current pediatric neuroimaging, is therefore limited to the assessment of acute trauma, potential cerebral calcifications, or when MRI is not available. CT is fast and easily performed in patients dependent on monitoring. For evaluation of osseous structures, acute intracranial hemorrhage, and calcifications, CT is superior to MRI. Cranial CT may be performed with a traditional sequential technique or helical (spiral) technique. Standard sequential scan is performed with thick collimation and section thickness of 5 mm. Overall image quality of thinly collimated spiral CT with image combining is better compared to thickly collimated sequential CT.[9] In addition, better visualization of brain tissue near the skull and reduction of image artifacts in the skull base can be achieved.

Primary indications for cranial CT include acute head trauma, acute intracranial bleeding, acute stroke, detection or evaluation of calcifications, immediate post-operative evaluation, and certain congenital or acquired skull lesions (Figs. 11.1 and 11.2). Furthermore, cranial CT is a useful screening tool for acute metal

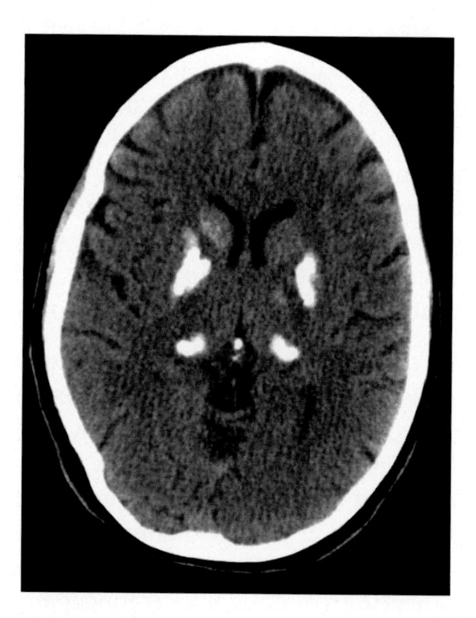

FIGURE 11.1 Axial noncontrast cranial computed tomography (CT) image shows symmetric hyperdense calcifications involving bilateral basal ganglia and thalami in a patient with the diagnosis of Fahr disease.

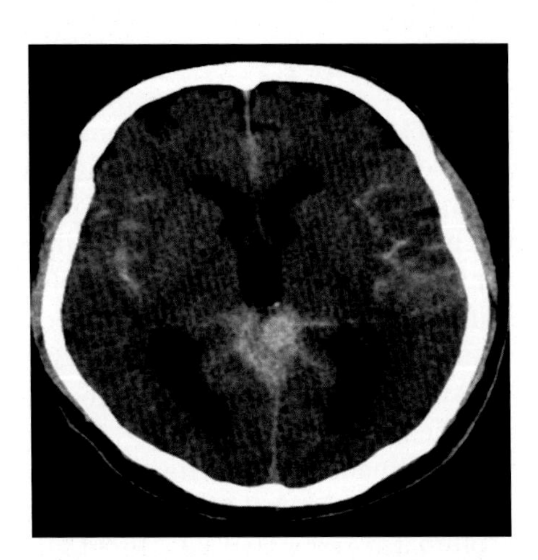

FIGURE 11.2 Axial noncontrast cranial CT image shows hyperdense subarachnoid hemorrhage in quadrigeminal cistern, bilateral Sylvian fissures, and interhemispheric fissure in a patient with acute hemorrhagic stroke.

status change, seizure, acute neurologic deficit, acute headache, and nonacute headache with neurologic findings. In addition, cranial CT with intravenous contrast can be performed in suspected neoplastic, infectious, or inflammatory conditions, but it is not sufficient for the evaluation of subtle changes.

Early diagnosis and imaging evaluation of acute stroke is crucial to direct patients to intravenous or intraarterial thrombolysis. Nonenhanced CT must be performed as soon as possible to rule out hemorrhage (a contraindication for thrombolytic therapy) and other mimics of stroke, such as infection or neoplasm. In addition, CT can identify early signs of acute ischemia, such as subtle hypoattenuation, obscuration loss of gray matter–white matter differentiation, cortical sulcal effacement, loss of the insular ribbon, and hyperdense vessel caused by hyperattenuation of acute intravascular thrombus.[10,11]

CT angiography and CT perfusion imaging are easy to perform with most spiral CT scanners, and increasingly included in many stroke protocols (Fig. 11.3). CT angiography can direct interventional therapy by demonstrating intravascular thrombi. In addition, CT angiography helps to: reveal arterial dissection, aneurisms, and vascular malformations; grade collateral blood flow; and evaluate atherosclerotic disease.[10,11] Image quality of CT angiography has been improved significantly with the introduction of multisection spiral CT. Multisection CT with multidetector row designs offer longer anatomic coverage due to simultaneous registration of multiple sections (data acquisition channels) during each gantry rotation.[12] With additional increase in gantry speed, multisection CT enables much faster scanning than conventional single-section spiral CT. Faster scanning results in improved temporal and contrast resolution

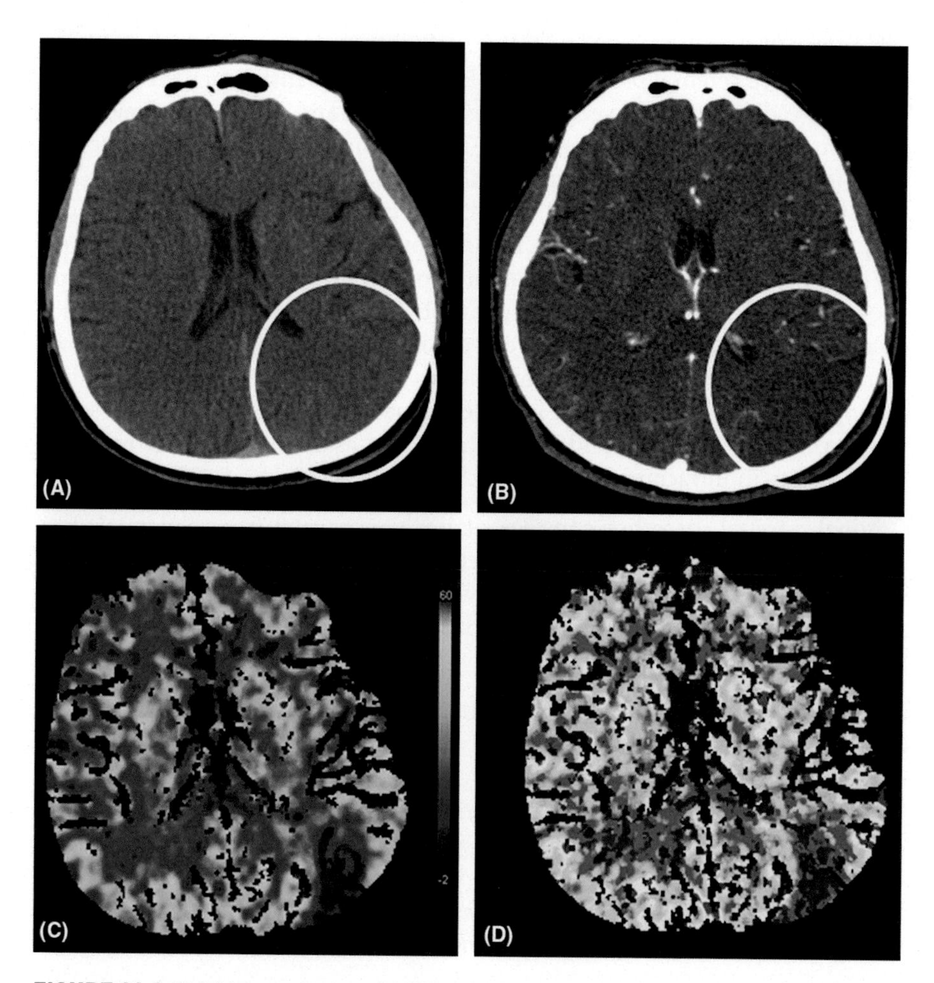

FIGURE 11.3 (A) Axial noncontrast cranial CT image of a patient with acute ischemic stroke reveals loss of gray–white matter differentiation in left parietooccipital region *(circle)*. (B) Axial CT angiography image shows parietooccipital hypoattenuation and focal loss of vascularity. (C) Perfusion CT images reveal (C) low cerebral blood volume (CBV) and (D) cerebral blood flow (CBF) demonstrated with *purple color code* in infarcted parietooccipital parenchyma.

with reduced motion artifacts.[12,13] Multisection CT also allows the selection of very thin section thickness, resulting in isotropic data acquisition and high-quality postprocessing.[12–14] CT angiography is preferably performed with injection of highly concentrated contrast medium (iodine, 350–370 mmol/mL) with rates of 4–5 mL/s. The most widely used postprocessing techniques in CT angiography are multiplanar reformation (MPR), maximum intensity projection (MIP), and 3D reconstruction with volume rendering.[13] MPR images perpendicular to the vessel lumen should be created for precise measurements of stenosis. MIP and/or volume rendering images help display vessel course.

Perfusion CT allows qualitative and quantitative evaluation of cerebral blood volume (CBV), cerebral blood flow (CBF), and mean transit time (MTT), which is the time difference between the arterial inflow and venous outflow.[10,11,15] Dynamic contrast enhanced perfusion imaging involves continuous cine imaging for 45 s over the same segment of tissue (1–32 sections) during the administration of a small (50 mL) high-flow contrast material bolus (injection rate, 4–5 mL/s).[10] Color-coded CT perfusion maps showing CBV, CBF, and MTT are obtained in a short time at an appropriate workstation.[10,11] In acute stroke, ischemic-salvageable tissue (penumbra) manifests with increased MTT, decreased CBF, and normal or increased CBV, whereas infracted tissue shows markedly decreased CBF (<30%) and decreased CBV (<40%).[10,11] Other clinical applications of perfusion CT include evaluation of cerebrovascular reserve, vasospasm after subarachnoid hemorrhage, and assessment of microvascular permeability in intracranial neoplasms.[15]

Coronal, sagittal, or other complex planes constructed from the axial data set are also major diagnostic tools used for paranasal, temporal, and spinal CT examinations. 3D reconstruction with shaded surface display or volume rendering can be performed to visualize cranium, spine, and postoperative instruments.

Mini-, micro-, and nano-CTs are CT-imaging modalities used for in vivo imaging of small laboratory animals or tissues. Information regarding anatomy, disease status, and disease progression can be achieved with high spatial resolution in animal models. Mini-CT describes CT systems with voxel resolutions of 50–200 μm, micro-CTs and nano-CTs are scanners with voxel resolutions of 1–50 μm and 0.1–1 μm, respectively. High-resolution imaging demands for high X-ray doses and long scan times, which make the use of extravascular contrast agents difficult. Flat-panel detector-based mini-CT systems (resolution~200 μm) offer a decrease in scan time (0.5 s) and applied X-ray dose.[16,17] This allows for angiographic scans with iodinated contrast agents, as well as dynamic examinations, such as perfusion studies and retrospective motion gating.

4 MAGNETIC RESONANCE IMAGING

MRI uses radiofrequency (RF) electromagnetic fields to produce high-quality multiplanar images. It provides a high degree of soft tissue contrast due to variations in the magnetic relaxation and susceptibility properties (T1, T2, and T2*) and proton density of different tissues.[18] Repetition time (TR) and echo time (TE) are two key parameters for creating image contrast. T1, T2, and proton density–weighted (W) images can be created with the use of different TR and TE values. T1W images are superior in revealing the anatomy. T2W images provide the best depiction of disease processes, most of which have a higher water content than normal and appear bright on T2W. Proton density–weighted MR images usually reveal both the anatomy and the disease entity. MRI is a powerful diagnostic tool for the evaluation, assessment of severity, and follow-up of diseases of the brain and spine. The basic imaging protocols for MRI of brain include a T1W sequence in the sagittal plane, and T2W and fluid-attenuated inversion recovery (FLAIR) sequences in the axial plane. FLAIR sequences are created with the use of an inversion recovery pulse to null the signal from the cerebrospinal fluid. Diffusion imaging, if available, is helpful in many indications. IV contrast material [gadolinium (Gd) chelates] may be administered when there is suspicion of breakdown of the blood–brain barrier (BBB). Postcontrast images can be obtained in T1W sequence with different planes.

Magnetic resonance angiography (*MRA*) is one of the most commonly used noninvasive techniques to evaluate neurovascular system. MRA with or without the administration of contrast material can be performed. Noncontrast MRA involves time-of-flight (TOF) or phase-contrast techniques. TOF images are gradient echo T1W images and produced by the application of multiple RF pulses with short TRs to saturate the spins in stationary tissues.[18,19] Inflowing blood is not affected and appears hyperintense compared to stationary tissue. Phase-contrast MR angiography uses opposite gradient pulses of the same strength and duration to rephase stationary spins.[18,19] Unlike TOF, phase-contrast images are not affected by the T1 values of stationary spins and can give information about the flow direction. Therefore, phase-contrast imaging can also be performed to evaluate venous and slow cerebrospinal fluid flow.

Perfusion specifically refers to the capillary blood supply providing oxygen and nutrients to the brain tissue, quantified in terms of the flow rate (milliliters per minute) normalized to the tissue mass (typically per 100 g). The assessment of the microcirculation is based on the use of a tracer of the vascular compartment, and investigating the distribution of this tracer over time within the tissue of interest.[20] The earliest and currently most frequently used method of *perfusion MR imaging* is dynamic susceptibility contrast (DSC) imaging, which utilizes Gd-based contrast as the tracer to estimate perfusion-related data. Following intravenous bolus administration of Gd contrast medium, rapid sequential images of the same brain tissue volume are acquired during the first pass of the contrast through the intracranial circulation over a fixed period of time. To be able to sample the first-pass transit of the contrast bolus, a fast-scan technique is required, such as echo planar imaging (EPI). Gd in the vasculature causes signal loss of T2 and T2* via spin dephasing and susceptibility-related effects, respectively, which is reflected as a dramatic reduction in signal intensity on T2-weighted or T2*-weighted images known as "negative enhancement." The result is a time course of signal change recorded for every voxel in the region of interest. Following data acquisition, this signal intensity versus time plot is analyzed using dedicated software functioning based on various MRI and pharmacokinetic models to estimate several important vascular parameters, such as CBV, CBF, time-to-peak concentration (TTP), and MTT. Based on the fact that the change in relaxation rate is proportional to the concentration of Gd contrast agent, the signal intensity–time curve is converted to concentration–time curve. The area under the concentration–time curve is derived to obtain relative cerebral blood volume (rCBV) map.

Parameters obtained via *MR perfusion imaging* may indicate the extent of irreversibly injured brain in the ischemic core and potentially salvageable but hypoperfused ischemic penumbra in patients with stroke, similar to CT perfusion[11,21] (Fig. 11.4). This may guide optimal patient selection for thrombolytic treatments including intraarterial endovascular therapy. In patients with brain tumor, it has been shown that rCBV measurements correlate reliably with tumor grade and histologic findings of increased tumor vascularity[22] (Fig. 11.5).

An additional dynamic MR perfusion imaging technique, utilizing Gd bolus contrast administration as a tracer method, is dynamic contrast enhanced (DCE) imaging, which is T1 weighted where the contrast agent causes signal enhancement, as opposed to DSC imaging. Typically, a 3D fast gradient echo sequence is used for DCE imaging, and unlike DSC imaging, DCE investigates contrast agent uptake and equilibrium in the brain tissue. Conventional Gd-based contrast agents used in clinical MRI are diffusible low–molecular weight extracellular agents that remain intravascular when the BBB is intact. Disruption of BBB due to a variety of pathological processes causes leakage of Gd-contrast agent into the extravascular extracellular space (EES) through the capillary endothelium. During DCE imaging, contrast accumulation within the EES results in T1 shortening and positive enhancement.[23] This phenomenon establishes the basis of steady-state, T1-weighted MR permeability imaging.

Via pharmacokinetic modeling of the DCE data, various permeability-related parameters can be determined, including the most frequently utilized metric "K^{trans}." K^{trans} is the transfer constant between plasma and EES and describes the rate of Gd flux into the EES. Permeability data can also be obtained using the T2* first-pass DSC MR technique.[24] However, this technique only measures the permeability in the first pass, which is likely to be different than the permeability measured in the steady state, where bidirectional exchange between plasma and EES occurs. In cases where there is very high vascular permeability, such as with most brain neoplasms, K^{trans} is highly flow related, and can be characterized in the first pass. On the other hand, in the setting of very low permeability, such as in inflammation and ischemia, K^{trans} is not necessarily flow limited, more proportional to the surface area product, and may be characterized using steady-state techniques.

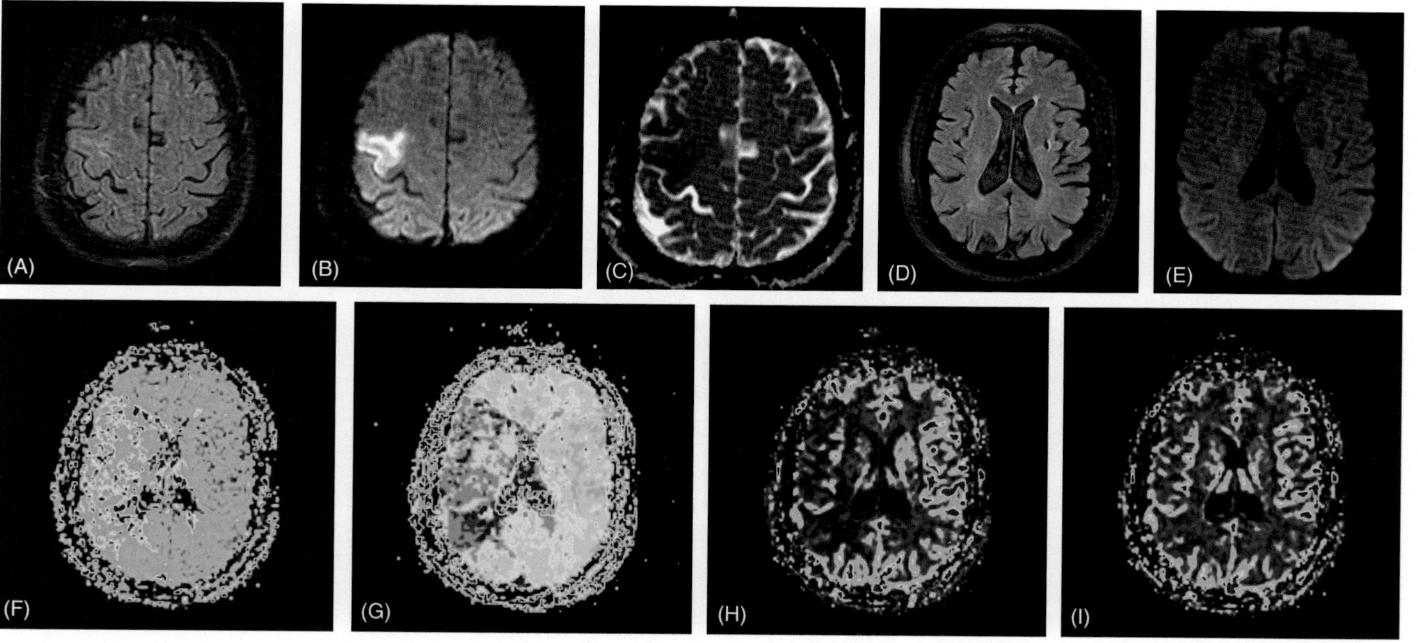

FIGURE 11.4 A 65-year-old male patient presented with left upper extremity weakness. (A) T2 fluid-attenuated inversion recovery (FLAIR) images show subtle increased signal intensity involving the right precentral gyrus. (B) Diffusion-weighted imaging (DWI) TRACE images and (C) apparent diffusion coefficient (ADC) map demonstrate diffusion restriction along the precentral gyrus consistent with acute infarction. A 49-year-old male patient with a history of diabetes mellitus presenting with vertigo. (D) T2 FLAIR images show a small left putaminal chronic lacunar infarct with no signs of acute infarction on (E) DWI images. However, magnetic resonance (MR) perfusion study demonstrates remarkable perfusion abnormality involving the right middle cerebral artery territory with markedly increased (F) mean transit time (MTT) and (G) time-to-peak concentration (TTP). (H) CBF also appears to be decreased moderately compared to contralateral hemisphere, with relative preservation of (I) CBV. On angiography, this patient has been shown to have high-grade stenosis of right ICA.

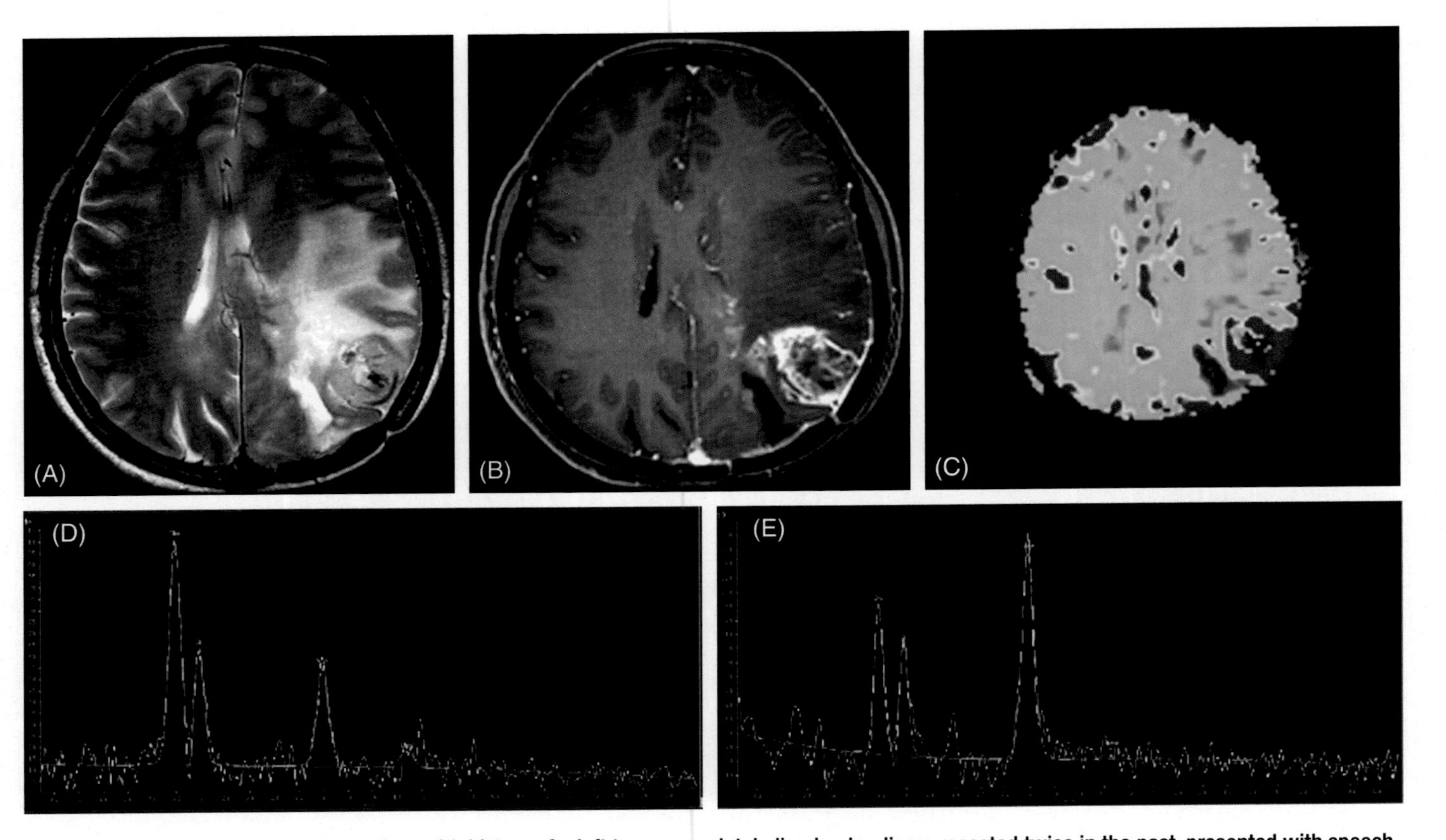

FIGURE 11.5 A 52-year-old female patient with history of a left temporoparietal oligodendroglioma resected twice in the past, presented with speech difficulty.

(A) T2-weighted and (B) postcontrast T1-weighted images demonstrated a left temporoparietal enhancing lesion surrounding the resection cavity. (C) Relative cerebral blood volume (rCBV) map showed increased CBV corresponding to the lesion. Multivoxel magnetic resonance spectroscopy (MRS) obtained using an echo time of 144 ms, there was (D) increased choline (Cho) and decreased *N*-acetyl aspartate (NAA) corresponding to the lesion compared to (E) normal-appearing contralateral white matter, with presence of a lactate peak. Patient had repeating surgery and pathological examination, and yielded grade 3 anaplastic oligodendroglioma.

There is another distinct category within perfusion MR imaging techniques, known as arterial spin labeling (ASL), which does not require any contrast agent, but makes use of endogenous water molecules within the in-flowing blood, which are magnetically labeled by pulsed RF energy, as a diffusible tracer.[25] ASL can be used in all patients where the administration of Gd contrast is restricted, such as in cases with renal insufficiency or in pediatric populations.

Diffusion-weighted imaging (*DWI*) is currently widely used in most brain-imaging protocols, being of utmost importance in the evaluation of acute stroke, as it allows the rapid diagnosis of brain ischemia. It is also essential for the assessment of brain tumors and intracranial infections. Molecular diffusion refers to random, microscopic translational movement of water and other small molecules in a tissue, known as the Brownian motion. In MR imaging, the motion of water molecules by diffusion through a magnetic gradient results in an irreversible signal through intravoxel dephasing. With increased diffusion within a tissue, there is greater signal attenuation in the MR image. The most common method for producing diffusion weighting is application of strong pulsed gradients along three orthogonal diffusion directions.[26] The *b*-value denotes the amount of diffusion weighting, which depends on the strength, duration, and spacing of the pulsed gradients. By combined analysis of at least two image sets obtained with different *b*-values, an apparent diffusion coefficient (ADC) map can be created which is an image of the calculated ADCs voxel by voxel.

The diffusion of water in the brain tissue is modulated by the presence of cell membranes, white matter axons, and myelin sheaths. The diffusion may be restricted by either cellular swelling or increased cellular density, as in the case of acute ischemia and tumor involvement, respectively. Conversely, necrosis, which results in breakdown of cellular membranes, increases diffusivity. In abscesses, diffusion is restricted most likely due to high viscosity and cellularity of the pus.

Diffusion tensor imaging (*DTI*) adds the directionality of the diffusion.[27] The tensor model is based on the anisotropic nature of diffusion, which is the consequence of complex microstructural architecture of the brain, facilitating diffusion in one direction and limiting in another. In particular, diffusion is predominantly along the direction parallel to the long axis of the white matter tracts, and limited in the direction perpendicular to the tract.[28] This anisotropic diffusion can be graphically represented as an ellipsoid, defined by eigenvalues and eigenvectors. By tensor analysis of each voxel, several quantitative parameters related to tissue microstructure can be obtained. Mean diffusivity (MD) is the tensor equivalent of ADC, reflecting the degree of diffusivity. Fractional anisotropy (FA) is defined as the ratio of the anisotropic component of the diffusion tensor to the whole diffusion tensor, and is a rotationally invariant scalar quantifying anisotropy between zero and one. The largest (major) eigenvalue is often called axial diffusivity, reflecting the magnitude of diffusion parallel to axons, and the average of the two lesser (medium and minor) eigenvalues is called radial diffusivity, reflecting the magnitude of diffusion perpendicular to axons. By combining the directional information and magnitude of anisotropic diffusion of the individual voxels, the trajectory of white matter tracts can be reconstructed, which is known as tractography. DTI has been reported in a broad spectrum of clinical applications, including analysis of white matter pathology (ischemia, axonal damage, and demyelination), tumor characterization and infiltrative features, and preoperative planning.[29] High-resolution, multitensor imaging (greater than six diffusion directions) may be helpful to discriminate between tumor-related white matter tract displacement versus infiltration, and identification of compensatory rearrangement of functional centers affected by adjacent brain tumors.

Magnetic resonance spectroscopy (*MRS*) involves the use of gradients to selectively excite a small volume of brain tissue, then recording of the free induction decay, and production of a spectrum from a voxel rather than an image.[30] MR spectrum consists of resonances or peaks of brain metabolites that represent signal intensities as a function of frequency (expressed as parts per million or ppm). MRS is based on the principle of a chemical shift occurring due to electron clouds of neighboring atoms shielding the nucleus from the magnetic field. In vivo MRS focuses on carbon-bound protons in the 1–5 ppm range of the chemical shift scale. The peak from the water has to be suppressed, so that the lower-concentration metabolites can be seen. The most important peaks are as follows: lactate (1.3 ppm; its elevation is indicative of anaerobic glycolysis), *N*-acetyl aspartate (NAA; 2.0 ppm; biomarker of neuronal integrity), creatine (Cr; 3.0 ppm; taken as a reference level and relatively constant throughout the brain), and choline (Cho; 3.2 ppm; may be a marker for cellular proliferation, increased membrane turnover, or inflammation). MRS is used as an adjunct to MR imaging in several common neurological disorders, including brain neoplasms, inherited metabolic disorders, demyelinating diseases, and infective focal lesions, with an aim to make an early noninvasive diagnosis or to increase confidence in a suspected diagnosis[31] (Fig. 11.5). Furthermore, it helps in differentiating tumor from other nonspecific lesions, identifying the optimal biopsy sites in heterogeneous gliomas, monitoring treatment response, and may aid in differentiating between treatment-induced changes and recurrent tumors that is often challenging. Cho/NAA ratio has been shown to reliably differentiate recurrent glioma from postradiation injury,[32] the accuracy appears to increase when combined with DSC MR imaging.

5 POSITRON EMISSION TOMOGRAPHY

Positron emission tomography (PET) with fluorine-18 fluorodeoxyglucose (^{18}F-FDG) allows noninvasive in vivo quantification of local cerebral metabolism and may demonstrate pathological conditions before morphological features are discernible.[33] PET metabolic imaging has significant usefulness in certain clinical conditions, such as refractory seizure disorders, dementia, and recurrent brain tumors. In dementia, different metabolic distribution patterns demonstrated on ^{18}F-FDG PET scans may aid in differentiating distinct disorders causing dementia, such as Alzheimer's disease, dementia with Lewy bodies, frontotemporal dementia, and PD.[34] PET imaging may also help distinguishing high-grade brain tumors from lower-grade tumors, improve the accuracy of stereotactic biopsy of lesions by targeting areas of hypermetabolism, and following treatment, it can help differentiate recurrent tumor from radiation necrosis.[35]

6 CONCLUSIONS

To provide evidence for a technological development in vivo, there has been an increased need for high-resolution MR imaging at the microscopic tissue level. To detect microscopic tissue changes in response to physiological processes (as in the case of newly developed pharmaceuticals and technical procedures, such as stem cell implantations for targeted treatments), the basic needs are dedicated as MR scanners for experimental animals equipped with proper coils. Besides Gd-based contrast materials, use of superparamagnetic iron oxide provides additional negative contrast effects best determined by T2* scans. Altogether, an interdisciplinary approach is mandatory in this experimental field.

Abbreviations

ADC Apparent diffusion coefficient
ASL Arterial spin labeling
BBB Blood–brain barrier
CBF Cerebral blood flow
CBV Cerebral blood volume
Cho Choline
CNS Central nervous system
Cr Creatine
CT Computed tomography
3D Three dimensional
DCE Dynamic contrast enhanced
DS Doppler sonography
DSC Dynamic susceptibility contrast
DTI Diffusion tensor imaging
DWI Diffusion-weighted imaging
EES Extravascular extracellular space
EPI Echo planar imaging
FA Fractional anisotropy

FLAIR	Fluid-attenuated inversion recovery
^{18}F-FDG	Fluorine-18 fluorodeoxyglucose
Gd	Gadolinium
GM	Germinal matrix
K^{trans}	Transfer constant between plasma and EES
MD	Mean diffusivity
MIP	Maximum intensity projection
MPR	Multiplanar reformation
MR	Magnetic resonance
MRA	Magnetic resonance angiography
MRI	Magnetic resonance imaging
MRS	Magnetic resonance spectroscopy
MTT	Mean transit time
NAA	*N*-acetyl aspartate
PD	Parkinson's disease
PET	Positron emission tomography
rCBV	Relative cerebral blood volume
RF	Radiofrequency
TCD	Transcranial doppler
TE	Echo time
TOF	Time of flight
TR	Repetition time
TTP	Time to peak
US	Ultrasonography

References

1. Karl K, Kainer F, Heling KS, Chaoui R. Fetal neurosonography: extended examination of the CNS in the fetus. *Ultraschall in der Medizin*. 2011;32(4):342–361.
2. Lowe LH, Bailey Z. State-of-the-art cranial sonography: Part 1, modern techniques and image interpretation. *Am J Roentgenol*. 2011;196(5):1028–1033.
3. Riccabona M. Neonatal neurosonography. *Eur J Radiol*. 2014;83(9):1495–1506.
4. Simbrunner J, Riccabona M. Imaging of the neonatal CNS. *Eur J Radiol*. 2006;60(2):133–151.
5. Fritz J, Polansky SM, O'Connor SC. Neonatal neurosonography. *Semin Ultrasound CT MR*. 2014;35(4):349–364.
6. Purkayastha S, Sorond F. Transcranial doppler ultrasound: technique and application. *Semin Neurol*. 2012;32(4):411–420.
7. Naqvi J, Yap KH, Ahmad G, Ghosh J. Transcranial Doppler ultrasound: a review of the physical principles and major applications in critical care. *Int J Vasc Med*. 2013;2013:629378.
8. Berg D, Godau J, Walter U. Transcranial sonography in movement disorders. *Lancet Neurol*. 2008;7(11):1044–1055.
9. van Straten M, Venema HW, Majoie CB, Freling NJ, Grimbergen CA, den Heeten GJ. Image quality of multisection CT of the brain: thickly collimated sequential scanning versus thinly collimated spiral scanning with image combining. *Am J Neuroradiol*. 2007;28(3):421–427.
10. de Lucas EM, Sanchez E, Gutierrez A, Mandly AG, Ruiz E, Florez AF, et al. CT protocol for acute stroke: tips and tricks for general radiologists. *Radiographics*. 2008;28(6):1673–1687.

11. Srinivasan A, Goyal M, Al Azri F, Lum C. State-of-the-art imaging of acute stroke. *Radiographics*. 2006;26(suppl 1):S75–S95.

12. Rydberg J, Buckwalter KA, Caldemeyer KS, Phillips MD, Conces Jr DJ, Aisen AM, et al. Multisection CT: scanning techniques and clinical applications. *Radiographics*. 2000;20(6):1787–1806.

13. Lell MM, Anders K, Uder M, Klotz E, Ditt H, Vega-Higuera F, et al. New techniques in CT angiography. *Radiographics*. 2006;26(suppl 1):S45–S62.

14. Cody DD. AAPM/RSNA physics tutorial for residents: topics in CT. Image processing in CT. *Radiographics*. 2002;22(5):1255–1268.

15. Hoeffner EG, Case I, Jain R, Gujar SK, Shah GV, Deveikis JP, et al. Cerebral perfusion CT: technique and clinical applications. *Radiology*. 2004;231(3):632–644.

16. Bartling SHSW, Semmler W, Kiessling F. Small animal computed tomography imaging. *Curr Med Imag Rev*. 2007;3:45–59.

17. Ritman EL. Current status of developments and applications of micro-CT. *Annu Rev Biomed Eng*. 2011;13:531–552.

18. Bitar R, Leung G, Perng R, Tadros S, Moody AR, Sarrazin J, et al. MR pulse sequences: what every radiologist wants to know but is afraid to ask. *Radiographics*. 2006;26(2):513–537.

19. Pandey S, Hakky M, Kwak E, Jara H, Geyer CA, Erbay SH. Application of basic principles of physics to head and neck MR angiography: troubleshooting for artifacts. *Radiographics*. 2013;33(3):E113–E123.

20. Viallon M, Cuvinciuc V, Delattre B, Merlini L, Barnaure-Nachbar I, Toso-Patel S, et al. State-of-the-art MRI techniques in neuroradiology: principles, pitfalls, and clinical applications. *Neuroradiology*. 2015;57(5):441–467.

21. Copen WA, Schaefer PW, Wu O. MR perfusion imaging in acute ischemic stroke. *Neuroimaging Clin N Am*. 2011;21(2):259–283.

22. Law M, Yang S, Wang H, Babb JS, Johnson G, Cha S, et al. Glioma grading: sensitivity, specificity, and predictive values of perfusion MR imaging and proton MR spectroscopic imaging compared with conventional MR imaging. *Am J Neuroradiol*. 2003;24(10):1989–1998.

23. Shiroishi MS, Lacerda S, Tang X, Muradyan N, Roberts TPL, Law M. Physical principles of MR perfusion and permeability imaging: gadolinium bolus technique. In: Faro SH, Mohamed FB, Law M, Ulmer JT, eds. *Functional Neuroradiology: Principles and Clinical Applications*. United States: Springer; 2012.

24. Lacerda S, Law M. Magnetic resonance perfusion and permeability imaging in brain tumors. *Neuroimaging Clin N Am*. 2009;19(4):527–557.

25. Williams DS, Detre JA, Leigh JS, Koretsky AP. Magnetic resonance imaging of perfusion using spin inversion of arterial water. *Proc Natl Acad Sci USA*. 1992;89(1):212–216.

26. Schaefer PW, Grant PE, Gonzalez RG. Diffusion-weighted MR imaging of the brain. *Radiology*. 2000;217(2):331–345.

27. Le Bihan D, Mangin JF, Poupon C, Clark CA, Pappata S, Molko N, et al. Diffusion tensor imaging: concepts and applications. *J Magn Reson Imaging*. 2001;13(4):534–546.

28. Huisman TA. Diffusion-weighted and diffusion tensor imaging of the brain, made easy. *Cancer Imag*. 2010;10:S163–S171.

29. Alexander AL, Lee JE, Lazar M, Field AS. Diffusion tensor imaging of the brain. *Neurotherapeutics*. 2007;4(3):316–329.

30. McRobbie DW, Moore EA, Graves MJ, Prince MR. *MRI from Picture to Proton*. Cambridge, United Kingdom: Cambridge University Press; 2004.

31. Oz G, Alger JR, Barker PB, Bartha R, Bizzi A, Boesch C, et al. Clinical proton MR spectroscopy in central nervous system disorders. *Radiology*. 2014;270(3):658–679.

32. Fink JR, Carr RB, Matsusue E, Iyer RS, Rockhill JK, Haynor DR, et al. Comparison of 3 Tesla proton MR spectroscopy, MR perfusion and MR diffusion for distinguishing glioma recurrence from posttreatment effects. *J Magn Reson Imaging*. 2012;35(1):56–63.

33. Mettler FAJ, Guiberteau MJ. Positron emission tomography (PET) imaging. In: Mettler FAJ, Guiberteau MJ, eds. *Essentials of Nuclear Medicine Imaging*. United States: Elsevier; 2006.

34. Bohnen NI, Djang DS, Herholz K, Anzai Y, Minoshima S. Effectiveness and safety of 18F-FDG PET in the evaluation of dementia: a review of the recent literature. *J Nucl Med*. 2012;53(1):59–71.

35. Basu S, Alavi A. Molecular imaging (PET) of brain tumors. *Neuroimaging Clin N Am*. 2009;19(4):625–646.

Targeted Treatment Strategies for Neurological Diseases

CHAPTER 12

Alzheimer

Qizhi Zhang, PhD*, Chi Zhang, MSc**
*Fudan University, Shanghai, China; **Renji Hospital, Shanghai Jiao Tong University School of Medicine, Shanghai, China

CONTENTS

1 INTRODUCTION

Alzheimer's disease (AD) is a complex neurodegenerative disease, which has become the most common form of dementia in the elderly. AD currently affects about 36 million people worldwide, and the estimated prevalence is expected to be over 115 million by 2050.[1] The official deaths of AD in the United States were recorded to be 84,767 in 2013, making AD the sixth leading cause of death in America.[2] The total payments for AD patients, including health care, long-term care, and hospice services, are expected to be 226 million dollars in 2015.[2] Thus AD has severe epidemiologic and economic impacts on the whole society, and effective therapeutic measures are urgently needed.

It is reported that more than 95% of AD cases are sporadic, which might result from a combination of environmental factors and risk genes.[3] The major genetic risk factor for the sporadic form of AD is the ε4 allele of apolipoprotein E (ApoE ε4), which influences AD pathogenesis to a great extent.[4] Symptoms of AD are characterized by a progressive decline of cognitive and functional abilities. The identified neuropathological hallmarks of AD are the deposition of the protein fragment amyloid-β (Aβ) (plaques), twisted stands of the hyperphosphorylated tau protein (neurofibrillary tangles), and evidence of nerve cell damage and death in the brains of patients. Other hallmarks associated with AD pathology are neuroinflammation and oxidative stress in the brain, which also play detrimental roles in AD.[3]

Nowadays, large numbers of drugs are available or under development for AD treatment. However, clinical drugs, such as cholinesterase inhibitors or glutamate receptor antagonists, are only symptomatic and do not cure the pathology. Moreover, the presence of the blood–brain barrier (BBB) becomes a formidable obstacle for AD drugs to enter the brain and reach the lesions. These challenges generally lead to the failure of AD treatment. Therefore, it is urgent

Nanotechnology Methods for Neurological Diseases and Brain Tumors. http://dx.doi.org/10.1016/B978-0-12-803796-6.00012-5

to develop strategies to improve the drug efficacy, as well as to increase drug delivery to the lesion to achieve better treatments.

Nanodelivery systems are now considered to be a promising tool, potentially able to enhance the drug transport to the brain lesion area, particularly with the functionalization of their surface with BBB-targeted or AD lesion–targeted agents. In Section 2, several nanodelivery systems, including single-targeted and dual-targeted ones delivered via intravenous (i.v.) or nasal route, are reviewed for their use in targeted treatment strategies for AD.

2 STRATEGIES FOR BRAIN DRUG DELIVERY IN AD TREATMENT

2.1 Intravenous Drug Delivery System

2.1.1 BBB Crossing

Although some aspects, such as P-gp function, cerebral blood, and cerebrospinal fluid reabsorption,[5] of the BBB of AD patients may differ from the normal BBB, it still remains a formidable obstacle for most drugs to enter the brain. Therefore, transport across the BBB is essential for AD drugs to achieve therapeutic effects. The use of surface-engineered drug delivery systems to enable transport across the BBB is now considered as a strategic approach, with receptor-mediated transcytosis (RMT) and adsorptive-mediated transcytosis (AMT) being the two most popular pathways.

2.1.2 Receptor-Mediated Transcytosis

RMT is the transport through specialized ligand-specific receptors expressed on the BBB, such as low-density lipoprotein receptors (LDLR), transferrin receptors (TfR), lactoferrin receptors (LfR), and acetylcholine receptors (AchR).

Low-density lipoprotein receptor–related protein (LRP) is a member of the LDLR family. LRP can bind numerous ligands, including proteinases, proteinase inhibitor complexes, and certain ApoE- and lipoprotein lipase–enriched lipoproteins, and subsequently mediate the cellular internalization of these ligands and their transport across the BBB.[6] Nonionic surfactant polysorbate 80 modified on polymeric nanoparticles (NPs) can mimic LDL, interact with LDLR, and enable NPs transport into the brain.[7] Polybutylcyanoacrylate (PBCA) NPs coated with polysorbate 80 were the first reported delivery system for the brain drug delivery, and improved the bioavailablity of AD drugs, rivastigmine and tacrine, in the brain.[8,9] Jose et al.[10] encapsulated bacoside-A into poly(lactic-*co*-glycolic acid) (PLGA) NPs modified with polysorbate 80 to evaluate the brain accumulation of the delivery system in rats. Bacoside-A is a plant extract that was reported to significantly improve acquisition, consolidation, and retention of memory.[11] Polysorbate 80–coated PLGA NPs were able to deliver 10-fold more bacoside-A into the brain compared to the free drug solution (23.95 ± 1.74 μg/g tissue vs. 2.56 ± 1.23 μg/g tissue), which highlighted the potential of polysorbate 80–modified PLGA NPs for the treatment of central nervous system (CNS) diseases, such as AD.[10]

ApoE also binds to lipoprotein receptors, including LDLR and LRP.[12] ApoE-coupled NPs can mimic lipoprotein particles that are endocytosed into and then transcytosed through the BBB endothelium into the brain.[13] Song et al.[14] constructed a nanostructure with ApoE3 (an isoform of ApoE) and reconstituted high-density lipoprotein (rHDL) (ApoE3–rHDL) for AD treatment. Utilizing the high binding affinity to Aβ of ApoE, ApoE3–rHDL not only possessed the ability to cross the BBB, but also facilitated Aβ degradation. About 0.4%ID/g of ApoE3–rHDL gained access to the mouse brain, 1 h after i.v. administration. Moreover, the daily treatment with ApoE3–rHDL for 4 weeks decreased Aβ deposition, attenuated microgliosis, ameliorated neurologic changes, and rescued memory deficits in senescence-accelerated mouse, P8 strain (SAMP8) mice. These results indicated that ApoE3–rHDL could serve as a novel brain-targeted nanomedicine for AD therapy.

Transferrin (Tf) is a naturally occurring TfR ligand, which assists iron uptake into the vertebrate cells through a cycle of endo- and exocytosis of Tf.[15] TfR is highly expressed in BBB endothelial cells and this TfR system is considered to be a pathway that increases the uptake of biologics into the brain.[15] Liposomes (LIP) modified with the whole Tf protein were prepared to target BBB.[16] After a 2-h incubation of brain capillary endothelial cells (BCECs) with LIP at 37°C, the cell uptake of Tf-modified LIP was found to be 1–3 times higher than the unmodified LIP. OX26 is a monoclonal antibody (mAb) against TfR on BCECs, which could be used to avoid the antigenicity of Tf. When injected into the femoral vein of rats, OX26-modified poly(ethyleneglycol)–poly(ε-caprolactone) (PEG–PCL) polymersomes (OX26-PO) with higher surface OX26 density accumulated greatly in rats' brain tissues.[17] OX26-PO loaded with AD therapeutic peptide NC-1900 also presented more significant learning and memory improvements in an AD rat model in a water maze task. However, using large proteins, such as Tf and OX26, in the formulations can easily cause problems in the synthesis procedure, formulation stability, and immunological response.[18] Therefore, a short peptide with high specificity, low cytotoxicity, and low immunological response is desired for the construction of the targeting delivery system. B6 peptide (CGHKAKGPRK), obtained from a phage display, showed high affinity to TfR and can therefore mediate brain drug delivery as a substitute for Tf.[19] Conjugating B6 to the surface of PEG–poly(lactic acid) block copolymer (PEG–PLA) NPs (B6-NPs) enhanced the entry of encapsulated neuroprotective peptide NAP (NAPVSIPQ) into the brain in BALB/c nude mice, when compared with the unmodified NPs.[20] After injected with NAP-loaded

B6-NPs into the tail vein, AD model mice revealed excellent amelioration in learning impairments, cholinergic disruption, and loss of hippocampal neurons even at lower dose (0.02 μg/day) than unmodified NPs. In contrast, the free NAP solution at concentrations up to 0.08 μg/day failed to produce any significant enhancement.[20]

Lactoferrin (Lf) is a natural single-chain iron-binding cationic glycoprotein of the Tf family, expressed in various tissues and involved in various physiological processes.[21] LfR was found to be highly expressed in the brain cells, such as endothelial cells and neurons, and also overexpressed in the CNS in patients with age-related neurodegenerative diseases, such as AD and Parkinson' disease.[22,23] Therefore, Lf might be a promising brain-targeting ligand for drug delivery system for CNS diseases. To increase the brain delivery of a neuroprotective peptide S14G-humainin for the treatment of AD, Yu et al.[24] built Lf-functionalized PEG–PLGA polymersomes (Lf-PO). Lf-PO achieved the greatest BBB permeability surface area and percentage of injected dose per gram (%ID/g) compared with Tf-modified PO and cationic bovine serum albumin–modified PO. AD model mice injected with Lf-PO achieved far more excellent learning ability and spatial memory than unmodified PO, which indicated that the modification of Lf on the PO effectively increased the drug concentration in the brain and its efficacy.

AChR is expressed on both BCECs and neuronal cells. Rabies virus glycoprotein (RVG), a 29–amino acid peptide, can specifically bind to AChR, thus enabling the viral entry into neuronal cells.[25] RVG has been conjugated to the exosomes to develop a brain-targeted siRNA delivery system for AD treatment.[26] After being intravenously injected into mice, RVG-modified exosomes delivered siRNA to the neurons, microglia, and oligodendrocytes in the mouse brain and resulted in a specific gene knockdown. When RVG-modified exosomes loaded with siRNA were administered, aiming at the β-site amyloid precursor protein–cleaving enzyme 1 (BACE1), a strong knockdown of BACE1 mRNA (60%) and protein (62%) was found in mice, as well as a significant decrease in the total Aβ42 levels was observed when compared with RVG-conjugated siRNA and RVG-modified exosomes without encapsulated siRNA.[26] Therefore, RVG-fused exosomes specifically delivered AD drugs to the brain after systemic administration, and exerted a therapeutic effect.

There are still some peptides obtained from the phage display technique that can interact with particular proteins or cells and have special affinity with the brain. TGN (TGNYKALHPHNG) is a peptide selected after four rounds of in vivo screening from a random 12-mer peptide library displayed on the surface of filamentous phage M13, which has a high potential for transport into the brain.[27] TGN conjugation on the surface of PEG–PLGA NPs or PEG–poly(2-(*N*,*N*-dimethylamino)ethyl methacrylate (PDMAEMA) polyplexes enhanced cellular uptake by BCECs.[28,29] Moreover, TGN modification can also significantly increase the amount of delivery system entering the brain by about twofold.[28,29] Encapsulation of NAP, a highly active fragment of activity-dependent neuroprotective protein, into TGN-modified PEG–PLGA NPs (TGN-NPs) produced better neuroprotective effects than NAP alone or NAP in unmodified NPs, with the greatest performance in Morris water maze test in AD model mice, the lowest activity of acetylcholine esterase (AChE), and the highest choline acteyl transferase activity, as well as the smallest Aβ plaque deposits.[30]

2.1.3 Adsorptive-Mediated Transcytosis

AMT is induced by some positively charged macromolecules, such as histone, avidine, and cationized albumin, that interact with the negatively charged cell surface and subsequently trigger transcytosis and exocytosis. Therefore, AMT pathway could be considered as a promising approach for brain drug delivery.

HIV-1 transactivating transcriptor (TAT), a small basic peptide containing six arginine and two lysine residues, is one of the most widely used cell-penetrating peptide, which can induce the cellular uptake through certain routes, such as clathrin-dependent endocytosis, micropinocytosis, and fluid-phase endocytosis.[31] When coupled to the surface of nanoliposomes (NLs), TAT increased the cellular uptake and permeability of NLs in hCMEC/D3 cells. Surface plasmon resonance (SPR) analysis showed that the K_D values of TAT-attached NLs to Aβ peptide were in the magnitude of nanomolar range, suggesting a high Aβ-binding affinity.[31] Therefore, TAT-functionalized NLs might become a promising tool for brain Aβ targeting.

Trimethyl chitosan (TMC) is a partially quaternized derivative of chitosan (CS), possessing mucoadhesive properties like CS, and having improved cationic characteristics. As a cationic ligand, TMC facilitated the brain transport of NPs through AMT pathway to enhance the drug delivery into the CNS. After being injected into the tail vein of mice, TMC-modified PLGA NPs loaded with coumarin-6 showed a higher accumulation in the cortex, paracoele, the third ventricle, and choroid plexus epithelium, while reduced brain uptake of unmodified PLGA NPs was observed.[32] Coenzyme Q10 was chosen as the AD model drug for the evaluation of the neuroprotective effects of TMC-modified PLGA NPs in AD transgenic mice. Mice administrated with TMC-modified PLGA NPs presented greater improvement in spatial memory and decreased number of senile plaques than unmodified NPs.[32] Thus TMC modification enabled NPs to transport across the BBB and effectively deliver the drug into the brain.

However, AMT strategy has a lack of selectivity to some degree and might cause side effects of drugs in nontargeted organs, which could limit its application in drug delivery system for AD therapy.

2.1.4 Brain Aβ Targeting

The drug distribution in normal brain tissues after transport across the BBB might cause unexpected side effects. Therefore, improving the drug delivery to the brain lesions is crucial for AD treatment. The extracellular aggregation of amyloid plaque is one of the primary histopathological characteristics of AD, which is caused by the increased production, accumulation, and aggregation of Aβ peptide. Accordingly, brain Aβ could be an ideal target for precise drug delivery in AD treatment. Several researches reported that brain Aβ–targeted drug delivery systems were able to effectively transport across the BBB and then precisely deliver drugs to the plaque deposits to exert the therapeutic effects.

The pF(ab′)$_2$ fragment is a polyamine-modified mAb against human Aβ (IgG4.1), which has high BBB permeability and extensive targeting to amyloid plaques throughout the brain parenchyma.[33] Smart nanovehicles (SNVs) consisting of a CS polymeric core and pF(ab′)$_2$ fragments coated on the surface were developed.[34] After being intravenously administrated into the vein of mice, ^{125}I-labeled SNVs (^{125}I-SNVs) achieved about 8–11 times higher uptake in various brain regions compared with the control nanovehicles (CNVs) (^{125}I-CNVs, CS nanovehicles coated with bovine serum albumin), and plasma clearance was 9 times higher than that of ^{125}I-CNVs.[34] The study indicated that SNVs were capable of delivering therapeutic agents across the BBB to target amyloid deposits in the brains of AD transgenic mice.[34]

As it is difficult to realize a single ligand crossing the BBB and targeting amyloid plaques simultaneously, in the past 5 years, a "dual-targeting" strategy (NPs modified with dual functional ligands) has been proposed for AD treatment.

Anti-Aβ1–42 monoclonal antibody (Aβ-mAb) has a very high affinity for Aβ1–42 peptides, and can be used as a targeting ligand for the plaque deposits.[35] OX26 (mAb against TfR of the BBB) and Aβ-mAb were both coupled to the surface of NLs to construct the dually decorated NLs (dd-NLs). dd-NLs, similar to the OX26-NLs, could be transferred intact through the BBB model built with hCMEC/D3 cells. The cellular uptake of dd-NLs increased significantly when cells were preincubated with Aβ1–42 peptides, while OX26-NLs uptake was not affected.[35]

Recently, Bana et al.[36] synthesized LIP bifunctionalized with phosphatidic acid (PA), which had high binding affinity to Aβ peptide, and with a modified peptide derived from the receptor-binding domain of human ApoE (MW 1698.18 g/mol, mApoE) as a BBB ligand. This bifunctionalization enhanced the amount of LIP transported across the BBB in vitro and in vivo, and the brain/blood ratio of mApoE-PA-LIP was fivefold higher than that of PA-LIP after i.v. administration into healthy mice. Moreover, the bifunctionalized LIP could strongly bind to the Aβ peptide (K_D = 0.6 μM), inhibit Aβ peptide aggregation (70% inhibition after 72 h), and trigger the disaggregation of preformed aggregates (60% decrease after 120-h incubation) according to the results of SPR, Thioflavin-T, and SDS-PAGE/WB assays.[36] After i.v. administration of mApoE-PA-LIP to APP/PS1 transgenic mice, a reduction of both amyloid plaques and the total amount of brain Aβ levels was found, suggesting that mApoE-PA-LIP could destabilize brain Aβ aggregates and promote Aβ removal. The results in novel-object recognition (NOR) test also showed increased memory improvement in AD model mice.[37] Although the bifunctionalized LIP do not eliminate the cause of Aβ overproduction, it could slow down the neurodegeneration process, and be a potential delivery system for chronic AD treatments to minimize disease progression.[37]

The density evaluation of both ligands on the surface of NPs is essential for the bifunctionalized NPs to obtain the optimal dual-targeting effects. The best targeting effects could be achieved without excessive modification; which leads to aggregation of NPs. It is important to know whether the two ligands influence each other. Zhang et al.[38] developed bifunctionalized NPs targeting amyloid plaques in the brain of AD model mice with two peptides, TGN and QSH, for BBB transport and Aβ targeting, respectively. The modification densities of both peptides were evaluated in vitro and in vivo, respectively. Different molar ratios between maleimides on the NPs and TGN or QSH were studied. The results showed that the best targeting effects were obtained when the maleimide/peptide molar ratio was 3 for both TGN and QSH, namely T3Q3-NPs.[38] Then H102, a β-sheet breaker peptide, was encapsulated into T3Q3-NPs for AD treatment.[39] A higher uptake of H102 was observed in the mice hippocampi of the T3Q3-NPs/H102 group than those of nonmodified NPs (NP/H102) and TGN-modified NPs (T3-NPs/H102) 1 h after administration. The neuroprotective effects of T3Q3-NPs/H102 group were significantly improved compared with other groups at the same dosage. These results demonstrated that T3Q3-NPs could be a promising carrier for drug entry into the brain and targeting of the AD lesions, thus offering a precise therapy strategy for AD.[39]

2.2 Intranasal Drug Delivery System

Although brain-targeted i.v. drug delivery system makes drug transport across the BBB possible, it still remains an invasive administration, which is not practical for chronic use. The noninvasive intranasal (i.n.) delivery seems to be a promising alternative route for AD treatment. It is a method for bypassing the BBB and allowing direct drug deliver to the brain through olfactory and trigeminal neural pathways.[40] Encapsulating drugs into NPs could also protect drugs from enzymatic degradation and facilitate their transport across the mucosal barrier. Some investigations pertaining to CNS drug delivery through nasal administration of NPs for AD treatment have been reported in the past 20 years.

LIP loaded with H102 peptide were constructed for i.n. administration by Zheng et al[41] for the treatment of AD. Prepared LIP effectively penetrated the Calu-3 cell monolayers and were delivered into the brain. The area under the curve from 0 to 90 min ($AUC_{0\rightarrow 90\ min}$) of H102 in LIP was 1.67- to 2.92-fold greater in each region of the brain compared to H102 solution; no H102 was detected in the brain after i.v. administration. Moreover, H102 LIP excellently ameliorated the spatial memory impairment of AD model rats and showed no toxicity in the nasal mucosa.[41] Besides, galantamine hydrobromide (GH), a selective and reversible AChE inhibitor, was also encapsulated into LIP for i.n. AD treatment.[42] Compared with orally administrated GH, GH loaded in LIP achieved 3.52 and 3.36 times higher C_{max} and $AUC_{0\rightarrow 10\ min}$, respectively, and greatly enhanced the efficiency of AChE inhibition.[42]

To prolong the contact time between the formulation and nasal mucosa, decrease the mucociliary clearance, thus increase the amount of drug reaching the brain, mucoadhesive NPs, such as chitosan nanoparticles (CS-NPs), gained attention. CS-NPs are biodegradable, biocompatible, stable, present low toxicity, and can be prepared by simple and cheap preparation methods, such as ionic gelation.[43] CS-NPs significantly enhanced the uptake of rivastigmine, a cholinesterase inhibitor, into the brain via i.n. delivery. The brain concentration achieved from i.n. administration of CS-NPs (966 ± 20.66 ng/mL) was significantly higher than those achieved after i.v. administration of rivastigmine solution (387 ± 29.51 ng/mL), and i.n. administration of rivastigmine solution (508.66 ± 22.50 ng/mL).[44] CS-NPs also significantly improved estradiol transport into CNS with the drug targeting index (DTI) of nasal route of 3.2, which is beneficial for AD management.[45]

To enhance nasal adsorption of NPs, modifying them with biorecognitive ligands on the surface might be a promising approach. Lectins are proteins or glycoproteins, which specifically recognize sugar molecules on cell surface, thus initiating vesicular transport processes in cells.[46] Several lectins, such as wheat germ agglutinin (WGA), *Solanum tuberosum* lectin (STL), and odorranalectin (OL), have been conjugated to the NPs for AD treatment through i.n. pathway.[47–49] Gao et al.[50] developed WGA-conjugated PLA NPs encapsulated with vasoactive intestinal peptide (VIP). WGA specifically binds to *N*-acetylglucosamine and sialic acid, both of which are abundant on the nasal epithelial membrane.[49] Following i.n. administration, the concentrations of intact VIP were found to increase in the olfactory bulb, cerebrum, and cerebellum by 5.66- to 7.74-fold and 1.58- to 1.82-fold when compared with i.n. solution and unmodified NPs, respectively. The increase in VIP concentrations also corresponded to improved memory function, which was determined by the water maze behavioral test.[50] Another lectin, STL, similar to WGA, was conjugated to PLGA NPs (STL-NPs). The in vitro uptake significantly enhanced endocytosis of STL-NPs compared to NPs in Calu-3 cells.[47] After encapsulating basic fibroblast growth factor (bFGF) into NPs, the AUCs of STL-NPs in the olfactory bulb, cerebrum, and cerebellum were enhanced, and the neuroprotective effects on AD model rats were markedly improved compared with NPs.[51] Therefore, the selective affinity of lectins might be useful in the direct nose-to-brain drug delivery.

3 FUTURE PERSPECTIVES

The targeted delivery of drugs to the lesions in the brain in AD patients is expected to be one promising approach for future AD treatment. However, to make it available for clinical application, many more studies are still needed to confirm the efficiency and safety of the targeted delivery system. First, the mechanism of NPs transport across the BBB and delivery to the lesions should be clearly understood. The illustration of the mechanism is necessary for the optimization and the validation of the targeting effects, which are crucial for the construction of the delivery system. Second, researchers should dig deep into the safety and toxicity issues and consider the benefit–risk ratio. The delivery system should not be used in humans if it is toxic to any organ, tissue, or even cell. Besides, attention should be paid to the preparation processes and the manufacturing techniques. The production at an industrial scale and the reasonable cost should be realized. The reproducibility of batches of NPs needs to be refined, and the quality of formulations must meet clinical standards.

For the future development of targeted NPs in AD, several aspects, such as diagnosis, and vaccination could be further explored. NPs targeting the neuropathological hallmarks of AD can be of great potential in early diagnosis, having the ability to accurately reach brain lesions of AD. With encapsulated probes, NPs could clearly reflect the brain focal areas even at an early stage. Moreover, vaccination should be more effective before disease initiation or during early stages of disease progression.[52] Targeted NPs could also exert immunotherapeutic effects, as more aspects of immunopharmacology are understood.

Abbreviations

AD Alzheimer's disease
Aβ Amyloid-β
Aβ-mAb Aβ1–42 monoclonal antibody
AChE Acetylcholine esterase

AChR	Acetylcholine receptor
AMT	Adsorptive-mediated transcytosis
ApoE	Apolipoprotein E
AUC	Area under the curve
BACE1	β-Site amyloid precursor protein–cleaving enzyme 1
BBB	Blood–brain barrier
BCECs	Brain capillary endothelial cells
bFGF	Basic fibroblast growth factor
B6-NPs	B6-conjugated poly(ethyleneglycol)–poly(lactic acid) block copolymer nanoparticles
CNS	Central nervous system
CNVs	Control nanovehicles
CS	Chitosan
dd-NLs	Dually decorated nanoliposomes
DTI	Drug targeting index
GH	Galantamine hydrobromide
HDL	High-density lipoprotein
i.n.	Intranasal
i.v.	Intravenous
LDLR	Low-density lipoprotein receptors
Lf	Lactoferrin
Lf-PO	Lactoferrin-functionalized poly(ethyleneglycol)–poly(D,L-lactic-*co*-glycolic acid) polymersomes
LfR	Lactoferrin receptor
LIP	Liposomes
LRP	Low-density lipoprotein receptor–related protein
mAb	Monoclonal antibody
NL	Nanoliposomes
NOR	Novel-object recognition
NP	Nanoparticles
OL	Odorranalectin
OX26-PO	OX26-modified poly(ethyleneglycol)–poly(ε-caprolactone) polymersomes
OX26-NLs	OX26 nanoliposomes
PA	Phosphatidic acid
PBCA	Polybutylcyanoacrylate
PCL	Poly(ε-caprolactone)
PDMAEMA	Poly(2-(*N*,*N*-dimethylamino)ethyl methacrylate
PEG	Poly(ethyleneglycol)
PLA	Poly(lactic acid)
PLGA	Poly(lactic-*co*-glycolic acid)
PO	Polymersomes
RMT	Receptor-mediated transcytosis
rHDL	Reconstituted high-density lipoprotein
RVG	Rabies virus glycoprotein
SAMP8	Senescence-accelerated mouse, P8 strain
SNVs	Smart nanovehicles
STL	*Solanum tuberosum* lectin
SPR	Surface plasmon resonance
TAT	Transactivating transcriptor
Tf	Transferrin

TfR	Transferrin receptor
TGN-NPs	TGNYKALHPHNG-modified poly(ethyleneglycol)–poly(lactic-*co*-glycolic acid) nanoparticles
TMC	Trimethyl chitosan
VIP	Vasoactive intestinal peptide
WGA	Wheat germ agglutinin

References

1. Gregori M, Masserini M, Mancini S. Nanomedicine for the treatment of Alzheimer's disease. *Nanomedicine*. 2015;10(7):1203–1218.
2. Alzheimer's Association2015 Alzheimer's disease facts and figures. *Alzheimer Dementia*. 2015;11(3):332–384.
3. Persson T, Popescu BO, Cedazo-Minguez A. Oxidative stress in Alzheimer's disease: why did antioxidant therapy fail?. *Oxid Med Cell Longev*. 2014;2014:427318.
4. Yu JT, Tan L, Hardy J. Apolipoprotein E in Alzheimer's disease: an update. *Annu Rev Neurosci*. 2014;37:79–100.
5. Farrall AJ, Wardlaw JM. Blood-brain barrier: ageing and microvascular disease—systematic review and meta-analysis. *Neurobiol Aging*. 2009;30(3):337–352.
6. Ke W, Shao K, Huang R, Han L, Liu Y, Li J, et al. Gene delivery targeted to the brain using an Angiopep-conjugated polyethyleneglycol-modified polyamidoamine dendrimer. *Biomaterials*. 2009;30(36):6976–6985.
7. Gelperina SE, Khalansky AS, Skidan IN, Smirnova ZS, Bobruskin AI, Severin SE, et al. Toxicological studies of doxorubicin bound to polysorbate 80-coated poly (butyl cyanoacrylate) nanoparticles in healthy rats and rats with intracranial glioblastoma. *Toxicol Lett*. 2002;126(2):131–141.
8. Rocha S. Targeted drug delivery across the blood brain barrier in Alzheimer's disease. *Curr Pharm Design*. 2013;19(37):6635–6646.
9. Wilson B, Samanta MK, Santhi K, Kumar KP, Paramakrishnan N, Suresh B. Targeted delivery of tacrine into the brain with polysorbate 80-coated poly (*n*-butylcyanoacrylate) nanoparticles. *Eur J Pharm Biopharm*. 2008;70(1):75–84.
10. Jose S, Sowmya S, Cinu TA, Aleykutty NA, Thomas S, Souto EB. Surface modified PLGA nanoparticles for brain targeting of bacoside-A. *Eur J Pharm Sci*. 2014;63:29–35.
11. Singh HK, Dhawan BN. Effect of *Bacopa monniera* Linn. (brahmi) extract on avoidance responses in rat. *J Ethnopharm*. 1982;5(2):205–214.
12. Kreuter J, Hekmatara T, Dreis S, Vogel T, Gelperina S, Langer K. Covalent attachment of apolipoprotein A-I and apolipoprotein B-100 to albumin nanoparticles enables drug transport into the brain. *J Control Release*. 2007;118(1):54–58.
13. Michaelis K, Hoffmann MM, Dreis S, Herbert E, Alyautdin RN, Michaelis M, et al. Covalent linkage of apolipoprotein E to albumin nanoparticles strongly enhances drug transport into the brain. *J Pharm Exp Ther*. 2006;317(3):1246–1253.
14. Song Q, Huang M, Yao L, Wang X, Gu X, Chen J, et al. Lipoprotein-based nanoparticles rescue the memory loss of mice with Alzheimer's disease by accelerating the clearance of amyloid-beta. *ACS Nano*. 2014;8(3):2345–2359.
15. Chang J, Betbeder D. Targeting drug delivery to the brain via transferrin anchored nanoparticles. Nanomedicine and the Nervous System. Jersey, British Isles: Science Publishers; 2012:250–263.

16. Visser CC, Stevanovic S, Voorwinden LH, van Bloois L, Gaillard PJ, Danhof M, et al. Targeting liposomes with protein drugs to the blood-brain barrier in vitro. *Eur J Pharm Sci.* 2005;25(2–3):299–305.
17. Pang Z, Lu W, Gao H, Hu K, Chen J, Zhang C, et al. Preparation and brain delivery property of biodegradable polymersomes conjugated with OX26. *J Control Release.* 2008;128(2):120–127.
18. Lo SL, Wang S. An endosomolytic Tat peptide produced by incorporation of histidine and cysteine residues as a nonviral vector for DNA transfection. *Biomaterials.* 2008;29(15):2408–2414.
19. Nie Y, Schaffert D, Rodl W, Ogris M, Wagner E, Gunther M. Dual-targeted polyplexes: one step towards a synthetic virus for cancer gene therapy. *J Control Release.* 2011;152(1):127–131.
20. Liu Z, Gao X, Kang T, Jiang M, Miao D, Gu G, et al. B6 peptide-modified PEG-PLA nanoparticles for enhanced brain delivery of neuroprotective peptide. *Bioconjug Chem.* 2013;24(6):997–1007.
21. Nuijens JH, van Berkel PH, Schanbacher FL. Structure and biological actions of lactoferrin. *J Mamm Gland Biol Neoplasia.* 1996;1(3):285–295.
22. Suzuki YA, Lopez V, Lonnerdal B. Mammalian lactoferrin receptors: structure and function. *Cell Mol Life Sci.* 2005;62(22):2560–2575.
23. Qian ZM, Wang Q. Expression of iron transport proteins and excessive iron accumulation in the brain in neurodegenerative disorders. *Brain Res Brain Res Rev.* 1998;27(3):257–267.
24. Yu Y, Jiang X, Gong S, Feng L, Zhong Y, Pang Z. The proton permeability of self-assembled polymersomes and their neuroprotection by enhancing a neuroprotective peptide across the blood-brain barrier after modification with lactoferrin. *Nanoscale.* 2014;6(6):3250–3258.
25. Liu Y, Huang R, Han L, Ke W, Shao K, Ye L, et al. Brain-targeting gene delivery and cellular internalization mechanisms for modified rabies virus glycoprotein RVG29 nanoparticles. *Biomaterials.* 2009;30(25):4195–4202.
26. Alvarez-Erviti L, Seow Y, Yin H, Betts C, Lakhal S, Wood MJ. Delivery of siRNA to the mouse brain by systemic injection of targeted exosomes. *Nat Biotechnol.* 2011;29(4):341–345.
27. Li J, Feng L, Fan L, Zha Y, Guo L, Zhang Q, et al. Targeting the brain with PEG-PLGA nanoparticles modified with phage-displayed peptides. *Biomaterials.* 2011;32(21):4943–4950.
28. Qian Y, Zha Y, Feng B, Pang Z, Zhang B, Sun X, et al. PEGylated poly(2-(dimethylamino) ethyl methacrylate)/DNA polyplex micelles decorated with phage-displayed TGN peptide for brain-targeted gene delivery. *Biomaterials.* 2013;34(8):2117–2129.
29. Li J, Zhang C, Li J, Fan L, Jiang X, Chen J, et al. Brain delivery of NAP with PEG-PLGA nanoparticles modified with phage display peptides. *Pharm Res.* 2013;30(7):1813–1823.
30. Brooks H, Lebleu B, Vives E. Tat peptide-mediated cellular delivery: back to basics. *Adv Drug Deliv Rev.* 2005;57(4):559–577.
31. Sancini G, Gregori M, Salvati E, Cambianica I, Re F, Ornaghi F, et al. Functionalization with TAT-peptide enhances blood-brain barrier crossing in vitro of nanoliposomes carrying a curcumin-derivative to bind amyloid-β peptide. *J Nanomed Nanotechnol.* 2013;4:2013.
32. Wang ZH, Wang ZY, Sun CS, Wang CY, Jiang TY, Wang SL. Trimethylated chitosan-conjugated PLGA nanoparticles for the delivery of drugs to the brain. *Biomaterials.* 2010;31(5):908–915.
33. Poduslo JF, Ramakrishnan M, Holasek SS, Ramirez-Alvarado M, Kandimalla KK, Gilles EJ, et al. In vivo targeting of antibody fragments to the nervous system for Alzheimer's disease immunotherapy and molecular imaging of amyloid plaques. *J Neurochem.* 2007;102(2):420–433.
34. Agyare EK, Curran GL, Ramakrishnan M, Yu CC, Poduslo JF, Kandimalla KK. Development of a smart nano-vehicle to target cerebrovascular amyloid deposits and brain parenchymal plaques observed in Alzheimer's disease and cerebral amyloid angiopathy. *Pharm Res.* 2008;25(11):2674–2684.
35. Markoutsa E, Papadia K, Clemente C, Flores O, Antimisiaris SG. Anti-Abeta-MAb and dually decorated nanoliposomes: effect of Abeta1-42 peptides on interaction with hCMEC/D3 cells. *Eur J Pharm.* 2012;81(1):49–56.

36. Bana L, Minniti S, Salvati E, Sesana S, Zambelli V, Cagnotto A, et al. Liposomes bi-functionalized with phosphatidic acid and an ApoE-derived peptide affect Abeta aggregation features and cross the blood-brain-barrier: implications for therapy of Alzheimer disease. *Nanomedicine.* 2014;10(7):1583–1590.

37. Balducci C, Mancini S, Minniti S, La Vitola P, Zotti M, Sancini G, et al. Multifunctional liposomes reduce brain beta-amyloid burden and ameliorate memory impairment in Alzheimer's disease mouse models. *J Neurosci.* 2014;34(42):14022–14031.

38. Zhang C, Wan X, Zheng X, Shao X, Liu Q, Zhang Q, et al. Dual-functional nanoparticles targeting amyloid plaques in the brains of Alzheimer's disease mice. *Biomaterials.* 2014;35(1):456–465.

39. Zhang C, Zheng X, Wan X, Shao X, Liu Q, Zhang Z, et al. The potential use of H102 peptide-loaded dual-functional nanoparticles in the treatment of Alzheimer's disease. *J Control Release.* 2014;192:317–324.

40. Sood S, Jain K, Gowthamarajan K. Intranasal therapeutic strategies for management of Alzheimer's disease. *J Drug Target.* 2014;22(4):79–94.

41. Zheng X, Shao X, Zhang C, Tan Y, Liu Q, Wan X, et al. Intranasal H102 peptide-loaded liposomes for brain delivery to treat Alzheimer's disease. *Pharm Res.* 2015;32(12):3837–3849.

42. Li W, Zhou Y, Zhao N, Hao B, Wang X, Kong P. Pharmacokinetic behavior and efficiency of acetylcholinesterase inhibition in rat brain after intranasal administration of galantamine hydrobromide loaded flexible liposomes. *Environ Toxicol Pharmacol.* 2012;34(2):272–279.

43. Hanafy AS, Farid RM, ElGamal SS. Complexation as an approach to entrap cationic drugs into cationic nanoparticles administered intranasally for Alzheimer's disease management: preparation and detection in rat brain. *Drug Dev Ind Pharm.* 2015;41(12):2055–2068.

44. Fazil M, Md S, Haque S, Kumar M, Baboota S, Sahni JK, et al. Development and evaluation of rivastigmine loaded chitosan nanoparticles for brain targeting. *Eur J Pharm Sci.* 2012;47(1):6–15.

45. Wang X, Chi N, Tang X. Preparation of estradiol chitosan nanoparticles for improving nasal absorption and brain targeting. *Eur J Pharm Biopharm.* 2008;70(3):735–740.

46. Liu Z, Jiang M, Kang T, Miao D, Gu G, Song Q, et al. Lactoferrin-modified PEG-co-PCL nanoparticles for enhanced brain delivery of NAP peptide following intranasal administration. *Biomaterials.* 2013;34(15):3870–3881.

47. Chen J, Zhang C, Liu Q, Shao X, Feng C, Shen Y, et al. Solanum tuberosum lectin-conjugated PLGA nanoparticles for nose-to-brain delivery: in vivo and in vitro evaluations. *J Drug Target.* 2012;20(2):174–184.

48. Wen Z, Yan Z, Hu K, Pang Z, Cheng X, Guo L, et al. Odorranalectin-conjugated nanoparticles: preparation, brain delivery and pharmacodynamic study on Parkinson's disease following intranasal administration. *J Control Release.* 2011;151(2):131–138.

49. Gao X, Tao W, Lu W, Zhang Q, Zhang Y, Jiang X, et al. Lectin-conjugated PEG-PLA nanoparticles: preparation and brain delivery after intranasal administration. *Biomaterials.* 2006;27(18):3482–3490.

50. Gao X, Wu B, Zhang Q, Chen J, Zhu J, Zhang W, et al. Brain delivery of vasoactive intestinal peptide enhanced with the nanoparticles conjugated with wheat germ agglutinin following intranasal administration. *J Control Release.* 2007;121(3):156–167.

51. Zhang C, Chen J, Feng C, Shao X, Liu Q, Zhang Q, et al. Intranasal nanoparticles of basic fibroblast growth factor for brain delivery to treat Alzheimer's disease. *Int J Pharm.* 2014;461(1–2):192–202.

52. Marciani DJ. Alzheimer's disease vaccine development: a new strategy focusing on immune modulation. *J Neuroimmunol.* 2015;287:54–63.

CHAPTER 13

Parkinson's Disease

Banu Cahide Tel, PhD, Gul Yalçin Çakmakli, MD
Hacettepe University, Ankara, Turkey

CONTENTS

1 INTRODUCTION

Parkinson's disease (PD) is the second most common neurodegenerative disease, after Alzheimer's disease, described as a neurological syndrome by James Parkinson in 1817. It is a slow progressive neurodegenerative disease with a prevalence of 1%, in people over 60 years of age.[1,2] The estimated number of individuals over age 50 with PD is 5.4 million in the world, and the burden of the disease shifts from western to eastern countries.[3] The disease is mainly characterized by the well-known motor features (e.g., bradykinesia, rigidity, tremor, and postural instability), but the nonmotor symptoms, including depression, hallucinations, sleep disorders, autonomic dysfunction, and, more importantly, dementia, are also prominent components of the disease, and may be as much disabling as motor symptoms. Although the pathogenesis of PD is not exactly known, pathological intracellular aggregation of alpha-synuclein plays a key role in the neurodegenerative process. Motor findings are mainly related to nigral dopaminergic neuronal loss associated with Lewy bodies and neuritis, which are composed of fibrillar alpha-synuclein aggregates in the remaining neurons. The cause of dopaminergic cell death remains unknown, but it is well accepted that environmental toxins and/or genetic predisposition may play a role in the etiology. Recently, considerable advances have been made to understand the complex interaction between genetics and environment. Most obvious cases of idiopathic PD are not genetic in origin, but genetic susceptibility may render individuals sensitive to environmental factors. Currently, PD treatment is focused on symptomatic relief, as the pathogenesis of the illness is not completely understood and there is still no effective neuroprotective treatment. A variety of pharmacological and nonpharmacological approaches are employed for the symptomatic treatment of PD. Nonpharmacological interventions include surgical interventions, such as thalamotomy, pallidotomy, and electrical stimulation of subthalamic nucleus (STN) or globus pallidus pars interna (GPi).[4] Fetal cell transplantation, intraventricular administration

Nanotechnology Methods for Neurological Diseases and Brain Tumors. http://dx.doi.org/10.1016/B978-0-12-803796-6.00013-7

of neurotrophic factors, the use of viral vectors, and stem cell implantation are currently being explored.[5–10] However, the majority of patients with PD are treated pharmacologically, which is based on dopamine replacement therapy, but over time the treatment leads to complications (such as motor fluctuations and dyskinesia), behavioral and psychiatric problems (such as hallucinations, psychosis, and impulse control disorders), and dopamine dysregulation syndrome. Furthermore, due to the progressive nature of the disease, both motor and nonmotor symptoms become resistant to treatment with time, and as a result, freezing of gait, postural instability, falls, dysphagia, constipation, and dementia are reported in more than half of the patients in advanced stages.[11] The chronic and progressive nature of the disease deteriorates the life quality of the patients, and puts a huge burden on their families, caregivers, healthcare system, and society. Despite intense efforts for discovering neuroprotective or neurorestorative therapies for PD, currently, we don't have any such effective drug in hand. However, it is crucial to find new symptomatic treatments without side effects for improving patient's quality of life and ultimately reducing the costs for the society.

2 CLINICAL FEATURES

PD is defined by a triad of symptoms comprising slowing down of voluntary movements (bradykinesia), muscular rigidity, and "pill-rolling" tremor. These cardinal symptoms are accompanied by postural and gait disturbances in the further stages of the disease. Bradykinesia signifies slowness and a reduction in the amplitude of movement, and is often accompanied by a lack of spontaneous movement (hypokinesia). Increase in passive muscle tone and resistance to passive joint displacement lead to rigidity, and it is usually associated with a cogwheel phenomenon. Postural disturbances are characterized by trunk and knee flexion, with a reduction in arm swing and stride length, accompanied by a festinating gait. Other features of PD include hypophonia, reduced facial expression, and reduced blinking. Nonmotor symptoms of PD include olfactory dysfunction, cognitive deficits leading to dementia, depression, and autonomic deficits (such as constipation, urinary incontinence, sweating, orthostatic hypotension, and sexual dysfunction). Daytime sleepiness, REM sleep behavior disorder, pain, and fatigue are also common nonmotor features. These may present even a decade before motor symptoms show up.[12] The neurodegenerative process leading to PD starts at least 20 years before the disease onset with motor symptoms, when 70% of the dopaminergic neurons in the substantia nigra pars compacta (SNpc) and 50% of striatal dopaminergic terminals are already lost. Therefore, the prodromal period is critical for both early diagnosis and as a target period for administration of possible disease-modifying therapies. Thus, development of neuroprotective and neuropreventive therapies is crucial to stop the progression of the disease before motor symptoms emerge.

It is commonly accepted that there are two subtypes of idiopathic PD, depending on the clinical observations and prognosis: tremor-dominant and akinetic-rigid (nontremor-dominant) forms. Tremor-dominant PD is often characterized by slow progression and less functional disability, while akinetic-rigid form presents with higher motor and cognitive deficits and progresses more rapidly.[13,14]

As the disease progresses, the motor symptoms worsen over the years. Currently, there is only symptomatic dopaminergic therapy available for the relief of motor and nonmotor symptoms. However, in the later stage of the illness, complications of the treatment emerge, such as motor and nonmotor fluctuations, dyskinesia, and psychosis. At the same time, as the disease progresses, both motor and nonmotor symptoms become resistant to treatment and as a result, freezing of gait, postural instability, falls, dysphagia, constipation, and dementia were reported on more than half of the patients[11]; therefore new therapy approaches are urgently needed.

3 ETIOLOGY AND PATHOGENESIS

The primary pathology of idiopathic PD involves a progressive degeneration of SNpc dopaminergic neurons, terminating in the caudate putamen (corpus striatum). This loss is predominantly in the ventrolateral part of SNpc, which is a region projecting mainly to the dorsal putamen. The loss of dopaminergic innervation is more marked in the putamen than the caudate, with the posterior putamen most severely affected. The deafferentation predominantly affects the sensorimotor striatum, which gives rise to the motor symptoms described previously. Lewy bodies and Lewy neurites accompany degenerative changes in all affected areas; these are eosinophilic globular perikaryal inclusions and thread-like inclusions in cellular processes, respectively. Lewy bodies or Lewy neurites consist of abnormally folded presynaptic protein, namely alpha-synuclein.[15] However, the degenerative process also affects other dopaminergic cell groups and nondopaminergic nuclei in the midbrain, such as the locus coeruleus, the dorsal raphe nucleus, the nucleus basalis of Meynert, and the pedunculopontine nucleus (PPN), as well as cerebral cortex and spinal cord or even the peripheral nervous system. Thus, it is hypothesized that the alpha-synuclein pathology first appears in the lower brainstem at the early stages of the disease during prodromal or asymptomatic phase, then gradually spreads upward until the cortex in the advanced stages of the disease.[16] The role of Lewy bodies and Lewy neurites in the neurodegenerative process remain unknown, but pathological intracellular aggregation of alpha-synuclein could be the reason of cell death or a compensatory mechanism to protect the neurons.

Alpha-synuclein is an abundant presynaptic protein expressed mainly in the neurons and to a lesser extent in the glia of the central, peripheral, and enteric nervous systems.[17] Alpha-synuclein is a 16-kDa small acidic protein and does not appear to be a physiologically essential protein in knockout mice.[18] However, alpha-synuclein is involved in various synaptic functions, such as vesicular trafficking, synaptic behavior, signal transduction, the regulation of oxidative stress, mitochondrial function, protein phosphorylation, or even fatty acid binding.[19–21] It has an important role in neurotransmitter release,[22] especially for dopamine.[23] There is also evidence that alpha-synuclein is involved in dopamine biosynthesis[24–26] and uptake.[27,28] Understanding the function of alpha-synuclein is essential to be able to understand neurodegenerative process in PD.

Recent studies showed that alpha-synuclein assists SNARE complex assembly as a chaperone[29] during synaptic transmission or the exocytosis, and maintains the vesicle recycling and the size of the presynaptic vesicular pool.[30] Alpha-synuclein is a highly mobile protein that rapidly disperses from the presynaptic terminal following neuronal stimulation; thus provides free and unobstructed vesicular trafficking for an efficient neurotransmitter release.[23] At the end of the synaptic neurotransmission, alpha-synuclein quickly returns to the presynaptic terminal. This is possibly to hinder the synaptic vesicle trafficking and presynaptic membrane docking, therefore inhibiting chemical transmission.[23]

The alpha-synuclein protein is structurally an unfolded, monomeric protein, which has three main domains, namely the N-terminal amphipathic, the central nonamyloid component (NAC), and the acidic C-terminal domains.[31,32] The N-terminal of the protein takes an alpha-helical form in the presence of lipid membranes or membranes with high curvature, where alpha-synuclein can interact with the membrane.[31,33–37] The helices have high affinity to bind the lipids and stabilize the membranes.[31] As a result of this interaction, unfolded alpha-synuclein rapidly changes to form a folded alpha-helical secondary structure, which is prone to forming different types of oligomers or multimers that are thought to cause synaptic dysfunction or neuronal death.[38] With self-assembly, alpha-synuclein can be found in many different forms, including monomeric, oligomeric, and fibrillar species. Fibrils are the result of further aggregation of oligomers, which is referred to as protofibrils. They are highly ordered oligomeric structures and rich in beta-sheets. In the end, fibrils form an insoluble and protease-resistant aggregate.

Alpha-synuclein, in the fibrillar form, is the main component of Lewy bodies. Recent studies showed that fibrillar forms of alpha-synuclein are the major toxic strains.[39] Therefore, abnormally folded alpha-synuclein protein can clog the presynaptic terminal and cause synaptic dysfunction before neuronal death. Genetic mutations are also shown to cause fibrillization of alpha-synuclein.[40–42]

3.1 Prion-Like Propagation

The progressive nature of the pathology suggests that misfolded alpha-synuclein could spread via neural circuits. The first evidence came from PD patients that had been transplanted with fetal mesencephalic neurons in the striatum. Lewy body–like inclusions were observed in young grafted neurons in these patients[43]; thus suggesting that misfolded or aggregated alpha-synuclein from the diseased brain had infiltrated into the fetal grafts due to exposure to the diseased neurons nearby. This observation of transmission was repeated in vitro and in a transgenic mouse model as well.[44] Additionally, synthetic alpha-synuclein fibril inoculation into nontransgenic mice led to development of Lewy body pathology in interconnected anatomical regions.[45] Similar results were found when the brain lysate from symptomatic transgenic mice expressing mutant alpha-synuclein was injected into young asymptomatic mice.[46] The prion-like propagation theory was proposed to explain cell-to-cell spread of misfolded alpha-synuclein, which serves as a template to initiate abnormal folding and accumulation of normal proteins in healthy neurons. However, there are still certain unknown aspects in this theory; continuing research in this area may reveal them soon.

3.2 Risk Factors

The cause of dopamine cell death remains unknown, but exposure to environmental toxins, genetic predisposition, or oxidative stress may play a role in its etiology. Age is a risk factor.

Genetic predisposition in PD is based on concordance studies, risk ratios associated with affected relatives, and rare familial case of PD.[47–50] Approximately 5%–10% of PD patients have a hereditary form of parkinsonism, and less than 10% of these can be explained by classical Mendelian autosomal recessive or dominant inheritance. Actually PD was considered as a nongenetic disorder until 1997, but in the past 20 years, 15 different gene mutations have been shown to be responsible for PD.[51] Furthermore, genomewide association studies (GWAS) showed that, other than the variants of these well-known PD genes, there are other risk/susceptibility genes, which may play role in the pathogenesis of sporadic PD cases.[1,52–55] Leucine-rich repeat kinase 2 (LRRK2 and PARK8), alpha-synuclein (SNCA), parkin (PARK 2), PINK1 (PARK 6), and DJ-1 (PARK 7) mutations have been studied more extensively among the others.[15,56–69] The findings of GWAS studies, based on the functions of the proteins of these gene products, also provide insights about the cellular pathways, which play a role in dopaminergic cell death. Thus understanding the genetic basis of PD can be helpful for understanding the pathogenetic mechanisms underlying sporadic PD and for developing individualized therapies according to the genetic workup of the patient. The protein products of PD genes defined until now are implicated in interconnected cellular pathways, namely, synaptic transmission (endocytosis and exocytosis), mitochondrial quality control and oxidative metabolism, lysosome-mediated autophagy, calcium homeostasis, and neuroinflammation.[70] The sporadic PD and genetic forms converge to these similar molecular pathways, so these cascades should be targeted for developing neuroprotective therapies.

3.3 Environmental Hypothesis in Parkinson's Disease

Recent epidemiological and experimental studies have linked increased prevalence of PD to industrialization and pesticide/herbicide exposure in rural environments.[71–78] Metals (transition metals, such as iron and copper), organic solvents (*n*-hexane), carbon monoxide, and manganese have been proposed as neurotoxins capable of causing PD.[79–83] At the cellular level, neuronal loss probably occurs by apoptosis and may involve oxidative stress, excitotoxicity, nitric oxide toxicity, inflammation, mitochondrial dysfunction, and altered proteolysis.[1,5,75,84–93] Although the triggering factors for apoptosis are not still clear, each of the pathogenic mechanisms listed previously may initiate the process of programmed cell death. On the other hand, many studies show a robust negative association for cigarette smoking with PD, which may relate to protective or therapeutic effects of nicotine.[13,94,95] Like smoking, a number of environmental factors have been implicated with the potential to reduce the risk for developing PD, for example, diet, caffeine, alcohol consumption, or use of nonsteroidal antiinflammatory drugs; although these findings appear to be inconclusive.[78,96–100] To date no single causative or protective factor has been elucidated.

Systemic mitochondrial complex I deficiency has long been considered to have an important role in the pathogenesis of idiopathic PD,[101] as it leads to additional oxidative stress in dopaminergic neurons in SNpc.[102] The cause of the decreased complex I activity in PD remains unclear. Systemic administration of MPTP and rotenone (herbicide) can induce nigral cell death with selective complex I inhibition.[77,103] Toxicity due to accumulated exogenous or endogenous compounds may give rise to complex I deficiency as a secondary event. The mutations in mtDNA may contribute the development of PD. Hybrid cell lines expressing mtDNA from PD patients are deficient in complex I activity, suggesting that mutated mtDNA may be responsible for the enzymatic defect.[104] In addition, in some familial cases it has been shown that complex I dysfunction is transmitted by descent through the female line.[105] The cause of PD is still unknown, but multiple factors, for example, proteosomal dysfunction or mitochondrial dysfunction, may contribute to cell death.

The primary biochemical change in PD is the loss of striatal dopamine. This loss also occurs in extrastriatal regions [including the nucleus accumbens, globus pallidus pars externa (GPe), GPi, substantia nigra pars reticulata (SNr)],

STN, and the motor cortex.[106] Following the loss of striatal dopamine, compensatory biochemical changes occur in the nonstriatal regions, such as increase in striatal γ-aminobutyric acid (GABA) levels, particularly in regions most severely depleted of dopamine. Compensatory increases in dopaminergic innervation of the GPi are present in early PD, but are lost as the disease progresses.[106] Other compensatory mechanisms include upregulated glutamatergic input to the SNc from the STN.[107,108] However, symptoms of PD do not present themselves until striatal dopamine deficits reach the 70% level, implying that the plasticity of the nigrostriatal dopamine system enables adaptation, which helps maintain the function despite extensive neuronal loss.[109,110]

4 NEUROANATOMY OF THE BASAL GANGLIA

The basal ganglia, a group of subcortical nuclei, receive cortical inputs, process, and send them back to the cortex via interconnections through the midbrain and thalamus. Basal ganglia play a major role in the execution of learned movements and responses to new stimuli[111–113] and psychomotor behavior, including procedural learning.[114,115] It consists of the striatum, globus pallidus, substantia nigra, and STN. The striatum is the main receptive nucleus of the basal ganglia and comprises the caudate nucleus and putamen, and receives inputs predominantly from the cerebral cortex, specific thalamic nuclei, and the SNc. The main output nuclei of the basal ganglia are the GPi and the SNr, two cytologically similar structures separated by the fibers of the internal capsula in a manner similar to the separation of the caudate nucleus and putamen. The GPi and SNr project to restricted areas of the cortex via thalamus.[116–118] The striatum receives robust glutamatergic inputs from motor cortices to the striatum.[119–122] The basal ganglia comprises parallel and largely closed circuits that play a role in processing of sensorimotor, oculomotor, prefrontal, and limbic inputs separately and independently.[116,123]

The striatum may be divided into two functional compartments, namely matrix and striosomes.[119] The matrix or matrisomes compartment receives the striatal inputs most directly related to sensorimotor, motor, and associative cortices.[114,124–126] By contrast, striosomes or the striatal bodies tend to receive inputs from the limbic system.

The dopaminergic innervation of the basal ganglia is an essential and central phenomenon for the execution of a wide variety of motor, cognitive, and emotional functions.[127,128] The SNc neurons that project to the sensorimotor striatum are calbindin negative and express high levels of the dopamine transporter. The associative, motor, and limbic territories of the striatum all receive afferents from the SNc. The nigrostriatal pathway is composed of several neuronal subsystems, each provided with widely distributed axonal arborization that allows a versatile influence on the striatum.[129] The dorsal tier of the

SNc principally projects to the matrix of the striatum, whereas the ventral tier mainly targets the striosomes,[130–132] although some neurons arising from both areas arborize and innervate both the matrix and striosomes.[129]

The striatum integrates cortical, thalamic, and nigral inputs and projects via "direct" or "indirect" pathways to the main output nuclei of the basal ganglia, the GPi and SNr. There is also the "hyperdirect" pathway, the third main circuit of the basal ganglia, which involves the STN. It receives input directly from the cortex and sends excitatory outputs to the GPi and SNr.[133] The striatal output nuclei tonically inhibit ventral tier thalamic nuclei, while the STN exerts strong excitatory effects on the output nuclei. Therefore, the net thalamic output sends excitatory projections to the cortex.[133]

Striatal spiny projection neurons (SPN) utilize GABA as the main neurotransmitter, and are traditionally divided into two distinct populations. Neurons of the direct pathway coexpress substance P and dynorphin, and project directly to the GPi and SNr. The indirect pathway utilizes enkephalin as a cotransmitter and influences GPi/SNr activity by synapsing in the GPe with GABAergic projections to the STN, which, in turn, sends a glutamatergic input to the GPi/SNr.[134]

In the execution of voluntary movement, at first the cortico-STN-pallidal "hyperdirect" pathway inhibits large areas of thalamus and cortex that are related to both selected and other competing motor programs, by activating GPi/SNr neurons. Then, at the same time the "direct" pathway sends an inhibitory signal to the GPi/SNr, which disinhibits their targets and releases only the selected motor program. Finally, via the "indirect" pathway an excitatory input is transmitted to the GPi/SNr, thereby suppressing their targets extensively by sending inhibitory projections to the thalamus for fine tuning of the selected program.[135]

Enkephalin, released from the indirect pathway neurons, attenuates GABA release in the GPe. Dynorphin preferentially binds κ-opioid receptors, located presynaptically on the terminals of D-1–mediated direct pathway nigrostriatal neurons, directly inhibiting dopamine release in the striatum presynaptically at the terminals of glutamatergic projections from the STN. Dynorphin, released from striatofugal neurons terminating in the GPi, directly inhibits glutamate release from subthalamopallidal neurons.[136] Neurons of the direct pathway utilize substance P as a cotransmitter also. In the striatum, substance P, released from SPN collaterals, stimulates the release of acetylcholine and enhances dopamine and enkephalin release directly or via its actions on cholinergic interneurons.

The two striatofugal populations are further differentiated by their receptor expression patterns. Dopamine differentially modulates the response of direct and indirect pathway neurons to glutamate, facilitating transmission in the direct pathway by its actions on D-1 receptors and inhibiting indirect pathway transmission by stimulation of D-2 receptors.[137]

4.1 Dopamine Receptors in Striatum

Dopamine exerts its effects through G-protein–coupled, seven transmembrane domain receptors. There are five dopamine receptors characterized and these are classified into two families, D-1 like (D-1 and D-5) and D-2–like (D-2, D-3, and D-4) based on sequence homology and pharmacology. D-1–like receptors couple with stimulatory G-proteins ($G_{s\alpha}/G_{olf\alpha}$) that activate adenylate cyclase (AC), but D-2–like receptors inhibit AC via ($G_{i\alpha}/G_{o\alpha}$) signaling.[138] AC promotes 3′,5′-cyclic adenosine monophosphate (cAMP) formation, and cAMP activates catalytic subunits of cAMP-dependent protein kinase (PKA) to amplify the signaling cascade.[139,140] There is also evidence for D-l–like receptors, which stimulate phospholipase C, thus activating phosphatidylinositol cascade and mobilizing calcium from intracellular stores.[141,142]

The D-1 receptor is expressed more widely and at higher levels than other dopamine receptor subtypes.[137] In the striatum, D-1 receptors are located postsynaptically to asymmetric synapses on the dendritic spines of GABAergic SPNs projecting to the GPi/SNr.[143,144] D-1 receptors are also localized in the GPi/SNr presynaptically, at the terminals of GABAergic striatal projection neurons.[145,146]

Striatal D-2 receptors are located predominantly on the spineheads and dendrites of SPNs projecting to the GPe and also, presynaptically, on the terminal of these neurons in the GPe.[145,147] Presynaptic D-2 receptors in the striatum are located on the terminals of nigrostriatal dopaminergic neurons and on glutamatergic striatal afferents, where they inhibit neurotransmitter release.[147] Presynaptic D-2 receptors are also located on the somata of dopaminergic neurons in the striatum. The D-2S and D-2L splice variants are thought to mediate the pre- and postsynaptic effects of D-2 receptors, respectively.[148] Although dopamine D-1 and D-2 receptors are predominantly segregated on populations of striatal direct and indirect pathway neurons, respectively,[145,149,150] there is increasing evidence for the colocalization of both receptor subtypes.[151–153]

Adenosine is also a neuromodulator that modifies the effects produced by neurotransmitters or other neuromodulators of the striatum. Adenosine acts on A-2a receptors, which are G-protein–coupled receptors, and activates AC by receptor stimulation.[154] In the striatum, A-2a receptors are restricted to indirect pathway neurons, and activation of A-2a receptors causes a decrease in dopamine D-2 receptor–mediated neurotransmission.[155–158] The coexistence of D-2/A-2a receptors in the indirect pathway neurons working in an antagonistic direction could provide a new target for therapeutic intervention of PD.

D-3 receptors have a specific distribution in the ventral striatum especially in the nucleus accumbens, where they are expressed in high levels. In contrast, it is poorly expressed in the dorsal striatum. In primates, most D-3 mRNA–expressing SPNs also express D-2 mRNA[145,159,160] The extensive colocalization of D-2 and D-3 receptors could be functional and exert local control over the efferents to the GPi/SNr and therefore influence locomotor activity.[161]

Low levels of D-4 and D-5 receptor mRNA have been found in striatum, although these receptors appear to be highly expressed outside the basal ganglia, that is, hippocampus and frontal cortex.[162] The functions of D-4 and D-5 in the basal ganglia are not clear yet, but it is unlikely that they contribute to movement control in PD.

4.2 Neurochemical Organization of the Striatum

The striatum is neurochemically and histologically heterogeneous. The striatum displays an area rich in acetylcholine esterase (AChE) staining (striosomes), which is embedded in an AChE-poor matrix. Almost all neurotransmitters and neuromodulators in the striatum are differentially expressed in these two compartments.[163] The striosome and matrix compartments not only receive different cortical afferents, but are also distinguishable by their efferent projections. The striatum is cellularly heterogeneous, containing both projection neurons and interneurons. The predominant cell type within the striatum is the SPN, constituting 90%–95% of striatal neurons.[164] The axons of these neurons emit collaterals that arborize either across the striatum or, more commonly, within the dendritic field. The SPNs all express GABA, but are traditionally divided into two populations according to cotransmitter and receptor expression patterns and target sites. Direct pathway SPNs coexpress D-1 dopamine receptor, substance P, and dynorphin and convey signals directly to the GPi and SNr. However, indirect pathway neurons express D-2 dopamine receptor and enkephalin. This segregated view of SPN populations is challenged by evidence of extensive collateralization by both populations of neurons in the GPe, GPi, and SNr[165–167] and dopamine receptor colocalization.[153] The SPN is the principal target for a majority of striatal afferents, and it provides the predominant striatal output pathway, thereby playing a key role in striatal function.

The interaction between the dopaminergic and glutamatergic striatal inputs is functionally significant.[168] Dopaminergic and glutamatergic cortical inputs converge on the same population of SPNs, and nigrostriatal neurons synapse on the neck of dendritic spines that receive cortical glutamatergic inputs.[169]Activation of dopamine receptors on the spinenecks of SPNs activates a kinase–phosphorylase cascade that affects the activation state of the adjacent glutamate receptors, and therefore the synaptic strength of cortico- and thalamostriatal inputs.[168,170] Dopamine therefore plays an important role in modulating the glutamatergic influence on these cells. Furthermore, dopamine can inhibit the release of striatal glutamate by its action on presynaptic dopamine D–2 receptors on corticostriatal afferents.[169,171]

5 TREATMENT OF PARKINSON'S DISEASE

Currently, the only treatment available for PD is symptomatic therapy. Development of neuroprotective and/or neurorestorative treatments is required to stop the neurodegeneration process. The underlying causes of PD are complex and possibly involve more than one cellular process; therefore a combination of new targets and therapeutic strategies may be needed to lead the radical treatment. Possible novel pharmacological targets for PD include alpha-synuclein accumulation, aggregation, and prion-like propagation, as well as neuromitochondrial dysfunction and oxidative stress, neuroinflammation, calcium channel activity, and LRRK2 activity.[172,173]

Surgical treatments include deep brain stimulation,[174] gene therapy,[8,175] and cell transplantation.[9,10] The symptomatic dopamine-mimetic therapies, namely levodopa (L-DOPA; L-3,4-dihydroxyphenylalanine) and the dopamine agonists, are the most commonly used antiparkinsonian treatments. The other medical therapies include monoamine oxidase-B (MAO-B) and catecholamine-O-methyl transferase (COMT) inhibitors, which enhance the dopaminergic activity. The nondopaminergic treatments include amantadine and cholinergic antagonists, among others.

5.1 Levodopa

The dopamine precursor levodopa has dominated the pharmacological treatment of PD for over 30 years. L-DOPA is metabolized in the periphery by decarboxylation to dopamine by L-aromatic amino acid decarboxylase (dopa decarboxylase), and consequently is used in conjunction with a peripheral decarboxylase inhibitor, carbidopa, or benserazide. This reduces the decarboxylation of L-DOPA, thus minimizing peripheral side effects of L-DOPA (e.g., nausea, emesis, and postural hypotension) and increasing L-DOPA bioavailability to the brain. The antiparkinsonian efficacy of L-DOPA is excellent at all stages of the disease, especially with regard to bradykinesia and rigidity. Although L-DOPA is still the gold standard treatment for PD, a number of problems limit its long-term utility. First, 4–6 years after the treatment, beneficial effects of L-DOPA are compromised by the appearance of dyskinesia and motor fluctuations, affecting approximately 40% of patients.[176–186] The earliest sign of decreased efficacy is the "wearing off" phenomenon in which each dose of L-DOPA lasts for progressively shorter periods. The "on–off" phenomenon, which is characterized by sudden unpredictable shifts between mobile, drug-responsive "on," and immobile "off" periods, is less commonly seen. The primary reasons for these fluctuations are related to the peripheral pharmacokinetics of L-DOPA, which has a short plasma half-life (0.75–1.5 h). Fluctuations in plasma L-DOPA levels are reflected in the synaptic cleft. L-DOPA therefore provides pulsatile stimulation of striatal dopamine receptors.

Second, many patients with severe fluctuations also develop cognitive, psychiatric, and emotional problems. The most common manifestations during the "off" state are anxiety, panic attacks, and depression, with hallucinations occurring during the "on" periods. The pathogenesis of drug-induced psychiatric complications in PD is not clearly understood, but it may be related to the indiscriminate stimulation of dopamine receptors throughout the central nervous system that is mesolimbic–mesocortical dopaminergic and serotonergic system.[13]

Finally, progression of the disease results in declined levels of DOPA decarboxylase. The decarboxylation of L-DOPA to dopamine and the metabolism of dopamine by MAO-B lead to formation of free radicals, which have a role in the cell death mechanisms in PD,[1,87] suggesting L-DOPA may exacerbate the degeneration of nigrostriatal neurons, although this remains controversial.[187]

5.2 Dopaminergic Agonists

The clinical deficiencies of L-DOPA led to the development of selective dopamine receptor agonist drugs that directly stimulate postsynaptic striatal dopamine receptors, restoring dopaminergic transmission and motor function. Unfortunately in clinical trials, all dopamine agonists have been found to be less effective than L-DOPA.[183]

Bromocriptine is a lysergic acid amide peptide containing an ergoline nucleus, and predominantly stimulates D-2/D-3 dopamine receptors, but it is also a partial agonist of dopamine D-1 receptors.[188] In clinical trials, long-term monotherapy with bromocriptine lowers the frequency of motor fluctuations and dyskinesia compared with L-DOPA. This is due to the duration of antiparkinsonian activity of bromocriptine, which is longer than that of L-DOPA (approximately 5 h compared to 1.5–2.0 h, respectively).[183,188] The other commonly used dopamine agonists are the D-1/D-2 agonists cabergoline and pergolide, which are also semisynthetic ergoline derivatives having the longest duration of action, 80–110 h and 20 h, respectively.[189,190] Both drugs also induce less dyskinesia than L-DOPA when used as a monotherapy in the treatment of PD.[191] Importantly, these studies show that drugs that have a short pulsatile effect on dopamine receptors are more likely to induce dyskinesia than compounds that produce prolonged or continuous receptor stimulation. Thus the pharmacokinetic profile of compounds may be more important for dyskinesia induction than their receptor selectivity. However, the use of ergoline dopamine agonist was associated with a considerable risk of developing cardiovalvular, pleuropulmonary, and retroperitoneal fibrosis,[189,192] hence limiting the use of this drug.

Ropinirole, pramipexole, and rotigotine are nonergoline dopamine receptor agonists with similar receptor profiles, with a high affinity for dopamine D-2-like receptors, although they are more selective for D-3 receptors. Indeed, clinical studies have shown that these drugs provide effective symptomatic treatment in PD, both as a monotherapy or adjunct therapy to L-DOPA.[193] Ropinirole, pramipexole, and rotigotine have longer half-lives (6–8 h, 8–12 h, and 7 h, respectively) and induce less dyskinesia compared to L-DOPA, mostly in lower doses.[194] In clinical trials, all of the three drugs showed an improvement of nonmotor symptoms generally; however, ropinirole and pramipexole improved anxiety and depression, but rotigotine did not.[195,196] There is also evidence that ropinirole and pramipexole have some neuroprotective properties.

Piribedil is also a nonergoline derivative dopamine D-2/D-3 receptor agonist with α_2-antagonistic effects.[197] Piribedil has equal affinity for D-3 and D-2 receptors, with no significant affinity for the dopamine D-1 receptor. Piribedil has a shorter half-life (2–5 h). It is also effective as a mono- or combination therapy in PD patients. In clinical trials, piribedil use showed an improvement in cognitive skills and was suggested to treat apathy in PD patients.[198]

Apomorphine is a dopamine D-1/D-2 receptor agonist with the shortest half-life (0.5–1.0 h) among all the dopaminergic agonists. The advantage of apomorphine is rapid onset of action and significant improvement of motor functions. However, subcutaneous infusion of apomorphine caused severe motor fluctuations, but induced less dyskinesia.[199]

MAO-B inhibitors, selegiline and rasagiline, selectively and irreversibly inhibit MAO, which is responsible for dopamine catabolism and increasing synaptic dopamine levels. Selegiline and rasagiline have comparable efficacy in improving motor symptoms in early stages of the disease.[200] They can be used in combination with L-DOPA therapy, improve motor fluctuations, and reduce off time.

Tolcapone and entacapone are peripheral reversible COMT inhibitors, although neither of them provides a complete peripheral COMT inhibition. COMT is the enzyme that catabolizes dopamine, L-DOPA, and other catecholamines, and thus COMT inhibitors increase the availability of L-DOPA. Therefore, tolcapone and entacapone are used in combination with L-DOPA. Tolcapone can cause hepatotoxicity, which may limit the use of the drug.[201]

5.3 Side Effects and Their Management

L-DOPA remains the best pharmacological therapy for PD, but in the majority of patients, the benefits are compromised by the appearance of motor response complications, including abnormal involuntary movements or dyskinesia. Furthermore, dyskinesia is a major source of disability in patients and the predominant dose-limiting factor in the use of L-DOPA. The most common form of dyskinesia following L-DOPA treatment is rapid, involuntary, random, and purposeless movements known as chorea, ranging from fleeting twitches to violent ballistic flinging movements. Sustained, abnormal muscle contraction, known as dystonia, is the second most commonly displayed form of L-DOPA–induced dyskinesia. Dystonic posturing is painful and often more disabling than chorea. Although patients display idiosyncratic patterns of abnormal involuntary movements, dyskinesia characteristically originates in the foot before progressing through the shin to the upper limbs. Dyskinesia can occur when L-DOPA or dopamine agonists reach a maximum level in the plasma. Peak-dose chorea, the most common form of L-DOPA–induced dyskinesia, occurs early in the treatment, and increases with continuing treatment. Peak-dose dystonia can appear in combination with, or instead of, peak-dose chorea. Initially, lowering the dose or L-DOPA attenuates peak-dose dyskinesia. However, the therapeutic window narrows with disease progression, and the minimal dose required for alleviating the symptoms produces dyskinesia.[178,179,202] Amantadine, an antiviral agent, is found to be very efficient in treating L-DOPA–induced dyskinesia.

The degree of nigrostriatal denervation is a risk factor for the development of motor side effects. Patients who have received high doses of L-DOPA for conditions not involving destruction of the nigrostriatal tract do not develop dyskinesia, and patients in the later stages of PD develop dyskinesia more rapidly than patients with less severe dopamine depletion.[203–205] However, studies have shown that chronic high-dose L-DOPA treatment produces dyskinesia in unlesioned primates.[206,207] Nigrostriatal denervation is not essential for the appearance of abnormal involuntary movements, but rather reduces the threshold for the induction of dyskinesia.

Dopaminergic stimulation is essential for the appearance of dyskinesia. Repeated administration of L-DOPA consistently elicits dyskinesia in patients with idiopathic PD and in MPTP-treated primates.[208–210] The incidence of dyskinesia increases with increasing dose and duration of L-DOPA treatment.[202,211,212] Dyskinesia is also provoked by some D-1 and D-2 dopamine agonists.[213] Therefore, it is unlikely that a single receptor subtype is responsible from the formation of dyskinesia alone.

Pharmacokinetics of the dopamine agonist is also an important factor for the development of dyskinesia. It has been shown with several clinical trials that pulsatile stimulation, rather than receptor selectivity, is responsible for the appearance of involuntary movements. In clinical trials, intraduodenal continuous L-DOPA infusion with a portable pump reduced side effects with no tolerance development, demonstrating that pulsatile, and not continuous, stimulation induces dyskinesia.[214–216] Therefore, developing new pharmaceutical forms of the current dopaminergic drugs with altered pharmacokinetics is also crucial. Altered activity in the indirect or direct pathway was also suggested to lead to dyskinesia with differential expression of neuromodulators.[217]

Dopamine agonists often cause nausea and in the beginning of the therapy. Nausea generally decreases with time and responds well to both antiemetic therapy with domperidone or complementary remedies. Somnolence is also a common adverse effect of dopamine agonist intake. For its treatment, the dose should be reduced, the drug discontinued, or drug switched to a MAO-B inhibitor. Modafinil can be used for severe somnolence cases. Hallucinations and orthostatic hypotension peripheral edema are the other adverse effects seen with dopamine agonist therapy.[218]

Recent reports in PD patients show that dopamine agonists are associated with the development of impulse control disorders, which are compulsive or pathological gambling, buying, sexual, and eating behaviors.[219] Thus, impulse control disorders can cause serious personal, familial, financial, and medical consequences. Furthermore, some patients with developed impulse control disorders are unable to discontinue dopamine agonist therapy as a result of motor worsening or dopamine agonist withdrawal syndrome.[220] During dopamine agonist therapy, it is crucial to monitor the patients; hence patients and caregivers should also be educated about the development of impulse control disorders. Adverse effects, such as addiction and withdrawal, are possible due to the activation of relatively intact mesocorticolimbic dopaminergic neurons.

6 TARGETED TREATMENT STRATEGIES FOR PARKINSON'S DISEASE

Although it is difficult to target drugs to reach the brain tissue, it is even harder to target the dopaminergic system, which remains another big challenge. There are several studies that have used in vivo disease models for brain targeting. For example, Md. et al. conducted an in vivo study and produced bromocriptine-loaded chitosan nanoparticles, which were successfully carried into the brain via the nose-to-brain application. These nanoparticles provided neuroprotection and hence improved haloperidol-induced PD model–induced behavioral abnormalities.[221]

An important part of the PD-related targeted drug treatment approaches is the improvement of L-DOPA carriage mechanism to the brain tissue. In conventional treatments, L-DOPA binds and passes to the brain via the blood–brain barrier (BBB). But the main obstacle for this treatment option is the peripheral degradation that leads to a need for a higher-dose titration and a combination with the peripheral DOPA decarboxylase inhibitor. To overcome this issue, researchers have developed and performed in vivo animal model studies with intranasal mucoadhesive nanoparticulates and thermoreversible gel formulations of levodopa for brain delivery.[222] Both of them resulted in efficient delivery into the brain tissue, and provided prolonged effect especially on motor symptoms. For intravenous route, using a poly(lactic-*co*-glycolic acid) (PLGA)–based L-DOPA delivery system proved to be an efficient way to bypass peripheral degradation, and hence a better penetration into the brain.[223]

As previously stated in this chapter, the main disease-causing mechanism for PD is neurodegenerative cell loss, especially in substantia nigra. To overcome this issue, Zhao et al.[224] applied one of the growth factors, namely GDNF, in a novel method; they genetically modified macrophages in such a way that they could secrete GDNF. Furthermore, they proposed that inflammation is part of PD pathophysiology, which leads to BBB opening and allows passage of genetically modified macrophages to the PD-affected area. Interestingly, the group demonstrated decrease in neurodegeneration and inflammation in substantia nigra, as well as a better motor performance in 6-hydroxydopamine (6-OHDA)–induced PD model.[224]

Another strategy to reach the brain in the PD model can be through the modification of conventional drugs to achieve a better penetration through the BBB. For example, Li et al.[225] developed a new compound *N*-3,4-bis(pivaloyloxy) dopamine-3-(dimethylamino)propanamide (PDDP), which reached very high brain concentrations compared to L-DOPA (270 times higher), and was transported via cationic transporters.

Targeted gene therapy is another approach. Zhang et al.[226] loaded tyrosine hydroxylase–expressing plasmids to PEGylated immunoliposomes and targeted brain tissue via transferrin receptors with the glial fibrillary acidic protein (GFAP) promoter. They intravenously applied this plasmid in 6-OHDA–induced mice PD model and observed normalization of striatal dopamine levels in a week, as well as behavioral improvement.

Targeted treatment options for early, as well as late phases of PD opens a new era in movement disorders. For developing drugs and understanding pathophysiology of PD, future studies are warranted, including targeted strategies.

7 CONCLUSIONS

PD is a complex disease in respect of its physiopathology and treatment. To obtain useful biomarkers for early diagnosis of PD in the prodromal phase or disease-modifying drugs after diagnosis, it is essential to better decipher the basic pathophysiological mechanisms underlying the neurodegenerative process.

With the help of the continuing research, we soon expect to have more effective disease-modifying drugs, which will be a component of an individualized therapy.

Abbreviations

AC Adenylate cyclase
AChE Acetylcholine esterase
BBB Blood–brain barrier
cAMP 3′,5′-Cyclic adenosine monophosphate
COMT Catecholamine-O-methyl transferase
GABA γ-Aminobutyric acid
GFAP Glial fibrillary acidic protein
GPe Globus pallidus pars externa
GPi Globus pallidus pars interna
GWAS Genomewide association studies
L-DOPA L-3,4-Dihydroxyphenylalanine
LRRK-2 Leucine-rich repeat kinase 2
MAO-B Monoamine oxidase-B
NAC Nonamyloid component
6-OHDA 6-Hydroxydopamine
PD Parkinson's disease
PDDP *N*-3,4-Bis(pivaloyloxy)dopamine-3-(dimethylamino)propanamide
PKA Protein kinase A or cAMP-dependent protein kinase
PLGA Poly(lactic-*co*-glycolic acid)
PPN Pedunculopontine nucleus
SNpc Substantia nigra pars compacta
SNr Substantia nigra pars reticulata
SPN Spiny projection neuron
STN Subthalamic nucleus

References

1. Olanow CW, Tatton WG. Etiology and pathogenesis of Parkinson's disease. *Annu Rev Neurosci*. 1999;22:123–144.
2. Schapira AH. Science, medicine, and the future: Parkinson's disease. *BMJ*. 1999;318(7179):311–314.
3. Dorsey ER, Constantinescu R, Thompson JP, Biglan KM, Holloway RG, Kieburtz K, et al. Projected number of people with Parkinson disease in the most populous nations, 2005 through 2030. *Neurology*. 2007;68(5):384–386.
4. Olanow CW, Watts RL, Koller WC. An algorithm (decision tree) for the management of Parkinson's disease (2001): treatment guidelines. *Neurology*. 2001;56(11 Suppl. 5):S1–S88.
5. Lang AE, Lozano AM. Parkinson's disease. First of two parts. *N Engl J Med*. 1998;339(15):1044–1053.
6. Kirik D, Bjorklund A, Modeling CNS. neurodegeneration by overexpression of disease-causing proteins using viral vectors. *Trends Neurosci*. 2003;26(7):386–392.

7. Lo Bianco C, Ridet JL, Schneider BL, Deglon N, Aebischer P. Alpha synucleinopathy and selective dopaminergic neuron loss in a rat lentiviral-based model of Parkinson's disease. *Proc Natl Acad Sci USA*. 2002;99(16):10813–10818.
8. Kordower JH, Bjorklund A. Trophic factor gene therapy for Parkinson's disease. *Mov Disord*. 2013;28(1):96–109.
9. Bjorklund A, Kordower JH. Cell therapy for Parkinson's disease: what next?. *Mov Disord*. 2013;28(1):110–115.
10. Lindvall O. Developing dopaminergic cell therapy for Parkinson's disease—give up or move forward?. *Mov Disord*. 2013;28(3):268–273.
11. Hely MA, Morris JG, Reid WG, Trafficante R. Sydney Multicenter Study of Parkinson's disease: non-L-dopa-responsive problems dominate at 15 years. *Mov Disord*. 2005;20(2):190–199.
12. Postuma RB, Aarsland D, Barone P, Burn DJ, Hawkes CH, Oertel W, et al. Identifying prodromal Parkinson's disease: pre-motor disorders in Parkinson's disease. *Mov Disord*. 2012;27(5):617–626.
13. Kalia LV, Lang AE. Parkinson's disease. *Lancet*. 2015;386(9996):896–912.
14. Post B, Muslimovic D, van Geloven N, Speelman JD, Schmand B, de Haan RJ, et al. Progression and prognostic factors of motor impairment, disability and quality of life in newly diagnosed Parkinson's disease. *Mov Disord*. 2011;26(3):449–456.
15. Spillantini MG, Schmidt ML, Lee VM, Trojanowski JQ, Jakes R, Goedert M. Alpha-synuclein in Lewy bodies. *Nature*. 1997;388(6645):839–840.
16. Braak H, Del Tredici K, Rub U, de Vos RA, Jansen Steur EN, Braak E. Staging of brain pathology related to sporadic Parkinson's disease. *Neurobiol Aging*. 2003;24(2):197–211.
17. Beyer K. Alpha-synuclein structure, posttranslational modification and alternative splicing as aggregation enhancers. *Acta Neuropathol*. 2006;112(3):237–251.
18. Abeliovich A, Schmitz Y, Farinas I, Choi-Lundberg D, Ho WH, Castillo PE, et al. Mice lacking alpha-synuclein display functional deficits in the nigrostriatal dopamine system. *Neuron*. 2000;25(1):239–252.
19. Sharon R, Goldberg MS, Bar-Josef I, Betensky RA, Shen J, Selkoe DJ. Alpha-synuclein occurs in lipid-rich high molecular weight complexes, binds fatty acids, and shows homology to the fatty acid-binding proteins. *Proc Natl Acad Sci USA*. 2001;98(16):9110–9115.
20. Ellis CE, Murphy EJ, Mitchell DC, Golovko MY, Scaglia F, Barcelo-Coblijn GC, et al. Mitochondrial lipid abnormality and electron transport chain impairment in mice lacking alpha-synuclein. *Mol Cell Biol*. 2005;25(22):10190–10201.
21. Bonini NM, Giasson BI. Snaring the function of alpha-synuclein. *Cell*. 2005;123(3):359–361.
22. Liu S, Ninan I, Antonova I, Battaglia F, Trinchese F, Narasanna A, et al. Alpha-synuclein produces a long-lasting increase in neurotransmitter release. *EMBO J*. 2004;23(22):4506–4516.
23. Fortin DL, Nemani VM, Voglmaier SM, Anthony MD, Ryan TA, Edwards RH. Neural activity controls the synaptic accumulation of alpha-synuclein. *J Neurosci*. 2005;25(47):10913–10921.
24. Yu S, Zuo X, Li Y, Zhang C, Zhou M, Zhang YA, et al. Inhibition of tyrosine hydroxylase expression in alpha-synuclein-transfected dopaminergic neuronal cells. *Neurosci Lett*. 2004;367(1):34–39.
25. Perez RG, Waymire JC, Lin E, Liu JJ, Guo F, Zigmond MJ. A role for alpha-synuclein in the regulation of dopamine biosynthesis. *J Neurosci*. 2002;22(8):3090–3099.
26. Zigmond MJ, Hastings TG, Perez RG. Increased dopamine turnover after partial loss of dopaminergic neurons: compensation or toxicity?. *Parkinsonism Relat Disord*. 2002;8(6):389–393.
27. Tehranian R, Montoya SE, Van Laar AD, Hastings TG, Perez RG. Alpha-synuclein inhibits aromatic amino acid decarboxylase activity in dopaminergic cells. *J Neurochem*. 2006;99(4):1188–1196.

28. Fountaine TM, Venda LL, Warrick N, Christian HC, Brundin P, Channon KM, et al. The effect of alpha-synuclein knockdown on MPP+ toxicity in models of human neurons. *Eur J Neurosci.* 2008;28(12):2459–2473.

29. Burre J, Sharma M, Tsetsenis T, Buchman V, Etherton MR, Sudhof TC. Alpha-synuclein promotes SNARE-complex assembly in vivo and in vitro. *Science.* 2010;329(5999):1663–1667.

30. Scott D, Roy S. Alpha-synuclein inhibits intersynaptic vesicle mobility and maintains recycling-pool homeostasis. *J Neurosci.* 2012;32(30):10129–10135.

31. Davidson WS, Jonas A, Clayton DF, George JM. Stabilization of alpha-synuclein secondary structure upon binding to synthetic membranes. *J Biol Chem.* 1998;273(16):9443–9449.

32. Perrin RJ, Woods WS, Clayton DF, George JM. Interaction of human alpha-synuclein and Parkinson's disease variants with phospholipids. Structural analysis using site-directed mutagenesis. *J Biol Chem.* 2000;275(44):34393–34398.

33. Chandra S, Chen X, Rizo J, Jahn R, Sudhof TC. A broken alpha -helix in folded alpha-synuclein. *J Biol Chem.* 2003;278(17):15313–15318.

34. Bussell Jr R, Eliezer D. A structural and functional role for 11-mer repeats in alpha-synuclein and other exchangeable lipid binding proteins. *J Mol Biol.* 2003;329(4):763–778.

35. Jao CC, Der-Sarkissian A, Chen J, Langen R. Structure of membrane-bound alpha-synuclein studied by site-directed spin labeling. *Proc Natl Acad Sci USA.* 2004;101(22):8331–8336.

36. Jensen MB, Bhatia VK, Jao CC, Rasmussen JE, Pedersen SL, Jensen KJ, et al. Membrane curvature sensing by amphipathic helices: a single liposome study using alpha-synuclein and annexin B12. *J Biol Chem.* 2011;286(49):42603–42614.

37. Middleton ER, Rhoades E. Effects of curvature and composition on alpha-synuclein binding to lipid vesicles. *Biophys J.* 2010;99(7):2279–2288.

38. Kim WS, Kagedal K, Halliday GM. Alpha-synuclein biology in Lewy body diseases. *Alzheimers Res Ther.* 2014;6(5):73.

39. Peelaerts W, Bousset L, Van der Perren A, Moskalyuk A, Pulizzi R, Giugliano M, et al. Alpha-synuclein strains cause distinct synucleinopathies after local and systemic administration. *Nature.* 2015;522(7556):340–344.

40. Narhi L, Wood SJ, Steavenson S, Jiang Y, Wu GM, Anafi D, et al. Both familial Parkinson's disease mutations accelerate alpha-synuclein aggregation. *J Biol Chem.* 1999;274(14):9843–9846.

41. Conway KA, Harper JD, Lansbury PT. Accelerated in vitro fibril formation by a mutant alpha-synuclein linked to early-onset Parkinson disease. *Nat Med.* 1998;4(11):1318–1320.

42. Fredenburg RA, Rospigliosi C, Meray RK, Kessler JC, Lashuel HA, Eliezer D, et al. The impact of the E46K mutation on the properties of alpha-synuclein in its monomeric and oligomeric states. *Biochemistry.* 2007;46(24):7107–7118.

43. Kordower JH, Chu Y, Hauser RA, Freeman TB, Olanow CW. Lewy body-like pathology in long-term embryonic nigral transplants in Parkinson's disease. *Nat Med.* 2008;14(5):504–506.

44. Desplats P, Lee HJ, Bae EJ, Patrick C, Rockenstein E, Crews L, et al. Inclusion formation and neuronal cell death through neuron-to-neuron transmission of alpha-synuclein. *Proc Natl Acad Sci USA.* 2009;106(31):13010–13015.

45. Luk KC, Kehm V, Carroll J, Zhang B, O'Brien P, Trojanowski JQ, et al. Pathological alpha-synuclein transmission initiates Parkinson-like neurodegeneration in nontransgenic mice. *Science.* 2012;338(6109):949–953.

46. Luk KC, Kehm VM, Zhang B, O'Brien P, Trojanowski JQ, Lee VM. Intracerebral inoculation of pathological alpha-synuclein initiates a rapidly progressive neurodegenerative alpha-synucleinopathy in mice. *J Exp Med.* 2012;209(5):975–986.

47. Tanner CM, Ottman R, Goldman SM, Ellenberg J, Chan P, Mayeux R, et al. Parkinson disease in twins: an etiologic study. *JAMA*. 1999;281(4):341–346.
48. Tanner CM. Is the cause of Parkinson's disease environmental or hereditary? Evidence from twin studies. *Adv Neurol*. 2003;91:133–142.
49. Sveinbjornsdottir S, Hicks AA, Jonsson T, Petursson H, Gugmundsson G, Frigge ML, et al. Familial aggregation of Parkinson's disease in Iceland. *N Engl J Med*. 2000;343(24):1765–1770.
50. Elbaz A, Grigoletto F, Baldereschi M, Breteler MM, Manubens-Bertran JM, Lopez-Pousa S, et al. Familial aggregation of Parkinson's disease: a population-based case-control study in Europe. EUROPARKINSON Study Group. *Neurology*. 1999;52(9):1876–1882.
51. Verstraeten A, Theuns J, Van Broeckhoven C. Progress in unraveling the genetic etiology of Parkinson disease in a genomic era. *Trends Genet*. 2015;31(3):140–149.
52. Valenti L, Conte D, Piperno A, Dongiovanni P, Fracanzani AL, Fraquelli M, et al. The mitochondrial superoxide dismutase A16V polymorphism in the cardiomyopathy associated with hereditary haemochromatosis. *J Med Genet*. 2004;41(12):946–950.
53. Maimone D, Dominici R, Grimaldi LM. Pharmacogenomics of neurodegenerative diseases. *Eur J Pharmacol*. 2001;413(1):11–29.
54. Grimes DA, Bulman DE. Parkinson's genetics—creating exciting new insights. *Parkinsonism Relat Disord*. 2002;8(6):459–464.
55. Spataro N, Calafell F, Cervera-Carles L, Casals F, Pagonabarraga J, Pascual-Sedano B, et al. Mendelian genes for Parkinson's disease contribute to the sporadic forms of the disease. *Hum Mol Genet*. 2015;24(7):2023–2034.
56. Zimprich A, Muller-Myhsok B, Farrer M, Leitner P, Sharma M, Hulihan M, et al. The PARK8 locus in autosomal dominant parkinsonism: confirmation of linkage and further delineation of the disease-containing interval. *Am J Hum Genet*. 2004;74(1):11–19.
57. Zimprich A, Biskup S, Leitner P, Lichtner P, Farrer M, Lincoln S, et al. Mutations in LRRK2 cause autosomal-dominant parkinsonism with pleomorphic pathology. *Neuron*. 2004;44(4): 601–607.
58. Valente EM, Salvi S, Ialongo T, Marongiu R, Elia AE, Caputo V, et al. PINK1 mutations are associated with sporadic early-onset parkinsonism. *Ann Neurol*. 2004;56(3):336–341.
59. Valente EM, Bentivoglio AR, Dixon PH, Ferraris A, Ialongo T, Frontali M, et al. Localization of a novel locus for autosomal recessive early-onset parkinsonism, PARK6, on human chromosome 1p35-p36. *Am J Hum Genet*. 2001;68(4):895–900.
60. Polymeropoulos MH, Lavedan C, Leroy E, Ide SE, Dehejia A, Dutra A, et al. Mutation in the alpha-synuclein gene identified in families with Parkinson's disease. *Science*. 1997;276(5321):2045–2047.
61. Paisan-Ruiz C, Jain S, Evans EW, Gilks WP, Simon J, van der Brug M, et al. Cloning of the gene containing mutations that cause PARK8-linked Parkinson's disease. *Neuron*. 2004;44(4): 595–600.
62. Moore DJ, Zhang L, Dawson TM, Dawson VL. A missense mutation (L166P) in DJ-1, linked to familial Parkinson's disease, confers reduced protein stability and impairs homo-oligomerization. *J Neurochem*. 2003;87(6):1558–1567.
63. Mitsumoto A, Nakagawa Y, Takeuchi A, Okawa K, Iwamatsu A, Takanezawa Y. Oxidized forms of peroxiredoxins and DJ-1 on two-dimensional gels increased in response to sublethal levels of paraquat. *Free Radic Res*. 2001;35(3):301–310.
64. Mitsumoto A, Nakagawa Y. DJ-1 is an indicator for endogenous reactive oxygen species elicited by endotoxin. *Free Radic Res*. 2001;35(6):885–893.
65. Matsumine H, Saito M, Shimoda-Matsubayashi S, Tanaka H, Ishikawa A, Nakagawa-Hattori Y, et al. Localization of a gene for an autosomal recessive form of juvenile Parkinsonism to chromosome 6q25.2-27. *Am J Hum Genet*. 1997;60(3):588–596.

66. Leroy E, Anastasopoulos D, Konitsiotis S, Lavedan C, Polymeropoulos MH. Deletions in the Parkin gene and genetic heterogeneity in a Greek family with early onset Parkinson's disease. *Hum Genet.* 1998;103(4):424–427.

67. Kitada T, Asakawa S, Hattori N, Matsumine H, Yamamura Y, Minoshima S, et al. Mutations in the parkin gene cause autosomal recessive juvenile parkinsonism. *Nature.* 1998;392(6676):605–608.

68. Bonifati V, Lucking CB, Fabrizio E, Periquet M, Meco G, Brice A. Three parkin gene mutations in a sibship with autosomal recessive early onset parkinsonism. *J Neurol Neurosurg Psych.* 2001;71(4):531–534.

69. Abou-Sleiman PM, Healy DG, Quinn N, Lees AJ, Wood NW. The role of pathogenic DJ-1 mutations in Parkinson's disease. *Ann Neurol.* 2003;54(3):283–286.

70. Hirsch EC, Jenner P, Przedborski S. Pathogenesis of Parkinson's disease. *Mov Disord.* 2013;28(1):24–30.

71. Warner TT, Schapira AH. Genetic and environmental factors in the cause of Parkinson's disease. *Ann Neurol.* 2003;53(Suppl. 3):S16–S23.

72. Tanner CM, Goldman SM. Epidemiology of Parkinson's disease. *Neurol Clin.* 1996;14(2):317.

73. Petrovitch H, Ross GW, Abbott RD, Sanderson WT, Sharp DS, Tanner CM, et al. Plantation work and risk of Parkinson disease in a population-based longitudinal study. *Arch Neurol.* 2002;59(11):1787–1792.

74. Liu B, Gao HM, Hong JS. Parkinson's disease and exposure to infectious agents and pesticides and the occurrence of brain injuries: role of neuroinflammation. *Environ Health Perspect.* 2003;111(8):1065–1073.

75. Jenner P, Dexter DT, Sian J, Schapira AH, Marsden CD. Oxidative stress as a cause of nigral cell death in Parkinson's disease and incidental Lewy body disease. The Royal Kings and Queens Parkinson's Disease Research Group. *Ann Neurol.* 1992;32 Suppl:S82–S87.

76. Di Monte DA. The environment and Parkinson's disease: is the nigrostriatal system preferentially targeted by neurotoxins?. *Lancet Neurol.* 2003;2(9):531–538.

77. Betarbet R, Sherer TB, MacKenzie G, Garcia-Osuna M, Panov AV, Greenamyre JT. Chronic systemic pesticide exposure reproduces features of Parkinson's disease. *Nat Neurosci.* 2000;3(12):1301–1306.

78. Noyce AJ, Bestwick JP, Silveira-Moriyama L, Hawkes CH, Giovannoni G, Lees AJ, et al. Meta-analysis of early nonmotor features and risk factors for Parkinson disease. *Ann Neurol.* 2012;72(6):893–901.

79. Pezzoli G, Ricciardi S, Masotto C, Mariani CB, Carenzi A. *n*-Hexane induces parkinsonism in rodents. *Brain Res.* 1990;531(1–2):355–357.

80. Meco G, Bonifati V, Vanacore N, Fabrizio E. Parkinsonism after chronic exposure to the fungicide maneb (manganese ethylene-bis-dithiocarbamate). *Scand J Work Environ Health.* 1994;20(4):301–305.

81. Gorell JM, Johnson CC, Rybicki BA, Peterson EL, Kortsha GX, Brown GG, et al. Occupational exposures to metals as risk factors for Parkinson's disease. *Neurology.* 1997;48(3):650–658.

82. Good PF, Olanow CW, Perl DP. Neuromelanin-containing neurons of the substantia nigra accumulate iron and aluminum in Parkinson's disease: a LAMMA study. *Brain Res.* 1992;593(2):343–346.

83. Choi IS. Parkinsonism after carbon monoxide poisoning. *Eur Neurol.* 2002;48(1):30–33.

84. Schapira AH, Gu M, Taanman JW, Tabrizi SJ, Seaton T, Cleeter M, et al. Mitochondria in the etiology and pathogenesis of Parkinson's disease. *Ann Neurol.* 1998;44(3 Suppl. 1):S89–S98.

85. McGeer PL, Itagaki S, Akiyama H, McGeer EG. Rate of cell death in parkinsonism indicates active neuropathological process. *Ann Neurol.* 1988;24(4):574–576.

86. Langston JW, Forno LS, Tetrud J, Reeves AG, Kaplan JA, Karluk D. Evidence of active nerve cell degeneration in the substantia nigra of humans years after 1-methyl-4-phenyl-1,2,3,6-tetrahydropyridine exposure. *Ann Neurol.* 1999;46(4):598–605.

87. Jenner P, Olanow CW. Oxidative stress and the pathogenesis of Parkinson's disease. *Neurology.* 1996;47(6 Suppl. 3):S161–S170.

88. Jenner P. Oxidative stress in Parkinson's disease. *Ann Neurol.* 2003;53(Suppl. 3):S26–S36.

89. Hirsch EC, Hunot S, Faucheux B, Agid Y, Mizuno Y, Mochizuki H, et al. Dopaminergic neurons degenerate by apoptosis in Parkinson's disease. *Mov Disord.* 1999;14(2):383–385.

90. Hirsch EC, Faucheux BA. Iron metabolism and Parkinson's disease. *Mov Disord.* 1998;13 (Suppl. 1):39–45.

91. Cooper JM, Schapira AH. Mitochondrial dysfunction in neurodegeneration. *J Bioenerg Biomembr.* 1997;29(2):175–183.

92. Beal MF. Mitochondria, oxidative damage, and inflammation in Parkinson's disease. *Ann NY Acad Sci.* 2003;991:120–131.

93. Beal MF. Excitotoxicity and nitric oxide in Parkinson's disease pathogenesis. *Ann Neurol.* 1998;44(3 Suppl. 1):S110–S114.

94. Quik M. Smoking, nicotine and Parkinson's disease. *Trends Neurosci.* 2004;27(9):561–568.

95. Maggio R, Riva M, Vaglini F, Fornai F, Molteni R, Armogida M, et al. Nicotine prevents experimental parkinsonism in rodents and induces striatal increase of neurotrophic factors. *J Neurochem.* 1998;71(6):2439–2446.

96. Tan EK, Tan C, Fook-Chong SM, Lum SY, Chai A, Chung H, et al. Dose-dependent protective effect of coffee, tea, and smoking in Parkinson's disease: a study in ethnic Chinese. *J Neurol Sci.* 2003;216(1):163–167.

97. Hernan MA, Chen H, Schwarzschild MA, Ascherio A. Alcohol consumption and the incidence of Parkinson's disease. *Ann Neurol.* 2003;54(2):170–175.

98. Hellenbrand W, Seidler A, Boeing H, Robra BP, Vieregge P, Nischan P, et al. Diet and Parkinson's disease. I: a possible role for the past intake of specific foods and food groups. Results from a self-administered food-frequency questionnaire in a case-control study. *Neurology.* 1996;47(3):636–643.

99. Gorell JM, Rybicki BA, Johnson CC, Peterson EL. Smoking and Parkinson's disease: a dose-response relationship. *Neurology.* 1999;52(1):115–119.

100. Chen H, Zhang SM, Hernan MA, Schwarzschild MA, Willett WC, Colditz GA, et al. Nonsteroidal anti-inflammatory drugs and the risk of Parkinson disease. *Arch Neurol.* 2003;60(8): 1059–1064.

101. Schapira AH, Marsden CD. Platelet mitochondrial DNA in Parkinson's disease. *Mov Disord.* 1994;9(1):119–121.

102. Wenning GK, Quinn NP. Parkinsonism. Multiple system atrophy. *Baillieres Clin Neurol.* 1997;6(1):187–204.

103. Nicklas WJ, Vyas I, Heikkila RE. Inhibition of NADH-linked oxidation in brain mitochondria by 1-methyl-4-phenyl-pyridine, a metabolite of the neurotoxin, 1-methyl-4-phenyl-1,2,5,6-tetrahydropyridine. *Life Sci.* 1985;36(26):2503–2508.

104. Swerdlow RH, Parks JK, Miller SW, Tuttle JB, Trimmer PA, Sheehan JP, et al. Origin and functional consequences of the complex I defect in Parkinson's disease. *Ann Neurol.* 1996;40(4):663–671.

105. Swerdlow RH, Parks JK, Davis 2nd JN, Cassarino DS, Trimmer PA, Currie LJ, et al. Matrilineal inheritance of complex I dysfunction in a multigenerational Parkinson's disease family. *Ann Neurol.* 1998;44(6):873–881.

106. Whone AL, Moore RY, Piccini P, Brooks DJ. Compensatory changes in the globus pallidus in early Parkinson's disease: an F-18-DOPA PET study. *Neurology*. 2001;56(8):A72–A73.

107. Bezard E, Boraud T, Bioulac B, Gross CE. Involvement of the subthalamic nucleus in glutamatergic compensatory mechanisms. *Eur J Neurosci*. 1999;11(6):2167–2170.

108. Bezard E, Boraud T, Bioulac B, Gross CE. Compensatory effects of glutamatergic inputs to the substantia nigra pars compacta in experimental parkinsonism. *Neuroscience*. 1997;81(2):399–404.

109. Bernheimer H, Birkmayer W, Hornykiewicz O, Jellinger K, Seitelberger F. Brain dopamine and the syndromes of Parkinson and Huntington. Clinical, morphological and neurochemical correlations. *J Neurol Sci*. 1973;20(4):415–455.

110. Jellinger K. *Pathology of Parkinsonism*. New York, NY: Raven press; 1996:33–66.

111. Martin KE, Phillips JG, Iansek R, Bradshaw JL. Inaccuracy and instability of sequential movements in Parkinson's disease. *Exp Brain Res*. 1994;102(1):131–140.

112. Marsden CD, Obeso JA. The functions of the basal ganglia and the paradox of stereotaxic surgery in Parkinson's disease. *Brain*. 1994;117(pt 4):877–897.

113. Alexander GE, Crutcher MD. Functional architecture of basal ganglia circuits: neural substrates of parallel processing. *Trends Neurosci*. 1990;13(7):266–271.

114. Graybiel AM, Canales JJ, Capper-Loup C. Levodopa-induced dyskinesias and dopamine-dependent stereotypies: a new hypothesis. *Trends Neurosci*. 2000;23(10 Suppl.):S71–S77.

115. Graybiel AM. Building action repertoires: memory and learning functions of the basal ganglia. *Curr Opin Neurobiol*. 1995;5(6):733–741.

116. Parent A, Hazrati LN. Functional anatomy of the basal ganglia. I. The cortico-basal ganglia-thalamo-cortical loop. *Brain Res Brain Res Rev*. 1995;20(1):91–127.

117. Graybiel AM, Aosaki T, Flaherty AW, Kimura M. The basal ganglia and adaptive motor control. *Science*. 1994;265(5180):1826–1831.

118. Albin RL, Young AB, Penney JB. The functional anatomy of basal ganglia disorders. *Trends Neurosci*. 1989;12(10):366–375.

119. Jones EG. Laminar distribution of cortical efferent cells. In: Jones EG, Peters A, eds. *Cerebral Cortex Vol. I Cellular Components of the Cerebral Cortex*. New York, NY: Plenum Press; 1984:521–553.

120. Selemon LD, Goldman-Rakic PS. Longitudinal topography and interdigitation of corticostriatal projections in the rhesus monkey. *J Neurosci*. 1985;5(3):776–794.

121. Goldman-Rakic PS, Selemon LD, Topography of corticostriatal projections in non-human primates and implications for functional parcellation of the neostriatum. Jones EG, Peters A, eds. *Cerebral Cortex*, vol. 5. New York, NY: Plenum Press; 1986:447–466.

122. Flaherty AW, Graybiel AM. Motor and somatosensory corticostriatal projection magnifications in the squirrel monkey. *J Neurophysiol*. 1995;74(6):2638–2648.

123. Parent A. Extrinsic connections of the basal ganglia. *Trends Neurosci*. 1990;13(7):254–258.

124. Parthasarathy HB, Graybiel AM. Cortically driven immediate-early gene expression reflects modular influence of sensorimotor cortex on identified striatal neurons in the squirrel monkey. *J Neurosci*. 1997;17(7):2477–2491.

125. Gerfen CR. The neostriatal mosaic: compartmentalization of corticostriatal input and striatonigral output systems. *Nature*. 1984;311(5985):461–464.

126. Flaherty AW, Graybiel AM. Input-output organization of the sensorimotor striatum in the squirrel monkey. *J Neurosci*. 1994;14(2):599–610.

127. Smith Y, Kieval JZ. Anatomy of the dopamine system in the basal ganglia. *Trends Neurosci*. 2000;23(10 Suppl.):S28–S33.

128. Joel D, Weiner I. The connections of the dopaminergic system with the striatum in rats and primates: an analysis with respect to the functional and compartmental organization of the striatum. *Neuroscience*. 2000;96(3):451–474.

129. Prensa L, Parent A. The nigrostriatal pathway in the rat: a single-axon study of the relationship between dorsal and ventral tier nigral neurons and the striosome/matrix striatal compartments. *J Neurosci*. 2001;21(18):7247–7260.

130. Song DD, Haber SN. Striatal responses to partial dopaminergic lesion: evidence for compensatory sprouting. *J Neurosci*. 2000;20(13):5102–5114.

131. Jimenez-Castellanos J, Graybiel AM. Subdivisions of the dopamine-containing A8-A9-A10 complex identified by their differential mesostriatal innervation of striosomes and extrastriosomal matrix. *Neuroscience*. 1987;23(1):223–242.

132. Gerfen CR, Baimbridge KG, Miller JJ. The neostriatal mosaic: compartmental distribution of calcium-binding protein and parvalbumin in the basal ganglia of the rat and monkey. *Proc Natl Acad Sci USA*. 1985;82(24):8780–8784.

133. Nambu A, Tokuno H, Takada M. Functional significance of the cortico-subthalamo-pallidal 'hyperdirect' pathway. *Neurosci Res*. 2002;43(2):111–117.

134. Kawaguchi Y, Wilson CJ, Emson PC. Projection subtypes of rat neostriatal matrix cells revealed by intracellular injection of biocytin. *J Neurosci*. 1990;10(10):3421–3438.

135. Takada M, Inoue K-i, Koketsu D, Kato S, Kobayashi K, Nambu A. Elucidating information processing in primate basal ganglia circuitry: a novel technique for pathway-selective ablation mediated by immunotoxin. *Front Neural Circuits*. 2013;7:140.

136. Maneuf YP, Mitchell IJ, Crossman AR, Woodruff GN, Brotchie JM. Functional implications of kappa opioid receptor-mediated modulation of glutamate transmission in the output regions of the basal ganglia in rodent and primate models of Parkinson's disease. *Brain Res*. 1995;683(1):102–108.

137. Strange PG. Dopamine receptors in the basal ganglia: relevance to Parkinson's disease. *Mov Disord*. 1993;8(3):263–270.

138. Sibley DR, Monsma Jr FJ, Shen Y. Molecular neurobiology of dopaminergic receptors. *Int Rev Neurobiol*. 1993;35:391–415.

139. Edelman AM, Blumenthal DK, Krebs EG. Protein serine/threonine kinases. *Annu Rev Biochem*. 1987;56:567–613.

140. Mellon PL, Clegg CH, Correll LA, McKnight GS. Regulation of transcription by cyclic AMP-dependent protein kinase. *Proc Natl Acad Sci USA*. 1989;86(13):4887–4891.

141. Undie AS, Weinstock J, Sarau HM, Friedman E. Evidence for a distinct D1-like dopamine receptor that couples to activation of phosphoinositide metabolism in brain. *J Neurochem*. 1994;62(5):2045–2048.

142. Undie AS, Berki AC, Beardsley K. Dopaminergic behaviors and signal transduction mediated through adenylate cyclase and phospholipase C pathways. *Neuropharmacology*. 2000;39(1):75–87.

143. Dearry A, Gingrich JA, Falardeau P, Fremeau Jr RT, Bates MD, Caron MG. Molecular cloning and expression of the gene for a human D1 dopamine receptor. *Nature*. 1990;347(6288): 72–76.

144. Smith Y, Bennett BD, Bolam JP, Parent A, Sadikot AF. Synaptic relationships between dopaminergic afferents and cortical or thalamic input in the sensorimotor territory of the striatum in monkey. *J Comp Neurol*. 1994;344(1):1–19.

145. Gerfen CR, Engber TM, Mahan LC, Susel Z, Chase TN, Monsma Jr FJ, et al. D1 and D2 dopamine receptor-regulated gene expression of striatonigral and striatopallidal neurons. *Science*. 1990;250(4986):1429–1432.

146. Fremeau Jr RT, Duncan GE, Fornaretto MG, Dearry A, Gingrich JA, Breese GR, et al. Localization of D1 dopamine receptor mRNA in brain supports a role in cognitive, affective, and neuroendocrine aspects of dopaminergic neurotransmission. *Proc Natl Acad Sci USA*. 1991;88(9):3772–3776.

147. Missale C, Nash SR, Robinson SW, Jaber M, Caron MG. Dopamine receptors: from structure to function. *Physiol Rev*. 1998;78(1):189–225.

148. Usiello A, Baik JH, Rouge-Pont F, Picetti R, Dierich A, LeMeur M, et al. Distinct functions of the two isoforms of dopamine D2 receptors. *Nature*. 2000;408(6809):199–203.

149. Le Moine C, Normand E, Bloch B. Phenotypical characterization of the rat striatal neurons expressing the D1 dopamine receptor gene. *Proc Natl Acad Sci USA*. 1991;88(10): 4205–4209.

150. Le Moine C, Bloch B. D1 and D2 dopamine receptor gene expression in the rat striatum: sensitive cRNA probes demonstrate prominent segregation of D1 and D2 mRNAs in distinct neuronal populations of the dorsal and ventral striatum. *J Comp Neurol*. 1995;355(3): 418–426.

151. Surmeier DJ, Song WJ, Yan Z. Coordinated expression of dopamine receptors in neostriatal medium spiny neurons. *J Neurosci*. 1996;16(20):6579–6591.

152. Surmeier DJ, Eberwine J, Wilson CJ, Cao Y, Stefani A, Kitai ST. Dopamine receptor subtypes colocalize in rat striatonigral neurons. *Proc Natl Acad Sci USA*. 1992;89(21):10178–10182.

153. Aizman O, Brismar H, Uhlen P, Zettergren E, Levey AI, Forssberg H, et al. Anatomical and physiological evidence for D1 and D2 dopamine receptor colocalization in neostriatal neurons. *Nat Neurosci*. 2000;3(3):226–230.

154. Sebastiao AM, Ribeiro JA. Fine-tuning neuromodulation by adenosine. *Trends Pharmacol Sci*. 2000;21(9):341–346.

155. Schiffmann SN, Jacobs O, Vanderhaeghen JJ. Striatal restricted adenosine-A2 receptor (Rdc8) is expressed by enkephalin but not by substance-P neurons—an situ hybridization histochemistry study. *J Neurochem*. 1991;57(3):1062–1067.

156. Fink JS, Weaver DR, Rivkees SA, Peterfreund RA, Pollack AE, Adler RS. Molecular cloning of the rat adenosine-A2 receptor—selective coexpression with D2-dopamine receptors in rat striatum. *Mol Brain Res*. 1992;14(3):186–195.

157. Augood SJ, Emson PC. Adenosine A(2a) receptor messenger-RNA is expressed by enkephalin cells but not by somatostatin cells in rat striatum—a study. *Mol Brain Res*. 1994;22(1–4): 204–210.

158. Svenningsson P, Le Moine C, Kull B, Sunahara R, Bloch B, Fredholm BB. Cellular expression of adenosine A2A receptor messenger RNA in the rat central nervous system with special reference to dopamine innervated areas. *Neuroscience*. 1997;80(4):1171–1185.

159. Le Moine C, Tison F, Bloch B. D2 dopamine receptor gene expression by cholinergic neurons in the rat striatum. *Neurosci Lett*. 1990;117(3):248–252.

160. Gurevich EV, Joyce JN. Distribution of dopamine D3 receptor expressing neurons in the human forebra: comparison with D2 receptor expressing neurons. *Neuropsychopharmacology*. 1999;20(1):60–80.

161. Joyce JN, Gurevich EV. D3 receptors and the actions of neuroleptics in the ventral striatopallidal system of schizophrenics. *Ann NY Acad Sci*. 1999;877:595–613.

162. Lidow MS, Wang F, Cao Y, Goldman-Rakic PS, Layer V. neurons bear the majority of mRNAs encoding the five distinct dopamine receptor subtypes in the primate prefrontal cortex. *Synapse*. 1998;28(1):10–20.

163. Graybiel AM. Neurotransmitters and neuromodulators in the basal ganglia. *Trends Neurosci*. 1990;13(7):244–254.

164. Wilson CJ, Groves PM. Fine structure and synaptic connections of the common spiny neuron of the rat neostriatum: a study employing intracellular inject of horseradish peroxidase. *J Comp Neurol*. 1980;194(3):599–615.

165. Wu Y, Richard S, Parent A. The organization of the striatal output system: a single-cell juxtacellular labeling study in the rat. *Neurosci Res*. 2000;38(1):49–62.

166. Parent A, Sato F, Wu Y, Gauthier J, Levesque M, Parent M. Organization of the basal ganglia: the importance of axonal collateralization. *Trends Neurosci*. 2000;23(10 Suppl.):S20–S27.

167. Loopuijt LD, van der Kooy D. Organization of the striatum: collateralization of its efferent axons. *Brain Res*. 1985;348(1):86–99.

168. Chase TN, Oh JD. Striatal dopamine- and glutamate-mediated dysregulation in experimental parkinsonism. *Trends Neurosci*. 2000;23(10 Suppl.):S86–S91.

169. Bouyer JJ, Park DH, Joh TH, Pickel VM. Chemical and structural analysis of the relation between cortical inputs and tyrosine hydroxylase-containing terminals in rat neostriatum. *Brain Res*. 1984;302(2):267–275.

170. Cepeda C, Buchwald NA, Levine MS. Neuromodulatory actions of dopamine in the neostriatum are dependent upon the excitatory amino acid receptor subtypes activated. *Proc Natl Acad Sci USA*. 1993;90(20):9576–9580.

171. Zheng P, Zhang XX, Bunney BS, Shi WX. Opposite modulation of cortical *N*-methyl-D-aspartate receptor-mediated responses by low and high concentrations of dopamine. *Neuroscience*. 1999;91(2):527–535.

172. AlDakheel A, Kalia LV, Lang AE. Pathogenesis-targeted, disease-modifying therapies in Parkinson disease. *Neurotherapeutics*. 2014;11(1):6–23.

173. Tran HT, Chung CH, Iba M, Zhang B, Trojanowski JQ, Luk KC, et al. Alpha-synuclein immunotherapy blocks uptake and templated propagation of misfolded alpha-synuclein and neurodegeneration. *Cell Rep*. 2014;7(6):2054–2065.

174. Charles D, Konrad PE, Neimat JS, Molinari AL, Tramontana MG, Finder SG, et al. Subthalamic nucleus deep brain stimulation in early stage Parkinson's disease. *Parkinsonism Relat Disord*. 2014;20(7):731–737.

175. Coune PG, Schneider BL, Aebischer P. Parkinson's disease: gene therapies. *Cold Spring Harb Perspect Med*. 2012;2(4):a009431.

176. Lesser RP, Fahn S, Snider SR, Cote LJ, Isgreen WP, Barrett RE. Analysis of the clinical problems in parkinsonism and the complications of long-term levodopa therapy. *Neurology*. 1979;29 (9 pt 1):1253–1260.

177. Fabbrini G, Mouradian MM, Juncos JL, Schlegel J, Mohr E, Chase TN. Motor fluctuations in Parkinsons disease—central mechanisms. *Ann Neurol*. 1988;24(3):366–371.

178. Fabbrini G, Juncos J, Mouradian MM, Serrati C, Chase TN. Levodopa pharmacokinetic mechanisms and motor fluctuations in Parkinsons disease. *Ann Neurol*. 1987;21(4):370–376.

179. Mouradian MM, Juncos JL, Fabbrini G, Schlegel J, Bartko JJ, Chase TN. Motor fluctuations in Parkinson's disease: central pathophysiological mechanisms. *Ann Neurol*. 1988;24(3):372–378.

180. Obeso JA, Grandas F, Vaamonde J, Luquin MR, Artieda J, Lera G, et al. Motor complications associated with chronic levodopa therapy in Parkinson's disease. *Neurology*. 1989;39 (11 Suppl. 2):11–19.

181. Sage JI, Mark MH. Basic mechanisms of motor fluctuations. *Neurology*. 1994;44 (7 Suppl. 6):S10–S14.

182. Nutt JG, Holford NH. The response to levodopa in Parkinson's disease: imposing pharmacological law and order. *Ann Neurol*. 1996;39(5):561–573.

183. Evidente VGH, Adler CH. Pharmacologic options for managing Parkinson's disease. *Formulary*. 1997;32(6):594.

184. Obeso JA, Linazasoro G, Gorospe A, Rodriguez M, Lera G. Complications associated with chronic L-DOPA therapy in Parkinson's disease. In: Olanow WC, Obeso JA, eds. *Beyond the Decade of Brain*. Kent: Royal Tunbridge Wells; 1997.

185. Schrag A, Quinn N. Dyskinesias and motor fluctuations in Parkinson's disease. A community-based study. *Brain*. 2000;123(pt 11):2297–2305.

186. Ahlskog JE, Muenter MD. Frequency of levodopa-related dyskinesias and motor fluctuations as estimated from the cumulative literature. *Mov Disord*. 2001;16(3):448–458.

187. Parkkinen L, O'Sullivan SS, Kuoppamaki M, Collins C, Kallis C, Holton JL, et al. Does levodopa accelerate the pathologic process in Parkinson disease brain?. *Neurology*. 2011;77(15): 1420–1426.

188. Lieberman AN, Goldstein M. Bromocriptine in Parkinson disease. *Pharmacol Rev*. 1985;37(2):217–227.

189. Dhawan V, Medcalf P, Stegie F, Jackson G, Basu S, Luce P, et al. Retrospective evaluation of cardio-pulmonary fibrotic side effects in symptomatic patients from a group of 234 Parkinson's disease patients treated with cabergoline. *J Neural Transm*. 2005;112(5):661–668.

190. Evidente VG, Gwinn KA. 80-Year-old woman with dementia and parkinsonism. *Mayo Clin Proc*. 1997;72(12):1171–1174.

191. Markham A, Benfield P. Pergolide: A review of its pharmacology and therapeutic use in Parkinson's disease. *CNS Drugs*. 1997;7(4):328–340.

192. Danoff SK, Grasso ME, Terry PB, Flynn JA. Pleuropulmonary disease due to pergolide use for restless legs syndrome. *Chest*. 2001;120(1):313–316.

193. Rascol O, Lees AJ, Senard JM, Pirtosek Z, Montastruc JL, Fuell D. Ropinirole in the treatment of levodopa-induced motor fluctuations in patients with Parkinson's disease. *Clin Neuropharmacol*. 1996;19(3):234–245.

194. Blandini F, Armentero MT. Dopamine receptor agonists for Parkinson's disease. *Exp Opin Invest Drugs*. 2014;23(3):387–410.

195. Harada T, Ishizaki F, Horie N, Nitta Y, Yamada T, Sasaki T, et al. New dopamine agonist pramipexole improves parkinsonism and depression in Parkinson's disease. *Hiroshima J Med Sci*. 2011;60(4):79–82.

196. Rektorova I, Balaz M, Svatova J, Zarubova K, Honig I, Dostal V, et al. Effects of ropinirole on nonmotor symptoms of Parkinson disease: a prospective multicenter study. *Clin Neuropharmacol*. 2008;31(5):261–266.

197. Millan MJ, Cussac D, Milligan G, Carr C, Audinot V, Gobert A, et al. Antiparkinsonian agent piribedil displays antagonist properties at native, rat, and cloned, human alpha(2)-adrenoceptors: cellular and functional characterization. *J Pharmacol Exp Ther*. 2001;297(3): 876–887.

198. Thobois S, Lhommee E, Klinger H, Ardouin C, Schmitt E, Bichon A, et al. Parkinsonian apathy responds to dopaminergic stimulation of D2/D3 receptors with piribedil. *Brain*. 2013;136(pt 5):1568–1577.

199. Martinez-Martin P, Reddy P, Antonini A, Henriksen T, Katzenschlager R, Odin P, et al. Chronic subcutaneous infusion therapy with apomorphine in advanced Parkinson's disease compared to conventional therapy: a real life study of non motor effect. *J Parkinsons Dis*. 2011;1(2): 197–203.

200. Marconi S, Zwingers T. Comparative efficacy of selegiline versus rasagiline in the treatment of early Parkinson's disease. *Eur Rev Med Pharmacol Sci*. 2014;18(13):1879–1882.

201. Eggert K, Oertel WH, Lees AJ. German Competence Network on Parkinson's Disease. Safety and efficacy of tolcapone in the long-term use in Parkinson disease: an observational study. *Clin Neuropharmacol*. 2014;37(1):1–5.

202. Nutt JG, Woodward WR, Carter JH, Gancher ST. Effect of long-term therapy on the pharmacodynamics of levodopa. Relation to on-off phenomenon. *Arch Neurol.* 1992;49(11):1123–1130.

203. Riley D. Lack of dyskinesias or parkinsonism after long-term exposure to L-DOPA. *Mov Disord.* 1997;12:322.

204. Jenner P. Factors influencing the onset and persistence of dyskinesia in MPTP-treated primates. *Ann Neurol.* 2000;47(4 Suppl. 1):S90–S99.

205. Nutt JG. Clinical pharmacology of levodopa-induced dyskinesia. *Ann Neurol.* 2000;47 (4 Suppl. 1):S160–S164.

206. Pearce RK, Heikkila M, Linden IB, Jenner P. L-Dopa induces dyskinesia in normal monkeys: behavioural and pharmacokinetic observations. *Psychopharmacology.* 2001;156(4):402–409.

207. Togasaki DM, Tan L, Protell P, Di Monte DA, Quik M, Langston JW. Levodopa induces dyskinesias in normal squirrel monkeys. *Ann Neurol.* 2001;50(2):254–257.

208. Bedard PJ, Di Paolo T, Falardeau P, Boucher R. Chronic treatment with L-DOPA, but not bromocriptine induces dyskinesia in MPTP-parkinsonian monkeys. Correlation with [3H] spiperone binding. *Brain Res.* 1986;379(2):294–299.

209. Clarke CE, Sambrook MA, Mitchell IJ, Crossman AR. Levodopa-induced dyskinesia and response fluctuations in primates rendered parkinsonian with 1-methyl-4-phenyl-1,2,3,6-tetrahydropyridine (MPTP). *J Neurol Sci.* 1987;78(3):273–280.

210. Pearce RK, Jackson M, Smith L, Jenner P, Marsden CD. Chronic L-DOPA administration induces dyskinesias in the 1-methyl-4- phenyl-1,2,3,6-tetrahydropyridine-treated common marmoset (*Callithrix jacchus*). *Mov Disord.* 1995;10(6):731–740.

211. Blin J, Bonnet AM, Agid Y. Does levodopa aggravate Parkinson's disease?. *Neurology.* 1988;38(9):1410–1416.

212. Hely MA, Morris JG, Reid WG, O'Sullivan DJ, Williamson PM, Broe GA, et al. Age at onset: the major determinant of outcome in Parkinson's disease. *Acta Neurol Scand.* 1995;92(6):455–463.

213. Cenci MA. Presynaptic mechanisms of L-DOPA-induced dyskinesia: the findings, the debate, and the therapeutic implications. *Front Neurol.* 2014;5:242.

214. Nyholm D, Lewander T, Johansson A, Lewitt PA, Lundqvist C, Aquilonius SM. Enteral levodopa/carbidopa infusion in advanced Parkinson disease: long-term exposure. *Clin Neuropharmacol.* 2008;31(2):63–73.

215. Eggert K, Schrader C, Hahn M, Stamelou M, Russmann A, Dengler R, et al. Continuous jejunal levodopa infusion in patients with advanced parkinson disease: practical aspects and outcome of motor and non-motor complications. *Clin Neuropharmacol.* 2008;31(3):151–166.

216. Antonini A, Isaias IU, Canesi M, Zibetti M, Mancini F, Manfredi L, et al. Duodenal levodopa infusion for advanced Parkinson's disease: 12-month treatment outcome. *Mov Disord.* 2007;22(8):1145–1149.

217. Tel BC, Zeng BY, Cannizzaro C, Pearce RK, Rose S, Jenner P. Alterations in striatal neuropeptide mRNA produced by repeated administration of L-DOPA, ropinirole or bromocriptine correlate with dyskinesia induction in MPTP-treated common marmosets. *Neuroscience.* 2002;115(4):1047–1058.

218. Wood LD. Clinical review and treatment of select adverse effects of dopamine receptor agonists in Parkinson's disease. *Drugs Aging.* 2010;27(4):295–310.

219. Voon V, Fox SH. Medication-related impulse control and repetitive behaviors in Parkinson disease. *Arch Neurol.* 2007;64(8):1089–1096.

220. Weintraub D, David AS, Evans AH, Grant JE, Stacy M. Clinical spectrum of impulse control disorders in Parkinson's disease. *Mov Disord.* 2015;30(2):121–127.

221. Md S, Khan RA, Mustafa G, Chuttani K, Baboota S, Sahni JK, et al. Bromocriptine loaded chitosan nanoparticles intended for direct nose to brain delivery: pharmacodynamic, pharmacokinetic and scintigraphy study in mice model. *Eur J Pharm Sci.* 2013;48(3):393–405.

222. Sharma S, Lohan S, Murthy RSR. Formulation and characterization of intranasal mucoadhesive nanoparticulates and thermo-reversible gel of levodopa for brain delivery. *Drug Dev Ind Pharm.* 2014;40(7):869–878.

223. Gambaryan PY, Kondrasheva IG, Severin ES, Guseva AA, Kamensky AA. Increasing the efficiency of Parkinson's disease treatment using a poly(lactic-*co*-glycolic acid) (PLGA) based L-DOPA delivery system. *Exp Neurobiol.* 2014;23(3):246–252.

224. Zhao Y, Haney MJ, Gupta R, Bohnsack JP, He Z, Kabanov AV, et al. GDNF-transfected macrophages produce potent neuroprotective effects in Parkinson's disease mouse model. *PLoS One.* 2014;9(9):e106867.

225. Li Y, Zhou Y, Qi B, Gong T, Sun X, Fu Y, et al. Brain-specific delivery of dopamine mediated by *N,N*-dimethyl amino group for the treatment of Parkinson's disease. *Mol Pharm.* 2014;11(9):3174–3185.

226. Zhang Y, Schlachetzki F, Zhang Y-F, Boado RJ, Pardridge WM. Normalization of striatal tyrosine hydroxylase and reversal of motor impairment in experimental parkinsonism with intravenous nonviral gene therapy and a brain-specific promoter. *Human Gene Ther.* 2004;15(4):339–350.

CHAPTER 14

Stroke

Özgur Öztop-Çakmak, MD*, Ebru N. Vanli-Yavuz, PhD*, Yasemin Gürsoy-Özdemir, MD, PhD,†**

*Koç University Hospital, Istanbul, Turkey;
**Koç University School of Medicine, Istanbul, Turkey;
†Research Center for Translational Medicine, Istanbul, Turkey

CONTENTS

1 INTRODUCTION

Stroke, being a major health care problem, causes significant morbidity and mortality. It is one of the leading causes of global mortality, following coronary heart disease and cancer. The prevalence of stroke is expected to increase dramatically with increased life expectancy. Globally, stroke causes 9.5% of the total deaths each year, with developing countries having a greater burden. About 17 million strokes cases occur every year, and more than half of these stroke patients are left with a major disability, resulting in lifelong changes for both the survivors and the caregivers.[1–3]

Stroke is caused by either occlusion or rupture of the central nervous system blood vessels. In most of the countries, 73%–86% of stroke cases are ischemic due to an occlusion of a vessel with a blood clot, whereas the remaining 8%–18% cases are hemorrhagic.[4] The most frequent causes of ischemic stroke are atherosclerosis (50%), cardioembolism (20%), and microinfarcts (lacunar strokes, 25%).[5,6] Hypertensive vascular disease is the most common cause of hemorrhagic strokes, occurring due to the rupture of deep penetrating arteries supplying blood, especially to the basal ganglia. This devastating sudden insult in the vasculature leads to failure in the circulation and may also cause neuronal damage.

2 CURRENT TREATMENT STRATEGIES

Although it is a global health problem, there are very limited treatment strategies for stroke. For the past decade, several consensus reports regarding the need for development of new treatment strategies have been published.[7,8] Many laboratories focused on the development of novel therapeutic drugs, as well as the establishment of previously unrecognized mechanisms for stroke,

Nanotechnology Methods for Neurological Diseases and Brain Tumors. http://dx.doi.org/10.1016/B978-0-12-803796-6.00014-9

and have also tried to find new therapeutic approaches, such as novel rehabilitative strategies, for chronic problems.

Current strategies in the treatment of ischemic stroke for the acute phase aim to recanalise the occluded vessel by intravenous thrombolysis and mechanical thrombectomy. Intravenous thrombolysis has been approved as a treatment for acute ischemic stroke in most countries for more than 10 years. Mechanical thrombectomy, using new generation of endovascular tools (stent retrievers), is found to improve the functional outcome either alone or in combination with pharmacological thrombolysis, when indicated. The most important limitation for acute stroke treatment is the urgency caused by the narrow therapeutic window of intravenous thrombolysis. For the latter, intravenous recombinant tissue plasminogen activator (rtPA) for acute ischemic stroke should be used only within 4.5 h of stroke onset, which leads to better outcomes due to earlier treatments.[9] This means that only 2%–4% of stroke patients can be treated with rtPA. The reason for this urgent treatment is that neuronal death and brain infarction evolve progressively in a time-dependent fashion, determined by both the duration and severity of the ischemic insult. Therapeutic strategies designed to restore cerebral perfusion in a time-dependent manner have the potential to limit the cellular, biochemical, and metabolic consequences of cerebral ischemia that ultimately lead to irreversible brain injury. The ultimate goal of early reperfusion therapy is to reduce or prevent brain infarction, and thereby minimize the long-term disability, neurologic impairment, and stroke-related mortality.[10]

3 SAVING PENUMBRAL TISSUE

When a cerebral artery is occluded with a thrombus, blood flow drops dramatically. Depending on the amount of blood flow decrease, ischemic area is divided into two parts: ischemic core and ischemic penumbra region. In the core area, the blood flow drops below 20% of the normal level, whereas in ischemic penumbra, blood flow circulation is maintained at around 30%–70% of normal levels. Although there is a decrease in the blood flow, the energy level is enough to provide cell survival in the penumbral area. However, as time passes and amount of energy (ATP) becomes insufficient, it becomes more difficult to provide intracellular and extracellular ionic gradients leading to neuronal swelling and necrotic cell death. Moreover, calcium channels are activated, and this leads to release of excitatory amino acids and therefore to excitatoxicity. Intracellular increase in calcium amount also triggers apoptotic cell death pathways.This cascade of events finally leads to penumbral neuronal cell death either via necrosis or apoptosis, depending on the surrounding conditions. For this reason, the time window is important, so that energy and blood flow to the penumbral tissue can be maintained, and cell viability remains unaffected. Besides, in the ischemic core area, cells rapidly undergo irreversible necrosis, which underlines why treatment options for stroke focus on penumbral tissue survival. The two approaches established for tissue rescue in the penumbra are recanalization and neuroprotection.

4 NEUROPROTECTIVE STRATEGIES

For the last 2 decades, preclinical stroke studies have focused on novel neuroprotective agents targeting neuronal cell damage especially. Although recanalization, and hence reperfusion of ischemic tissue, is a good approach, it has a very limited effect on neuronal death. Targeting the brain parenchyma with pharmacological compounds (called neuroprotection) may also be a promising strategy to curb the spread of infarcted tissue. However so far, none of the investigated agents have demonstrated positive results in clinical trials.[11] The Stroke Treatment Academic Industry Roundtable (STAIR) has already made seven consensus meetings, and the final STAIR meeting even offered recommendations for neuroprotective strategies. They stated that "The strategy of delivering neuroprotective therapies before reperfusion treatments to extend penumbra survival is attractive and could potentiate the benefit of early reperfusion therapy and expand the time window for late reperfusion interventions" and "Selective cerebral delivery of neuroprotection interventions is a promising strategy to maximize local neuroprotection and minimize systemic toxicity."[8]

One of the main obstacles for effective neuroprotective treatment strategies is the presence of the blood–brain barrier (BBB). Although BBB is damaged after ischemic insult, it opens hours later after ischemia.[12,13] On the other hand, for a neuroprotective agent to be effective, it must be administered within the early hours after ischemia, when BBB is still functioning and can prevent entry of substances into the brain tissue. Hence, targeted drug delivery strategies across the BBB would open a new era for stroke treatment.

4.1 Nanotherapeutics

As we have learned about the mechanisms of cell death in the neurodegenerative process during the last decades, it is now possible to use that information to target several steps of the neuronal cell death process. Nanomaterials could be used for the generation of more effective therapeutics for these purposes. They can be used as tools to eliminate the high rate of drug metabolism, rapidly clear substances in central nervous system, and enable the delivery of substances across the BBB. Receptor- and adsorptive-mediated transcytosis seem to be the most promising mechanisms for enhancing transcellular transport of nanomedicine from the blood to the brain.[14,15]

Today, nanomedicine treatment studies for stroke focus on recanalization, neuroprotection, and imaging.[16,17] Recanalization strategies are based on improving the effects of the thrombolytic and antiaggregant drugs, such as rtPA and salicylic acid, in a relatively longer time window, as well as minimizing the risk for hemorrhagic complications and neurotoxicity, by lowering the amount of the drugs used. Korin et al.[18,19] developed shear stress–activated nanoparticles, which become activated in narrowed atherosclerotic areas of blood flow and inhibit thrombocyte activation. Under normal conditions, the physiology of blood flow is strictly controlled. Shear stress is a major determinant of vascular pathophysiology, which occurs due to atherosclerosis and narrowing of the vessel. Hemodynamic regulatory control mechanisms limit the shear stress in a certain range. Platelets can sense the abnormal shear stress, and react by sticking to the vascular wall and aggregate. This physiological feature inspired the creation of mechanically activated nanotherapeutics, which specifically target drugs to flow-obstructed vessels and concentrate thrombolytic agents at these sites.[18] Shear-activated nanotherapeutics coated with tPA were engineered in a micrometer-size range with the ability to disintegrate only at pathologic levels of shear stress. These nanoparticles were tested in acute ischemic stroke, acute myocardial infarction, and massive pulmonary embolism. The thrombolytic drug was distributed to the place it was most required by shear stress. The results of the studies suggest that the dose of the administrated drug can be decreased. Moreover, faster recanalization and lowering the risk of hemorrhage can beachieved.[18–20]

For a neuroprotective agent to be successful, ischemic cell damage mechanisms need to be targeted. Ischemia leads to the development of a cascade of pathological and biochemical reactions: over production of free radicals, activation of the immune system, gene overexpression, and cell death. Neuroprotection can be achieved by inhibiting proinflammatory cytokine production, apoptosis, and cell adhesion mechanisms and decreasing lipid peroxidation processes. In the light of this knowledge, many antioxidant molecules, as well as antiinflammatory agents became candidates for neuroprotection in stroke models. Moreover, many of the agents were assumed as neuroprotective, and both preclinical and clinical studies were performed. Although preclinical studies were promising, clinical trials were disappointing. One of the reasons for this ineffective treatment may be related to the inability of those reagents to achieve an adequate concentration in the infarct zone in the brain or to cross the BBB. Recent preclinical studies suggest that application of neuroprotective agents as brain-targeted nanoparticles or formulation optimization for better bioavailability will overcome these issues.

4.2 Improving Bioavailability

There are several studies performed with improved bioavailability and better outcomes in stroke models. In one of them, Yun et al.[21] studied the impact of various nanoparticles [liposomes, polybutylcyanoacrylate (PBCA), or poly(lactide-*co*-glycolide) (PLGA)] that contained the active superoxide dismutase (SOD) enzyme (4,000–20,000 U/kg) in a mouse model of cerebral ischemia and reperfusion injury. The nanoparticles were directed against *N*-methyl-D-aspartate receptor-1 (NMDA-1). The nanoparticles protected primary neurons from oxygen–glucose deprivation, and limited ongoing apoptosis in in vitro models. In addition when applied after injury, they reduced the infarct volume by 50%–60%, reduced levels of inflammatory markers, and improved behavioral outcomes.[21] Similarly, in a transient focal ischemia model in rats, Sopala et al.[22] demonstrated the neuroprotective activity of glycine B site NMDA receptor antagonist MRZ2/576. The formulated nanoparticles provided prolonged intravenous administration of relevant doses within the physiological range of pH. Normally this compound is not soluble at physiological pH, but the formulation provided an effective intravenous application for 6 h with minimal side effects, and a 53% reduction of total infarct size.[22] Adenosine is also a beneficial active nucleoside in several neurological diseases. Due to its short plasma half-life, moderate side effects, and the inability to pass the BBB, it's not possible to use adenosine in the treatment of neurodegenerative diseases. In a recent study, Gaudin et al.[23] designed a biocompatible lipid molecule by conjugating adenosine to squalene. This created molecule is an amphiphilic prodrug that spontaneously forms 120-nm nanoparticles, allowing prolonged drug interaction with the neurovascular unit, as well as minimizing systemic toxicity and side effects. In a mouse model of ischemia, systemic administration of this squalene–adenosine nanoassembly showed significant neuroprotection. These results suggest that alterations in drug bioavailability can enhance the neuroprotective effect.[23]

4.3 Reaching the Brain via BBB

Other than increasing bioavailability, directly targeting drugs to pass through the BBB is another strategy. Caspase activation takes place in cell death mechanisms in acute and also chronic neurodegenerative diseases. Growth factors suppress cell death at several steps in the pathway, and also promote regeneration, which therefore makes them therapeutic candidates for the treatment of stroke. Brain-targeted nanoparticles can transport large amounts of BBB-impermeable agents across the BBB via systemic administration. In a recent study, Yemisci et al.[24] prepared therapeutic nanopeptides that could pass from the blood to the brain and decrease infarct size in a stroke model of mice. They loaded chitosan particles with a caspase inhibitor, z-DEVD-FMK, and a growth factor, basic fibroblast growth factor (bFGF), and administered the brain-targeted nanoparticles systemically. These nanoparticles crossed the BBB via receptor-mediated transcytosis through transferrin receptor-1, which is found on the endothelial cells that form the BBB. In this study, it is shown that not only small peptides, but also larger peptides, such as bFGF, can be encapsulated into chitosan nanopeptides and targeted to the brain tissue. These studies also showed the possibility of encapsulating multiple agents together to achieve additive effects.[24] In another study, Karatas-Kursun et al.[25] demonstrated that those nanoparticles were rapidly transported across the

BBB, without being measurably taken up by liver and spleen, and reached the brain tissue to provide effective release for at least 3 h after the application. Pre- or posttreatment (2 h) with intravenously injected z-DEVD-FMK–loaded nanospheres (a caspase inhibitor), dose dependently decreased the infarct volume, neurological deficit, and ischemia-induced caspase-3 activity in mice subjected to 2 h of MCA occlusion and 24 h of reperfusion, suggesting that they released sufficient amount of peptide to inhibit caspase activity. They also clearly showed that nanospheres were able to inhibit physiological caspase-3 activity during development in the neonatal mouse cerebellum on postnatal day 17 after the closure of the BBB.[25]

5 IMAGING TECHNIQUES

Nanoparticles recently became an area of focus for imaging studies. For imaging purposes, nanoparticles must have a relatively long-circulating half-life, be sensitive and selective against the epitope of interest, have a prominent contrast-to-noise enhancement, be nontoxic, and ultimately have the capacity to be imaged with commercially available equipment, such as with magnetic resonance imaging (MRI) or computed tomography (CT).[17,26] For clinical success in the treatment of stroke, it would be ideal if the place and amount of intravascular thrombosis are documented. To achieve this issue, Yu et al. developed gadolinium-labeled fibrin–avid nanoparticles and labeled intravascular fibrin.[27] Similarly, recently McCarthy et al. used fluorescently labeled cross-linked iron oxide nanoparticles functionalized with GPR or FXIIIa-targeting peptides. In animal models, these particles successfully detected the thrombus formation tracked by MRI.[28,29]

In acute ischemic stroke, MRI is not the first choice of imaging, mostly due to time limitations and availability. Hence CT-compatible nanoparticles are required for thrombus detection. Kim et al.[30] recently developed a contrast agent utilizing glycol chitosan–gold nanoparticles. These nanoparticles accumulated in the thrombus and allowed CT visualization of both the presence and extent of primary and recurrent thrombi in mouse carotid arteries, without a single failure of detection. Besides, thrombus imaging with these nanoparticles was also effective in monitoring the therapeutic efficacy of thrombolysis.[30]

6 CONCLUSIONS

In line with these studies and new knowledge gained about stroke pathophysiology, targeted treatment strategies will have a great impact on stroke diagnosis and treatment. It is an exciting and novel area for stroke research. Both increasing the bioavailability and directly targeting drugs to brain tissue will increase success in clinical settings. These novel treatments can also be used as adjuvant treatments applied together with other stroke treatment options. One of the drawbacks is the limited number of in vivo studies. Although there are a lot of neuroprotective agents, only a limited number of them are thoroughly investigated in animal models. Future studies are needed for a better understanding of their clinical safety and feasibility. Careful screening of their toxicities and determination of their pharmacokinetic and pharmacodynamic profiles are necessary before they can be passed through clinical trials.

Abbreviations

BBB Blood–brain barrier
bFGF Basic fibroblast growth factor
CT Computed tomography
MRI Magnetic resonance imaging
NMDA *N*-Methyl-D-aspartate receptor
PBCA Polybutylcyanoacrylate
PLGA Poly(lactide-*co*-glycolide)
rtPA Recombinant tissue plasminogen activator
SOD Superoxide dismutase
STAIR Stroke Treatment Academic Industry Roundtable

References

1. Camak DJ. Addressing the burden of stroke caregivers: a literature review. *J Clin Nurs.* 2015;24(17–18):2376–2382.
2. Woimant F, Biteye Y, Chaine P, Crozier S. Severe stroke: which medicine for which results?. *Ann Fr Anesth Reanim.* 2014;33(2):102–109.
3. Adamson J, Beswick A, Ebrahim S. Is stroke the most common cause of disability?. *J Stroke Cerebrovasc Dis.* 2004;13(4):171–177.
4. Shiber JR, Fontane E, Adewale A. Stroke registry: hemorrhagic vs ischemic strokes. *Am J Emerg Med.* 2010;28(3):331–333.
5. Kolominsky-Rabas PL, Weber M, Gefeller O, Neundoerfer B, Heuschmann PU. Epidemiology of ischemic stroke subtypes according to TOAST criteria: incidence, recurrence, and long-term survival in ischemic stroke subtypes: a population-based study. *Stroke.* 2001;32(12): 2735–2740.
6. Kim BJ, Kim JS. Ischemic stroke subtype classification: an Asian viewpoint. *J Stroke.* 2014;16(1):8–17.
7. Stroke Therapy Academic Industry RoundtableRecommendations for standards regarding preclinical neuroprotective and restorative drug development. *Stroke.* 1999;30(12):2752–2758.
8. Albers GW, Goldstein LB, Hess DC, Wechsler LR, Furie KL, Gorelick PB, et al. Stroke Treatment Academic Industry Roundtable (STAIR) recommendations for maximizing the use of intravenous thrombolytics and expanding treatment options with intra-arterial and neuroprotective therapies. *Stroke.* 2011;42(9):2645–2650.
9. Marler JR, Tilley BC, Lu M, Brott TG, Lyden PC, Grotta JC, et al. Early stroke treatment associated with better outcome: the NINDS rt-PA stroke study. *Neurology.* 2000;55(11):1649–1655.
10. Albers GW, Amarenco P, Easton JD, Sacco RL, Teal P. Antithrombotic and thrombolytic therapy for ischemic stroke: American College of Chest Physicians Evidence-Based Clinical Practice Guidelines (8th Edition). American College of Chest Physicians. *Chest.* 2008;133 (6 Suppl.):630S–669S.

11. Sahota P, Savitz SI. Investigational therapies for ischemic stroke: neuroprotection and neurorecovery. *Neurotherapeutics*. 2011;8(3):434–451.
12. Gursoy-Ozdemir Y, Bolay H, Saribas O, Dalkara T. Role of endothelial nitric oxide generation and peroxynitrite formation in reperfusion injury after focal cerebral ischemia. *Stroke*. 2000;31(8):1974–1980.
13. Gursoy-Ozdemir Y, Can A, Dalkara T. Reperfusion-induced oxidative/nitrative injury to neurovascular unit after focal cerebral ischemia. *Stroke*. 2004;35(6):1449–1453.
14. Re F, Gregori M, Masserini M. Nanotechnology for neurodegenerative disorders. *Nanomedicine*. 2012;8(Suppl. 1):S51–S58.
15. Omidi Y, Barar J. Impacts of blood-brain barrier in drug delivery and targeting of brain tumors. *Bioimpacts*. 2012;2(1):5–22.
16. Shcharbina N, Shcharbin D, Bryszewska M. Nanomaterials in stroke treatment: perspectives. *Stroke*. 2013;44(8):2351–2355.
17. Kyle S, Saha S. Nanotechnology for the detection and therapy of stroke. *Adv Health Mater*. 2014;3(11):1703–1720.
18. Korin N, Kanapathipillai M, Matthews BD, Crescente M, Brill A, Mammoto T, et al. Shear-activated nanotherapeutics for drug targeting to obstructed blood vessels. *Science*. 2012;337(6095):738–742.
19. Korin N, Gounis MJ, Wakhloo AK, Ingber DE. Targeted drug delivery to flow-obstructed blood vessels using mechanically activated nanotherapeutics. *JAMA Neurol*. 2015;72(1):119–122.
20. IWootton DM, Alevriadou BR. The shear stress of busting blood clots. *N Engl J Med*. 2012;367(14):1361–1363.
21. Yun X, Maximov VD, Yu J, Zhu H, Vertegel AA, Kindy MS. Nanoparticles for targeted delivery of antioxidant enzymes to the brain after cerebral ischemia and reperfusion injury. *J Cereb Blood Flow Metab*. 2013;33(4):583–592.
22. Sopala M, Schweizer S, Schafer N, Nurnberg E, Kreuter J, Seiller E, et al. Neuroprotective activity of a nanoparticulate formulation of the glycineB site antagonist MRZ 2/576 in transient focal ischaemia in rats. *Arzneimittelforschung*. 2002;52(3):168–174.
23. Gaudin A, Yemisci M, Eroglu H, Lepetre-Mouelhi S, Turkoglu OF, Donmez-Demir B, et al. Squalenoyl adenosine nanoparticles provide neuroprotection after stroke and spinal cord injury. *Nat Nanotechnol*. 2014;9(12):1054–1062.
24. Yemisci M, Caban S, Gursoy-Ozdemir Y, Lule S, Novoa-Carballal R, Riguera R, et al. Systemically administered brain-targeted nanoparticles transport peptides across the blood-brain barrier and provide neuroprotection. *J Cereb Blood Flow Metab*. 2015;35(3):469–475.
25. Karatas H, Aktas Y, Gursoy-Ozdemir Y, Bodur E, Yemisci M, Caban S, et al. A nanomedicine transports a peptide caspase-3 inhibitor across the blood-brain barrier and provides neuroprotection. *J Neurosci*. 2009;29(44):13761–13769.
26. Feigin VL, Forouzanfar MH, Krishnamurthi R, Mensah GA, Connor M, Bennett DA, et al. Global and regional burden of stroke during 1990-2010: findings from the Global Burden of Disease Study 2010. *Lancet*. 2014;383(9913):245–254.
27. Yu X, Song SK, Chen J, Scott MJ, Fuhrhop RJ, Hall CS, High-resolution MRI, et al. characterization of human thrombus using a novel fibrin-targeted paramagnetic nanoparticle contrast agent. *Magn Reson Med*. 2000;44(6):867–872.
28. McCarthy JR, Patel P, Botnaru I, Haghayeghi P, Weissleder R, Jaffer FA. Multimodal nanoagents for the detection of intravascular thrombi. *Bioconjug Chem*. 2009;20(6):1251–1255.
29. Cicha I, Thrombosis:. Novel nanomedical concepts of diagnosis and treatment. *World J Cardiol*. 2015;7(8):434–441.
30. Kim DE, Kim JY, Sun IC, Schellingerhout D, Lee SK, Ahn CH, et al. Hyperacute direct thrombus imaging using computed tomography and gold nanoparticles. *Ann Neurol*. 2013;73(5):617–625.

CHAPTER 15

Nanotechnology-Based Management of Neurological Autoimmune Diseases

Erdem Tüzün, MD
Institute for Experimental Medicine, Istanbul University, Istanbul, Turkey

CONTENTS

1 GENERAL CONCEPTS

Nanotechnology has recently emerged as an exciting and prospective tool for treating neurological disorders. Advances in nanoengineering and methods for conjugation of therapeutic molecules to nanoparticles have enabled innovation of a wide variety of nanomaterials with the capability of crossing the blood–brain barrier (BBB), and thus gaining access into critical target central nervous system (CNS) structures. While the majority of these novel nanomolecules have been utilized to support cell survival in damaged neurons or interfere with intraneuronal molecular responses in brain tumors, many others have been designed to suppress inflammation in the CNS and to ameliorate symptoms of neurological disorders induced by autoimmunity.[1–3]

CNS inflammation is primarily determined by astrocytes and microglia that are activated upon exposure to pathogens, which leads to the release of proinflammatory cytokines (e.g., IL-1β and IL-6) and chemotactic molecules (e.g., CCL2, CXCL2, and CXCL12), altered expression of adhesion molecules, and simultaneous production of reactive oxygen species, which lead to neuronal cell death, disruption of the BBB, and recruitment of peripheral immune cells (e.g., neutrophils, eosinophils, monocytes, and T lymphocytes) to the CNS. This sequence of pathogenic events is not only restricted to autoimmune or infectious diseases, but also includes trauma, stroke, and neurodegenerative disorders.[4–6] In several preclinical studies, nanoparticles were effectively used to suppress inflammation cascades. Antiinflammatory reagents within nanoparticles that cross the BBB may be taken up by immune cells and subsequently interfere with intracellular inflammation pathways or interact with biomolecules present on cell membranes, and consequently alter mobility and adhesion of immune cells.[6,7]

The BBB is situated along the brain capillaries and is made up of endothelial cells that are connected by tight junctions, a thick basement membrane, pericytes, and astrocytic end feet. The BBB restricts the diffusion of microorganisms

Nanotechnology Methods for Neurological Diseases and Brain Tumors. http://dx.doi.org/10.1016/B978-0-12-803796-6.00015-0

and certain molecules with high molecular weight or hydrophilic features, whereas certain essential molecules, such as glucose, are actively transported through the BBB by specific transport mechanisms.[3,8] The BBB and pericytes, members of the reticuloendothelial system (RES) family, significantly reduce the likelihood of delivery of conventional drugs to active inflammation sites in the CNS. Utilization of nanoparticles with active targeting and transport strategies [e.g., coating liposomes with polyethylene glycol (PEG) or antibodies to the transferrin receptor] increases penetration, half-life, and transport capacity of drugs into the CNS, prevents their rapid RES clearance, and increases the bioavailability of antiinflammatory drugs in the brain.[9,10] In addition to treatment approaches, these novel nanoparticle targeting strategies have also enhanced the value of nanotechnology in imaging of neuroinflammation.[11]

2 A USEFUL MODEL FOR NANOPARTICLE EXPERIMENTS ON CNS AUTOIMMUNITY

Multiple sclerosis (MS) is a chronic autoimmune disease of the CNS, characterized by inflammation, demyelination, and neurodegeneration and by lesions that are distributed into white matter, optic nerve, brain stem, and spinal cord. These lesions are dominated by perivascular infiltrates of CD4+ and CD8+ T lymphocytes, myelin-loaded macrophages, monocytes, and rare B lymphocytes. Resident CNS microglia and infiltrating inflammatory cells contribute to myelin loss in earlier stages, and to irreversible axonal degeneration in successive progressive stages of the disease.[12]

MS, one of the most disabling neurological disorders of young adults, is a prototypical autoimmune disease of the CNS. Hereupon, nanoparticle-based experiments, focused on disease mechanisms, diagnosis, and treatment of autoimmune CNS disorders, have extensively utilized animal models of MS. The animal model used to study pathogenic and therapeutic aspects of MS is the experimental autoimmune encephalomyelitis (EAE). EAE is generally induced in mice and rats by immunization with myelin proteins or peptides in complete Freund's adjuvant (CFA).[13,14] The EAE model closely mimics the foremost pathological features of MS, such as BBB dysfunction, mononuclear cell infiltration, Th1- and Th17-type cytokine production, and neuronal tissue degradation[14,15] and is thus considered as a valuable model for MS.

3 IMAGING OF INFLAMMATORY NERVOUS SYSTEM DISORDERS BY NANOPARTICLES

Magnetic resonance imaging (MRI) techniques are the primary means for diagnosing CNS autoimmune diseases. MRI studies are also used in follow-up studies of patients with inflammatory brain disorders and in studies investigating the pathophysiology of autoimmune neurological diseases. However, MRI does not address specific pathological processes involved in disease mechanisms. With the support of nanotechnology, contrast agents can be conjugated to specific cells or antibodies to carry out imaging of specific infiltrating immune cells.[16]

The most commonly used contrast agents in nanoparticle-based imaging are the superparamagnetic iron oxide (SPIO) (50–180 nm diameter), ultrasmall superparamagnetic iron oxide (USPIO) (10–50 nm diameter), and very small superparamagnetic iron oxide (VSPIO) ($<$10 nm diameter) particles.[6,16] These reagents decrease the longitudinal and transverse relaxation times of water molecules, producing a hyperintense signal on T1-weighted images and a hypointense signal on T2-weighted images.[17] These superparamagnetic nanoparticles have enabled safe and continuous visualization of cell trafficking from the blood stream to the brain in living organisms, and thus have been proven useful not only in autoimmune neurological diseases, but also in the evaluation of CNS inflammation induced by stroke and traumatic nerve injury.[6,16,18,19]

Due to their larger size, iron oxide particles extravasate slower than gadolinium chelates, which are routinely used in MRI studies, and thus can more accurately assess the cerebral blood volume and blood flow rather than gadolinium-based contrast agents.[18,20] Furthermore, following intravenous administration, ferromagnetic nanoparticles are actively phagocytosed by circulating immune cells, enabling visualization and tracking of the extravasation of infiltrating macrophages, polymorphonuclear cells, and lymphocytes into inflammation zones in the CNS.[21] As SPIO particles have a larger diameter, they can be absorbed by RES cells, mostly located in liver, spleen, and lymph nodes. USPIO and VSPIO particles are therefore preferred in animal model studies.[22] An alternative to directly injecting iron oxide particles into the circulation is ex vivo labeling of peripheral blood monocytes, which includes the removal of peripheral blood, exposure of mononuclear cells to iron oxide nanoparticles, and injection of the labeled monocytes into the blood stream. An important limitation of this method is that iron oxide labels can leach or be released from injected cells and can be phagocytosed by other cell types, including the resident microglia,[22] reducing the specificity of the technique.

SPIO particles have also found their use in EAE. They enable cell-specific visualization of CNS-infiltrating immune cells after the establishment of clinical symptoms. Several studies have shown that USPIO and VSPIO are not necessarily more sensitive than gadolinium-based contrast agents in demonstrating established inflammatory demyelinating lesions in EAE.[23–25] However, they can be used to track the trafficking of immune cells to lesion sites and indicate the presence of lesions that occur before the disruption of the BBB, which are not visible by gadolinium-based agents.[25] In a recent study, VSPIO particles were used to identify CNS alterations that precede the clinical disease onset, by monitoring early inflammatory processes in EAE. In this study, long before the development of inflammatory plaques in the white matter, immune cells have been found to accumulate in the choroid plexus and perivascular areas, prior to infiltrating the brain parenchyma.[24] In a rat model of EAE, USPIO-positive MRI lesions were correlated with disease severity,[26] implying that nanotechnology can serve as a prognostic tool in MS patients.

An alternative method to pinpoint specific cell populations in living organisms is to use superparamagnetic particles bound to monoclonal antibodies. For instance, CD3+ T cells located in inflammatory white matter lesions have been shown in a murine model of EAE using iron oxide–conjugated monoclonal antibodies.[27] Likewise, expression of intercellular adhesion molecule 1 (ICAM-1), which is important in EAE and MS pathogenesis, has been demonstrated by antibody-conjugated paramagnetic liposomes.[28]

4 NANOTECHNOLOGY IN THE TREATMENT OF AUTOIMMUNE NEUROLOGICAL DISORDERS

Preclinical studies of nanotechnology-based treatment methods have been mostly performed by using the EAE model. The relapsing remitting phase of MS is characterized with CNS inflammation, and generally favorable responses to immunosuppressants. This phase is followed by a progressive stage characterized by neurodegeneration and absence of response to immunomodulation.[12] Significant progress has been achieved for the treatment of the earlier stages of MS, and most of the nanoparticle-based treatment methods address this initial phase. Nanoparticles have been used to deliver antiinflammatory compounds to specific lesion sites, to carry autoantigens to tolerogenic antigen-presenting cells and to inhibit specific immunological factors.[6,11,16]

5 ADMINISTRATION OF ANTIINFLAMMATORY AGENTS

PEG liposomes accumulate predominantly within sites of inflammation[29] and therefore have been used for specific delivery of antiinflammatory agents in autoimmune disease models. CNS delivery of methylprednisolone by the use of PEGylated liposomes has enhanced its efficacy in suppressing EAE, and made it superior not only to free methylprednisolone, but also to interferon-beta and glatiramer acetate, which are the currently used immunomodulators for MS treatment.[30–32]

PEG liposomes loaded with minocycline, a potent inhibitor of matrix metalloproteinase-9, have been administered to mice with EAE to prevent the entry of immune cells from the peripheral blood into the CNS. Injection of

PEG–minocycline liposomes every 5 days has been as effective, as daily injections of free minocycline, in ameliorating clinical and pathological features of EAE.[33] In an earlier study in a rat model, injection of liposome-encapsulated dichlormethylene diphosphonate, an antagonist of phagocytic cells, effectively suppressed the Guillain–Barré syndrome, an autoimmune disease of the peripheral nervous system.[34]

Nanotechnology can also be used for selective targeting of specific immune cell subpopulations. CD40L is mostly expressed by activated T and B lymphocytes. Liposomes loaded with the cytostatic drug methotrexate and conjugated with a CD40L-specific peptide ligand have significantly reduced the CD40L+ cell ratios in peripheral blood and spleen, and thus ameliorated clinical muscle weakness in a mouse EAE model.[35] Moreover, in three different studies, researchers have taken advantage of the lymphotoxic action of the diphtheria toxin to reduce the numbers of specific immune cell populations that are involved in EAE pathogenesis. Administration of liposome-embedded diphtheria toxin conjugated with IL-18, interferon gamma–inducible protein 10 (IP-10), or regulated by activating normal T cell expressed and secreted (RANTES) molecules has improved EAE symptoms of mice by reducing the proportions of infiltrating IL-18 receptor–positive, CXCR3-positive, and CCR5-positive cells, respectively.[36–38]

Notably, certain nanoparticles have been found to demonstrate direct anti-inflammatory effects. Iron oxide nanoparticles appear to suppress IL-1β production of lipopolysaccharide-stimulated microglia.[39] Similarly, myelin basic protein-reactive T cells cultured with antigen-presenting cells previously exposed to multiwalled carbon nanotubes show deficient IL-17 production due to the increased antigen-presenting cell expression of IL-27, a cytokine that inhibits Th17-type immune responses. Moreover, adaptive transfer of these cells to naïve mice causes less severe EAE when compared to T cells that are not exposed to multiwalled carbon nanotubes.[40]

6 TOLERANCE INDUCTION

Autoimmune inflammation is governed by cognate interactions between antigen-presenting cells and autoantigen-specific T cells. Alternatively, recognition of autoantigens by antigen-presenting cells in noninflammatory conditions might promote tolerance by expanding antigen-specific T regulatory cells and thus ameliorating tissue destruction induced by autoimmunity.[41–44]

Administration of autoantigens, together with immunoregulatory molecules in a nanoparticle form, enables their controlled release for several days to weeks. The uptake of myelin peptides and immunoregulatory molecules by lymphoid tissues is followed by T-cell anergy (a shift from the Th1/Th17-type immune responses to T regulatory–type immune responses), reduction in the number of autoreactive lymphocytes, and alleviation of clinical symptoms of EAE. For this purpose, different types of nanoparticles loaded with myelin peptides and immunoregulatory molecules have been used in immunization-based EAE models.[6] In a recent study, subcutaneous administration of the myelin oligodendrocyte glycoprotein (MOG_{35-55}) autoantigen and recombinant IL-10 in polymeric biodegradable lactic–glycolic acid (PLGA) nanoparticles significantly reduced clinical scores, demyelination, CD3+ T-cell infiltrates in the CNS, and secretion of IFN-γ (Th1) and IL-17 (Th17) by spleen lymphocytes in the MOG-induced mouse model of EAE.[45]

An alternative approach is to inject dendritic cells modified by nanoparticles containing myelin peptides and immunoregulatory molecules. This treatment causes a shift in dendritic cells toward an antiinflammatory profile. As an example, injection of dendritic cells modified with nanoparticles loaded with MOG and coinhibitory receptor B and T lymphocyte attenuator (BTLA) increased Foxp3+ CD4+ regulatory T cells, increased IL-10 and TGF-β production, and decreased IL-2 and IFN-γ secretion, leading ultimately to a reduction in CD4+ T-cell responses to MOG, cellular infiltrates in the CNS, and clinical severity of EAE.[46] Similar results have been obtained by treatment of MOG-immunized mice with dendritic cells modified by myelin peptides and the immunoregulatory molecule 2-(1′H-indole-3′-carbonyl)-thiazole-4-carboxylic acid methyl ester (ITE).[47]

7 NEUROPROTECTION

Ongoing inflammation in MS/EAE plaques brings forward oxidative stress with mitochondria impairment, which in due course enhances axonal loss and neuronal death caused by inflammation.[48] Nanoparticles loaded with antioxidant agents, such as tempamine, have been effectively used as a neuroprotective measure. Injection of tempamine-loaded liposomes has reduced clinical scores and cellular infiltrates in mouse EAE models induced by proteolipid protein or MOG immunization.[49] In another study, administration of cerium oxide nanoparticles improved muscle weakness, which was induced by MOG immunization, by avoiding free radical accumulation in the CNS.[50]

8 NANOTECHNOLOGY IN NONAUTOIMMUNE NEUROINFLAMMATION

Nanoparticles have been widely used in the diagnosis and/or treatment of inflammation induced by degenerative, traumatic, and cerebrovascular disorders of the nervous system.[51–54] In the rat model of familial amyotrophic lateral sclerosis (ALS), neuroimaging studies performed with USPIO nanoparticles disclosed CD4+ and CD8+ T-cell infiltrates in nonspinal CNS regions, such as brainstem and hippocampus. Simultaneous studies performed with gadolinium contrast also confirmed that the BBB was compromised in these regions.[51] In the rat model of Alzheimer's disease, administration of lipid-core nanocapsules loaded with the nonsteroidal antiinflammatory drug, indomethacin, ameliorated clinical symptoms, neuronal cell death, and inflammation, which were induced by amyloid beta 1–42.[52] Likewise, lipid-core nanocapsules loaded with resveratrol, an antiinflammatory and neuroprotective compound, have suppressed amyloid beta 1–42–induced inflammation and cell death in hippocampal cultures more effectively than free resveratrol.[55] In animal models of ischemic stroke, infiltrating inflammatory cells can be visualized by iron oxide particles. More importantly, USPIO-labeled antibodies to specific cell adhesion molecules, such as E-selectin and vascular cell adhesion molecule 1 (VCAM-1), have enabled imaging of inflamed cerebrovascular endothelial cells in a stroke model.[53]

9 NEUROINFLAMMATORY ADVERSE EFFECTS OF NANOPARTICLES

An important aspect of nanotechnology is the possible CNS inflammatory reaction in response to administered nanomaterials. Industrial nanoparticles (particles <100 nm in size) originating from various workplaces and combustion engines (e.g., diesel exhaust particles, welding fumes, carbon nanotubes, and titanium dioxide) have been shown to display neurotoxic effects. As clearly demonstrated in animal studies, toxic nanoparticles may enter the brain via the olfactory bulb after inhalation, and may also accumulate in the systemic circulation. In the brain, nanoparticles induce inflammation, apoptosis, and oxidative stress, ultimately leading to neurodegeneration. Nanoparticles appear to induce neuroinflammation, mainly by triggering and enhancing the expression of proinflammatory cytokines (e.g., IL-1β, IL-6 and TNF-α), chemokines (e.g., CCL2, CCL3, and CXCL9), prostaglandins, and the transcription factor NF-κB.[56–58] These findings put forward the possibility that nanoparticles produced for medical purposes might also induce inflammation.

In both cell culture and animal studies, iron oxide particles that are widely used in neuroimaging have been shown to accumulate in CNS glial cells, impair cell viability, induce apoptosis, and increase the expression levels of proinflammatory cytokines.[58] Likewise, upon intravenous injection to mice, lipid nanoparticles have been shown to accumulate in microglia for several months and induce apoptotic and inflammatory responses.[59]

10 CONCLUDING REMARKS

Recently established novel therapeutic and diagnostic techniques are promising, and relatively noninvasive methods might improve the management of autoimmune CNS disorders. Development and usage of iron oxide–labeled antibodies directed against specific molecules might provide a platform for the detection of disease markers that are not noticed by conventional imaging techniques.

Although nanotechnology has found massive use in treatment of neuroinflammation, treatment trials are restricted to animal models and therefore rapid and urgent translation of these results to patients is required. In this context, progress in research for innovation of treatment methods that will protect and repair injured axons and neurons or induce remyelination in the progressive phase of MS still appears to be slow.

Despite the well-known neurotoxic effects of nanoparticles, most preclinical studies evaluating the therapeutic effects of nanotechnology have failed to investigate the toxicity of administered molecules. It is imperative to calculate risk–benefit ratios and make treatment decisions based on these assessments. Future studies in animal models and human subjects, together with satisfactory assessment of safety issues associated with nanoparticle injection, will provide further opportunities for the efficacy of nanotechnology-based methods in preclinical and clinical neuroimmunology.

Abbreviations

ALS Amyotrophic lateral sclerosis
BBB Blood–brain barrier
BTLA B and T lymphocyte attenuator
CFA Complete Freund's adjuvant
CNS Central nervous system
EAE Experimental autoimmune encephalomyelitis
ICAM-1 Intercellular adhesion molecule 1
IP-10 Interferon gamma–inducible protein 10
ITE 2-(1′H-indole-3′-carbonyl)-thiazole-4-carboxylic acid methyl ester
MOG Myelin oligodendrocyte glycoprotein
MRI Magnetic resonance imaging
MS Multiple sclerosis
PEG Polyethylene glycol
PLGA Polymeric biodegradable lactic–glycolic acid
RANTES Regulated on activation normal T cells expressed and secreted
RES Reticuloendothelial system
SPIO Superparamagnetic iron oxide
USPIO Ultrasmall superparamagnetic iron oxide
VCAM-1 Vascular cell adhesion molecule 1
VSPIO Very small superparamagnetic iron oxide

References

1. Lin LN, Liu Q, Song L, Liu FF, Sha JX. Recent advances in nanotechnology based drug delivery to the brain. *Cytotechnology*. 2010;62:377–380.
2. Kanwar JR, Sun X, Punj V, Sriramoju B, Mohan RR, Zhou SF, Chauhan A, Kanwar RK. Nanoparticles in the treatment and diagnosis of neurological disorders: untamed dragon with fire power to heal. *Nanomedicine*. 2012;8:399–414.
3. Neuwelt EA, Bauer B, Fahlke C, Fricker G, Iadecola C, Janigro D, Leybaert L, Molnár Z, O'Donnell ME, Povlishock JT, Saunders NR, Sharp F, Stanimirovic D, Watts RJ, Drewes LR. Engaging neuroscience to advance translational research in brain barrier biology. *Nat Rev Neurosci*. 2011;12:169–182.
4. Zhang Q, Raoof M, Chen Y, Sumi Y, Sursal T, Junger W, Brohi K, Itagaki K, Hauser CJ. Circulating mitochondrial DAMPs cause inflammatory responses to injury. *Nature*. 2010;464:104–107.
5. Hanisch UK, Kettenmann H. Microglia: active sensor and versatile effector cells in the normal and pathologic brain. *Nat Neurosci*. 2007;10:1387–1394.
6. Ballerini C, Baldi G, Aldinucci A, Maggi P. Nanomaterial applications in multiple sclerosis inflamed brain. *J Neuroimmune Pharmacol*. 2015;10:1–13.
7. Miller IS, Lynch I, Dowling D, Dawson KA, Gallagher WM. Surface-induced cell signaling events control actin rearrangements and motility. *J Biomed Mater Res A*. 2010;93:493–504.
8. Saunders NR, Ek CJ, Habgood MD, Dziegielewska KM. Barriers in the brain: a renaissance?. *Trends Neurosci.*. 2008;31:279–286.
9. Siwak DR, Tari AM, Lopez-Berestein G. The potential of drug-carrying immunoliposomes as anticancer agents. Commentary re: J. W. Park et al., Anti-HER2 immunoliposomes: enhanced efficacy due to targeted delivery. *Clin Cancer Res*. 2002;8:955–956.
10. Patel T, Zhou J, Piepmeier JM, Saltzman WM. Polymeric nanoparticles for drug delivery to the central nervous system. *Adv Drug Deliv Rev*. 2012;64:701–705.
11. Singh AV, Khare M, Gade WN, Zamboni P. Theranostic implications of nanotechnology in multiple sclerosis: a future perspective. *Autoimmune Dis*. 2012;2012:160830.
12. Hauser SL, Chan JR, Oksenberg JR. Multiple sclerosis: prospects and promise. *Ann Neurol*. 2013;74:317–327.
13. Sisay S, Pryce G, Jackson SJ, Tanner C, Ross RA, Michael GJ, Selwood DL, Giovannoni G, Baker D. Genetic background can result in a marked or minimal effect of gene knockout (GPR55 and CB2 receptor) in experimental autoimmune encephalomyelitis models of multiple sclerosis. *PLoS One*. 2013;8:e76907.
14. Baxter AG. The origin and application of experimental autoimmune encephalomyelitis. *Nat Rev Immunol*. 2007;7:904–912.
15. Rausch M, Hiestand P, Baumann D, Cannet C, Rudin M. MRI-based monitoring of inflammation and tissue damage in acute and chronic relapsing EAE. *Magn Reson Med*. 2003;50:309–314.
16. Weinstein JS, Varallyay CG, Dosa E, Gahramanov S, Hamilton B, Rooney WD, Muldoon LL, Neuwelt EA. Superparamagnetic iron oxide nanoparticles: diagnostic magnetic resonance imaging and potential therapeutic applications in neurooncology and central nervous system inflammatory pathologies, a review. *J Cereb Blood Flow Metab*. 2010;30:15–35.
17. Absinta M, Sati P, Gaitán MI, Maggi P, Cortese IC, Filippi M, Reich DS. Seven-tesla phase imaging of acute multiple sclerosis lesions: a new window into the inflammatory process. *Ann Neurol*. 2013;74:669–678.
18. Yang YM, Feng X, Yin le K, Li CC, Jia J, Du ZG. Comparison of USPIO-enhanced MRI and Gd-DTPA enhancement during the subacute stage of focal cerebral ischemia in rats. *Acta Radiol*. 2014;55:864–873.

19. Weise G, Stoll G. Magnetic resonance imaging of blood brain/nerve barrier dysfunction and leukocyte infiltration: closely related or discordant?. *Front Neurol.* 2012;3:178.
20. Varallyay CG, Muldoon LL, Gahramanov S, Wu YJ, Goodman JA, Li X, Pike MM, Neuwelt EA, Dynamic MRI. using iron oxide nanoparticles to assess early vascular effects of antiangiogenic versus corticosteroid treatment in a glioma model. *J Cereb Blood Flow Metab.* 2009;29:853–860.
21. Saleh A, Wiedermann D, Schroeter M, Jonkmanns C, Jander S, Hoehn M. Central nervous system inflammatory response after cerebral infarction as detected by magnetic resonance imaging. *NMR Biomed.* 2004;17:163–169.
22. Venneti S, Lopresti BJ, Wiley CA. Molecular imaging of microglia/macrophages in the brain. *Glia.* 2013;61:10–23.
23. Hunger M, Budinger E, Zhong K, Angenstein F. Visualization of acute focal lesions in rats with experimental autoimmune encephalomyelitis by magnetic nanoparticles, comparing different MRI sequences including phase imaging. *J Magn Reson Imaging.* 2014;39:1126–1135.
24. Millward JM, Schnorr J, Taupitz M, Wagner S, Wuerfel JT, Infante-Duarte C. Iron oxide magnetic nanoparticles highlight early involvement of the choroid plexus in central nervous system inflammation. *ASN Neuro.* 2013;5:e00110.
25. Tysiak E, Asbach P, Aktas O, Waiczies H, Smyth M, Schnorr J, Taupitz M, Wuerfel J. Beyond blood brain barrier breakdown—in vivo detection of occult neuroinflammatory foci by magnetic nanoparticles in high field MRI. *J Neuroinflammation.* 2009;6:20.
26. Brochet B, Deloire MS, Touil T, Anne O, Caillé JM, Dousset V, Petry KG. Early macrophage MRI of inflammatory lesions predicts lesion severity and disease development in relapsing EAE. *Neuroimage.* 2006;32:266–274.
27. Luchetti A, Milani D, Ruffini F, Galli R, Falini A, Quattrini A, Scotti G, Comi G, Martino G, Furlan R, Politi LS. Monoclonal antibodies conjugated with superparamagnetic iron oxide particles allow magnetic resonance imaging detection of lymphocytes in the mouse brain. *Mol Imaging.* 2012;11:114–125.
28. Sipkins DA, Gijbels K, Tropper FD, Bednarski M, Li KC, Steinman L. ICAM-1 expression in autoimmune encephalitis visualized using magnetic resonance imaging. *J Neuroimmunol.* 2000;104:1–9.
29. Metselaar JM, Storm G. Liposomes in the treatment of inflammatory disorders. *Expert Opin Drug Deliv.* 2005;2:465–476.
30. Lee DH, Rötger C, Appeldoorn CC, Reijerkerk A, Gladdines W, Gaillard PJ, Linker RA. Glutathione PEGylated liposomal methylprednisolone (2B3-201) attenuates CNS inflammation and degeneration in murine myelin oligodendrocyte glycoprotein induced experimental autoimmune encephalomyelitis. *J Neuroimmunol.* 2014;274:96–101.
31. Gaillard PJ, Appeldoorn CC, Rip J, Dorland R, van der Pol SM, Kooij G, de Vries HE, Reijerkerk A. Enhanced brain delivery of liposomal methylprednisolone improved therapeutic efficacy in a model of neuroinflammation. *J Control Release.* 2012;164:364–369.
32. Avnir Y, Turjeman K, Tulchinsky D, Sigal A, Kizelsztein P, Tzemach D, Gabizon A, Barenholz Y. Fabrication principles and their contribution to the superior in vivo therapeutic efficacy of nano-liposomes remote loaded with glucocorticoids. *PLoS One.* 2011;6:e25721.
33. Hu W, Metselaar J, Ben LH, Cravens PD, Singh MP, Frohman EM, Eagar TN, Racke MK, Kieseier BC, Stüve O. PEG minocycline-liposomes ameliorate CNS autoimmune disease. *PLoS One.* 2009;4:e4151.
34. Jung S, Huitinga I, Schmidt B, Zielasek J, Dijkstra CD, Toyka KV, Hartung HP. Selective elimination of macrophages by dichlormethylene diphosphonate-containing liposomes suppresses experimental autoimmune neuritis. *J Neurol Sci.* 1993;119:195–202.
35. Ding Q, Si X, Liu D, Peng J, Tang H, Sun W, Rui M, Chen Q, Wu L, Xu Y. Targeting and liposomal drug delivery to CD40L expressing T cells for treatment of autoimmune diseases. *J Control Release.* 2015;207:86–92.

36. Jia J, Li H, Tai S, Lv M, Liao M, Yang Z, Zhang B, Zhou B, Zhang G, Zhang L. Construction and preliminary investigation of a plasmid containing a novel immunotoxin DT390-IL-18 gene for the prevention of murine experimental autoimmune encephalomyelitis. *DNA Cell Biol.* 2008;27:279–285.

37. Chen W, Li H, Jia Y, Lv M, Li M, Feng P, Hu H, Zhang L. In vivo administration of plasmid DNA encoding recombinant immunotoxin DT390-IP-10 attenuates experimental autoimmune encephalomyelitis. *J Autoimmun.* 2007;28:30–40.

38. Jia Y, Li H, Chen W, Li M, Lv M, Feng P, Hu H, Zhang L. Prevention of murine experimental autoimmune encephalomyelitis by in vivo expression of a novel recombinant immunotoxin DT390-RANTES. *Gene Ther.* 2006;13:1351–1359.

39. Wu HY, Chung MC, Wang CC, Huang CH, Liang HJ, Jan TR. Iron oxide nanoparticles suppress the production of IL-1beta via the secretory lysosomal pathway in murine microglial cells. *Part Fibre Toxicol.* 2013;10:46.

40. Moraes AS, Paula RF, Pradella F, Santos MP, Oliveira EC, von Glehn F, Camilo DS, Ceragioli H, Peterlevitz A, Baranauskas V, Volpini W, Farias AS, Santos LM. The suppressive effect of IL-27 on encephalitogenic Th17 cells induced by multiwalled carbon nanotubes reduces the severity of experimental autoimmune encephalomyelitis. *CNS Neurosci Ther.* 2013;19:682–687.

41. Probst HC, Muth S, Schild H. Regulation of the tolerogenic function of steady-state DCs. *Eur J Immunol.* 2014;44:927–933.

42. Bonifaz L, Bonnyay D, Mahnke K, Rivera M, Nussenzweig MC, Steinman RM. Efficient targeting of protein antigen to the dendritic cell receptor DEC-205 in the steady state leads to antigen presentation on major histocompatibility complex class I products and peripheral CD8+ T cell tolerance. *J Exp Med.* 2002;196:1627–1638.

43. Steinman RM, Hawiger D, Nussenzweig MC. Tolerogenic dendritic cells. *Annu Rev Immunol.* 2003;21:685–711.

44. Hawiger D, Inaba K, Dorsett Y, Guo M, Mahnke K, Rivera M, Ravetch JV, Steinman RM, Nussenzweig MC. Dendritic cells induce peripheral T cell unresponsiveness under steady state conditions in vivo. *J Exp Med.* 2001;194:769–779.

45. Cappellano G, Woldetsadik AD, Orilieri E, Shivakumar Y, Rizzi M, Carniato F, Gigliotti CL, Boggio E, Clemente N, Comi C, Dianzani C, Boldorini R, Chiocchetti A, Renò F, Dianzani U. Subcutaneous inverse vaccination with PLGA particles loaded with a MOG peptide and IL-10 decreases the severity of experimental autoimmune encephalomyelitis. *Vaccine.* 2014;32:5681–5889.

46. Yuan B, Zhao L, Fu F, Liu Y, Lin C, Wu X, Shen H, Yang Z. A novel nanoparticle containing MOG peptide with BTLA induces T cell tolerance and prevents multiple sclerosis. *Mol Immunol.* 2014;57:93–99.

47. Yeste A, Nadeau M, Burns EJ, Weiner HL, Quintana FJ. Nanoparticle-mediated codelivery of myelin antigen and a tolerogenic small molecule suppresses experimental autoimmune encephalomyelitis. *Proc Natl Acad Sci USA.* 2012;109:11270–11275.

48. Henderson AP, Barnett MH, Parratt JD, Prineas JW. Multiple sclerosis: distribution of inflammatory cells in newly forming lesions. *Ann Neurol.* 2009;66:739–753.

49. Kizelsztein P, Ovadia H, Garbuzenko O, Sigal A, Barenholz Y. Pegylated nanoliposomes remote-loaded with the antioxidant tempamine ameliorate experimental autoimmune encephalomyelitis. *J Neuroimmunol.* 2009;213:20–25.

50. Heckman KL, DeCoteau W, Estevez A, Reed KJ, Costanzo W, Sanford D, Leiter JC, Clauss J, Knapp K, Gomez C, Mullen P, Rathbun E, Prime K, Marini J, Patchefsky J, Patchefsky AS, Hailstone RK, Erlichman JS. Custom cerium oxide nanoparticles protect against a free radical mediated autoimmune degenerative disease in the brain. *ACS Nano.* 2013;7:10582–10596.

51. Bataveljić D, Stamenković S, Bačić G, Andjus PR. Imaging cellular markers of neuroinflammation in the brain of the rat model of amyotrophic lateral sclerosis. *Acta Physiol Hung.* 2011;98:27–31.

52. Bernardi A, Frozza RL, Meneghetti A, Hoppe JB, Battastini AM, Pohlmann AR, Guterres SS, Salbego CG. Indomethacin-loaded lipid-core nanocapsules reduce the damage triggered by Aβ1-42 in Alzheimer's disease models. *Int J Nanomed.* 2012;7:4927–4942.
53. Deddens LH, Van Tilborg GA, Mulder WJ, De Vries HE, Dijkhuizen RM. Imaging neuroinflammation after stroke: current status of cellular and molecular MRI strategies. *Cerebrovasc Dis.* 2012;33:392–402.
54. Rissiek B, Koch-Nolte F, Magnus T. Nanobodies as modulators of inflammation: potential applications for acute brain injury. *Front Cell Neurosci.* 2014;8:344.
55. Frozza RL, Bernardi A, Hoppe JB, Meneghetti AB, Battastini AM, Pohlmann AR, Guterres SS, Salbego C. Lipid-core nanocapsules improve the effects of resveratrol against Abeta-induced neuroinflammation. *J Biomed Nanotechnol.* 2013;9:2086–2104.
56. Win-Shwe TT, Fujimaki H. Nanoparticles and neurotoxicity. *Int J Mol Sci.* 2011;12:6267–6280.
57. Czajka M, Sawicki K, Sikorska K, Popek S, Kruszewski M, Kapka-Skrzypczak L. Toxicity of titanium dioxide nanoparticles in central nervous system. *Toxicol In Vitro.* 2015;29:1042–1052.
58. Migliore L, Uboldi C, Di Bucchianico S, Coppedè F. Nanomaterials and neurodegeneration. *Environ Mol Mutagen.* 2015;56:149–170.
59. Huang JY, Lu YM, Wang H, Liu J, Liao MH, Hong LJ, Tao RR, Ahmed MM, Liu P, Liu SS, Fukunaga K, Du YZ, Han F. The effect of lipid nanoparticle PEGylation on neuroinflammatory response in mouse brain. *Biomaterials.* 2013;34:7960–7970.

CHAPTER 16

Infectious Diseases of the Brain

Melike Ekizoğlu, PhD
Hacettepe University, Ankara, Turkey

CONTENTS

1 INTRODUCTION

Despite advances in antimicrobial treatment and critical care, infections in the brain and spinal cord continue to be diseases of concern globally. For example, bacterial meningitis is recognized as a significant cause of infection-related deaths, and the survivors suffer from long-term neurological sequelae with an incidence rate of approximately 30%–50%.[1,2] Mortality and morbidity rates vary and depend on several factors, such as age, etiologic agents, residence in developing countries, and the onset of treatment. The prognosis of central nervous system (CNS) infections mainly depends on rapid identification of the causative organisms and subsequent selection of the right drugs that can reach effective microbicidal concentrations in cerebrospinal fluid (CSF).[3] Three main obstacles need to be overcome to achieve this: (1) limited access of antimicrobial agents into the CNS, which is caused by the blood–brain barrier (BBB)/blood–cerebrospinal fluid barrier (BCSFB); (2) developed resistance against antibiotics during treatments; and (3) the effect of immunomodulators. Nowadays, insufficient knowledge of the microbe–host relations and the pathogenesis of brain infections is another contributing factor to the failure of treatments against infections. This chapter focuses on CNS infections and their treatments, factors creating obstacles for these treatments, and newly developed antimicrobial agents that can be targeted to the infection site.

2 BRAIN INFECTIONS

The CNS is especially sensitive to infections due to a lack of normal host defense mechanisms, including antibody and complement activities. A number of pathogenic microbes, including neuroinvasive viruses, bacteria, parasites, and fungi, can enter into the CNS and cause many clinical problems. Brain infections can be characterized according to the corresponding anatomical components; meningitis (meninges), encephalitis (brain parenchyma), abscess

Nanotechnology Methods for Neurological Diseases and Brain Tumors. http://dx.doi.org/10.1016/B978-0-12-803796-6.00016-2

(brain parenchyma and extradural), empyema (subdural), myelitis (spinal cord), or meningoencephalitis and encephalomyelitis (simultaneous infections in multiple regions).

2.1 Meningitis

Meningitis occurs when there is an inflammation of the meninges and/or CSF that normally surrounds and protects the brain and spinal cord. The characteristic changes of the CSF are caused by bacteria and viruses. Neurological side effects, such as hearing loss, developmental disorders, and neuropsychological impairment, are expected in nearly half of the surviving patients. The symptoms of viral meningitis are generally mild and recover after treatment with antiviral drugs without any significant long-term sequelae. Besides, microorganisms noninfectious factors, such as systemic and neoplastic diseases, and use of certain drugs can trigger meningeal inflammation. Thus, compared with bacterial and fungal infections, in most cases viral meningitis is considered to be the least deadly meningitis form.[4]

2.1.1 Bacterial Meningitis

Bacterial meningitis is described as a bacterial infection of the meninges. It is a medical urgency that needs early diagnosis and treatment, and a very common cause of infection-related deaths. The disease is responsible for over 1.2 million reported cases worldwide each year.[5] According to the region, country, pathogen, age group, vaccination programs, and immune status, the incidence and mortality rates for bacterial meningitis vary. According to recent data, acute bacterial meningitis annual incidence in Turkey decreased from 3, 5 to 0.9 cases per 100,000 population.[6] Furthermore, while treatment can decrease the case fatality rate levels to 10%–30%, without treatment this rate can even reach 70%. Around 10%–20% of survivors of bacterial meningitis suffer from neuron degeneration and permanent damage of the visual and hearing systems. The potential cause of persistent neuropsychological sequelae in survivors has been identified as the damage to neurons, particularly in hippocampal structures.[7–10]

Bacterial meningitis is characterized by four main processes of pathogenesis, namely nasopharyngeal colonization, bacterial invasion into the blood stream, persistance in the blood stream, and access to the subarachnoid space.[10–12] The common virulence factors of meningitis-causing bacteria are species-specific factors (such as antiphagocytic capsules, which ensure bacterial multiplication in the blood), complement resistance, and the ability to survive in phagocytic cells.[13–15] After typical meningeal pathogens cross the BBB, host defense mechanism in the subarachnoid space is not sufficient for controlling the infection due to lack of complement factors, polymorphonuclear leukocytes, and other plasma cells. When bacteria multiply in the CSF, fever and inflammation (with the release of cytokines, mainly IL-1beta, IL-6, and TNF-alpha; free radicals; and matrix metalloproteinases) result in BBB dysfunction and neuronal injury. Bacterial toxins and virulence factors also cause neuronal injury.[16] The number of white blood cells passing through the BBB into the CNS tissue increases. Furthermore, meningeal inflammation not only increases the permeability of the BBB, but also has an inhibitory effect on the elimination pump. With antibiotic treatment, inflammation subsides and the functioning of the BBB slowly normalizes.[17] The formation of brain edema, increased intracranial pressure, and alterations of cerebral blood flow are the main consequences of bacterial meningitis that cause hypoxic brain damage and may lead to death. Moreover, after subsequent use of antibiotics (e.g., beta-lactams), bacterial lysis leads to the release of proinflammatory/toxic bacterial products, such as lipopolysaccharide, lipoteichoic acid, and peptidoglicans, that may result in the overstimulation of immune cells and enhance the inflammatory response.[10,15,18]

2.1.1.1 Etiology of Bacterial Meningitis

The most common causative organisms of acute bacterial meningitis are *Streptococcus pneumoniae* (also called *Pneumococcus*), *Streptococcus agalactiae, Neisseria meningitidis* (also called *Meningococcus*), and *Listeria monocytogenes.*[19,20]

S. pneumoniae is the most common pathogen that causes meningitis in children and adults. It results in 20%–30% of the cases resulting in hospital mortality and up to 40% in intracranial complications (e.g., brain edema, hydrocephalus, and intracranial hemorrhage), with hearing loss and motor deficits.[21] Persons younger than age 5 and older than age 60 constitute the highest risk group. Some predisposing factors, such as acute and chronic otitis media, alcoholism, diabetes mellitus, head trauma, and CSF rhinorrhea, increase the risk of pneumococcal meningitis, as well as are involved in immunosuppression.[22] Vaccines, currently in practice, were developed using more than 90 antigenically different serotypes of *S. pneumoniae,* as determined by the polysaccharide capsule.[10]

Gram-negative anaerobs, *Haemophilus* sp., or Enterobacteriaceae sp. can be pathogens responsible for meningitis as secondary to sinusitis, otitis, mastoiditis, and *S. pneumoniae.* Moreover, following postneurosurgical procedure or head trauma, *Staphylococcus aureus* is the most common pathogen.

N. meningitidis has become the leading cause of meningitis and septicemia across the globe. The epidemiologic profile of meningococcal meningitis varies all over the world; however, 6 of 12 recognized serogroups, (Men) A, B, C, W135, X, and Y, are responsible for the majority of the cases throughout the world.[23] *N. meningitidis* serogroup B causes meningococcal meningitis in developed countries. In Turkey, serogroup W135, followed by serogroup B are reported to be responsible for the majority of cases. Infants seem to be at the highest risk for fatal meningitis caused by serogroup B.[6] Asplenia or terminal complement deficiency predisposes an individual to meningococcal meningitis.

L. monocytogenes, a Gram-positive bacillus, is commonly found in soil and fecal flora of human beings. Bacteremia and meningitis occur in <2 months of age, in individuals over 50 years of age, among persons with deficiencies in cell-mediated immunity, who have undergone organ transplantation, women who are pregnant, and persons with malignancy, chronic illness, or immunosuppressive therapy.[24] On the other hand, ingestion of a large inoculum may cause gastroenteritis in healthy individuals. After crossing the placental barrier, *L. monocytogenes* can cause a stillborn baby or the baby dies shortly after birth. As meningitis caused by *L. monocytogenes* is rare in young and healthy individuals, screening for HIV infection must be performed if *L. monocytogenes* meningitis occurs in these patient groups.[25]

Group B streptococcus (*S. agalactiae*) is another cause of bacterial meningitis, and it has been increasingly recognized in neonates, puerperal women, and patients with critical underlying diseases, such as diabetes mellitus.[26]

There are other pathogens that infrequently cause bacterial meningitis. *Haemophilus influenzae* type b was the most common cause of bacterial meningitis among children before *H. influenzae* type b conjugate vaccine was routinely used. Child and infant meningitis was largely prevented by the development of the vaccine. On the other hand, in adults there is limited improvement, as *H. influenzae* type B still causes bacterial meningitis in cases, such as immunologic deficiencies and illnesses (e.g., sickle cell anemia, leukemia, and chronic lung disease).[27] Further, children under age 2 to possess a complete primary *H. influenzae* type b vaccination course are at risk.

Escherichia coli is a pathogen that predominantly causes meningitis in newborns, infants, old, or immunosuppressed people. Other Gram-negative bacteria, such as *Pseudomonas aeruginosa*, *Klebsiella pneumoniae*, *Salmonella* sp., and *Serratia marcescens*, can occasionally be responsible for meningitis in infants, patients with head trauma or immunosuppression, neurosurgical patients, and those with underlying diseases, such as, cancer, diabetes, or congestive heart failure. *Mycobacterium tuberculosis* is a cause of chronic or subacute meningitis.

Bacterial meningitis can be classified in two different categories: community-onset and nosocomial infections. The incidence of community-acquired bacterial meningitis among children in developed countries has seen a profound decline with the development of effective vaccines against *H. influenzae* type b and *S. pneumoniae*, especially in children. Among adults, the annual incidence of community-onset bacterial meningitis, which is 3–6 cases per 100,000 persons, remained unchanged over the past decade. *S. pneumoniae*, *N. meningitidis*, and *L. monocytogenes* are among the common causes of community-onset bacterial meningitis in high-income countries. *Streptococcus pyogenes* is a rare cause of community-onset meningitis, often secondary to otitis media, with an incidence less than 0.5% of bacterial meningitis.[28,29]

Nosocomial bacterial meningitis has been a growing concern in critical care medicine. Invasive neurosurgical procedures, particularly CSF shunt placement, administration by lumbar puncture of intrathecal chemotherapy, or CSF leakage (e.g., recent head injury) predispose individuals to nosocomial meningitis. The incidence of nosocomial meningitis ranges from 1% to 6% among neurosurgical patients. Although *S. pneumoniae* is the most common cause of nosocomial meningitis, *S. aureus* and coagulase-negative staphylococci are frequently isolated in meningitis after neurosurgery (37%). Aerobic, Gram-negative bacilli (especially Enterobacteriaceae) cause up to 33% cases of nosocomial meningitis.[30,31]

2.1.1.2 Treatment

After both blood and CSF of the patient have been obtained for culture, the most critical point in bacterial meningitis is to immediately start with antibiotic treatment. The patient's age, predisposing factors of infection, type of pathogen, regional antibiotic resistance patterns, and BBB-permeating abilities and bactericidal properties of antibiotics in CSF determine the choice of the treatment antibiotic. Once the results of microbiological identification and antimicrobial susceptibility tests are known, therapy can be adjusted and narrowed. Standard empirical therapies for community-acquired infections consist of third-generation cephalosporin (ceftriaxone and cefotaxime) or fourth-generation cephalosporin (cefepime) and vancomycin and acyclovir. Acyclovir is recommended to cover herpes simplex virus (HSV). In neonates amoxicillin and cefotaxime should be used for empirical therapy, ceftriaxone is contraindicated in this group of patients. Ampicillin should be added if the patient is immunosuppressed, against the risk of *L. monocytogenes*.

Nosocomial meningitis should include a combination of vancomycin plus ceftazidime, cefepime, or meropenem.[25] The patient should be isolated when meningococcal meningitis is suspected or proven during the first 24 h treatment and also for people who had a close contact with the patient chemoprophylaxis should be used.

Due to its synergistic effect, rifampin can be added to vancomycin/ceftriaxone; however, it is not recommended to be used as a single dose, as incomplete drug course can give rise to resistance. Besides, pregnant women should not use rifampin.[32]

2.1.1.3 Novel Antibiotics

2.1.1.3.1 Cefepime. Cefepime is a fourth-generation cephalosporin with in vitro activity similar to that of cefotaxime or ceftriaxone against meningitis-causing bacteria, such as *H. influenzae, S. pneumoniae, N. meningitidis, P. aeruginosa,* and *Enterobacter* sp.[33]

2.1.1.3.2 Meropenem. Meropenem has a high in vitro activity against *L. monocytogenes,* multidrug-resistant Gram-negative bacteria and *P. aeruginosa.* It has been found in the experiments that meropenem had an activity similar to ceftriaxone and vancomycin combination used for *S. pneumaniae.*[34]

2.1.1.3.3 Daptomycin. Daptomycin, a cyclic lipopeptide, has a bactericidal effect on Gram-positive bacteria, including pneumococci. The penetration rate of daptomycin into the CSF of infected animals is comparable to that of ceftriaxone. Daptomycin is shown to be equally effective as vancomycin in the treatment of pneumococcal meningitis.

2.1.1.3.4 Televancin. Televancin is a derivative of vancomycin with bactericidal activity against Gram-positive bacteria, including methicillin-resistant *S. aureus,* vancomycin–intermediate resistant *S. aureus,* and linezolid-resistant *S. aureus.* Televancin seems to be more effective than the combination of vancomycin and ceftriaxone in eradicating bacteria from CSF.[35]

2.1.1.4 New Treatment Options

2.1.1.4.1 Adjunctive Treatment With Corticosteroid. The use of dexamethasone, the most commonly studied corticosteroid, as an adjunctive therapy to the standard therapy of bacterial meningitis is controversial. Some studies suggest that dexamethasone is only recommended in adjunctive therapy and used in community-acquired bacterial meningitis to prevent both the meningeal inflammation and neurologic sequelae like hearing loss.[4,10,19,25] The Infectious Diseases Society of America (IDSA) practice guidelines for the empiric treatment of suspected or proven pneumococcal meningitis recommend the dexamethasone use with a combination of vancomycin and ceftriaxone (or cefotaxime) in adults.[36] Nevertheless, by reducing meningeal inflammation in the treatment of bacterial meningitis with the use of dexamethasone could decrease the penetration of hydrophilic antibiotics like vancomycin and immune cells into the CSF, causing serious concern in the treatment with dexamethasone.[18,37] Besides, data regarding the effects of dexamethasone therapy in patient with both meningitis and septic shock are limited.[19]

A number of adjunctive treatments are under investigation. These include glycerol, melatonin, hypothermia, vitamin B6, and antipyretic treatments. Glycerol and hypothermia treatments are being considered for reducing the intracranial pressure and hypertension. Further controlled trials are needed for determining their benefits in meningitis management.[10,38,39] Hence, none of these options have been proven effective in patients.

2.1.1.4.2 Nonbacteriolytic Antibiotics. To prevent the overstimulation of microglial cells and to reduce neouronal injury by released bacterial components, an effective antibiotic treatment should be started rapidly, and compounds that do not release significant amount of pathogen products should be selected. There is a strong link between bacteriolytic and -cidal actions with regard to cell wall antibacterials, particularly beta-lactam antibiotics. Bactericidal antibiotics act by inhibiting the RNA or protein synthesis or DNA replication, they circumvent or at least prolong bacterial lysis. As compared to the usual standard treatment with beta-lactam antibiotics, rifampicin, clindamycin, and daptomycin decrease inflammation, mortality, neuronal injury, or/and neurological long-term residual effects in bacterial meningitis of animals.[15,18] Additional studies are needed to investigate whether nonlytic agents are effective for reducing neurophysiological deficits.

2.1.1.4.3 Vaccines. The implementation of vaccines resulted in a dramatic decrease in the incidence of bacterial meningitis in the past 2 decades, especially those caused by *H. influenzae* type b and *S. pneumoniae*. The introduction of Hib polysaccharide–protein conjugate vaccine, pneumococcal conjugate 7, and recently 13-valent vaccine has reduced the incidence of bacterial meningitis drastically, especially in infants and young children. Immunization against *N. meningitidis* with a conjugate meningococcal vaccine has been implemented in various parts of the world against high-incidence serotypes. While serotypes B and C are the leading causes of bacterial meningitis in Europe, serotype W135 (followed by serogroup B) is predominant in Turkey. The refugee population living in Turkey and pilgrims attending the Hajj are factors that affect the frequency of occurrence of the serogroups.[40,41] Although quadrivalent conjugate meningococcal vaccines that contain A, C, Y, and W135 serogroups (available since 2005) are not widely used, they will play an important role in the reduction of the incidence of this infection.[19]

2.1.2 Viral Meningitis

Viral meningitis is classically defined as nonbacterial inflammation of the tissues lining the brain. Any inflammation or pathology that also involves the brain parenchyma is referred to as meningoencephalitis or encephalitis. The vast majority of viral meningitis cases are caused by human enteroviruses (HEV) that mostly target children. However, many other viruses have the ability to cause meningitis, which are: arboviruses (e.g., California encephalitis virus group, St. Louis encephalitis virus, and WNV), bunyaviruses, mumps virus, lymphocytic choriomeningitis virus (LCMV), HSV-1 and -2, and human immunodeficiency virus (HIV)-1.[42]

HSV encephalitis commonly occurs in patients younger than 20 and older than 50 years of age. HSV-2 is an important cause of viral meningitis, accounting for 17% of cases of aseptic meningitis. Varicella zoster virus accounted for 8% of viral meningitis cases among adults and was not recognized as a cause of meningitis, until polymerase chain reaction was widely available.

2.1.2.1 Treatment

Currently, there are no treatments with proven efficiency available for viral CNS infections, except for herpesviruses and HIV. Acyclovir remains the most widely used drug for HSV. But other drugs, including valaciclovir, famciclovir, and foscarnet, have been used for several years.

2.2 Encephalitis

Viral inflammation of the brain parenchyma, encephalitis, is caused by a direct viral invasion. When a hypersensitivity reaction to a virus occurs, an acute disseminated encephalomyelitis in brain and spinal cord inflammation is developed. Thus viruses can indirectly cause both inflammations.

2.2.1 Etiology of Encephalitis

In encephalitis incidents, viruses were the main cause in 68.5% of the cases with identified etiology, which constitutes 15.8% of the total cases. In the developed world, HSV continues to be the main cause of encephalitis.[43] Other causes of encephalitis include enteroviruses, varicella zoster virus, Epstein–Barr virus, measles virus, and arboviruses, such as Japanese encephalitis virus (JEV), West Nile virus (WNV), and Murray Valley encephalitis virus (MVEV).[44] A list of other significant viral causes of encephalitis provided by Tyler.[45] *L. monocytogenes* is another possible cause of meningoencephalitis, which exhibits features of meningitis and encephalitis. Encephalitis is generally a primary disease, sometimes being a postinfectious immunologic complication after a viral infection. These viruses in original encephalitis go directly to the brain. There might be two types of viral infections: epidemic or sporadic. Epidemic infections could be caused by arbovirus, echovirus, coxsackievirus, poliovirus (in some underdeveloped areas and in sporadic infections), HSV, rabies, varicella zoster virus, or mumps virus.[31] Treatment is complementary and in certain cases, and requires antiviral drugs.

2.3 Brain Abscesses

Brain abscesses commonly caused by bacterial, fungal, and parasitic pathogens are considered to be a focal infection of the brain parenchyma. The incidence of brain abscess is 1%–2% in developed countries, while in nondeveloping countries they continue to be a significant health problem, with an incidence of 8%.[46] The prevalence of brain abscesses is higher in immunocompromised persons compared to individuals with normal immune functions. A brain abscess generally occurs in males within the age range of 30–50 years.[47] Achievements in microbiological methods of diagnosis (better anaerobic culture techniques and neuroradiological imaging procedures), detection (modern neurosurgical techniques), and treatment (broad-spectrum antibiotics) have led to a decrease in the mortality rates of patients. Consequently the mortality rate of patients with a brain abscess has decreased to less than 5%–15%; however, intraventricular rupture of a brain abscess is associated with a high mortality rate up to

80%.[48] Despite these improvements, brain abscess continues to be a serious, life-threatening disease. The frequency of neurological squealae varies from 20% to 79% in patients who survive the infection, and is predicated on how quickly the diagnosis is reached and antibiotics administered.[49]

2.3.1 Etiology of Brain Abscess

Brain abscesses can be attributed to three different causes: cranial infections (e.g., osteomyelitis, sinusitis, and otitis) and cranial trauma, recent neurosurgical interventions, and hematogenous spread (e.g., in bacterial endocarditis).

Despite bacterial and viral meningitis, the geographical distribution of causative microorganisms of brain abscess over continents was similar and did not substantially change over the past 60 years.[50]

Generally, bacterial brain abscesses are caused by a contiguous spread of cranial infection from the oropharynx, middle ear, and paranasal sinuses. Although it still is not completely clear how microorganisms spread into the brain from these sources, it is possible that valveless emissary veins allow microorganisms to flow into the venous system of the brain from these sites.[51] *Bacteroides, Peptostreptococcus*, and *Streptococcus* are mostly identified in brain abscesses due to contiguous spread. Cranial trauma is another cause of brain abscess by contiguous means. The prevalence of brain abscesses ranges from 2% to 14% after trauma or neurosurgical procedures,[49,52] *S. aureus* being the dominant species. Another significant cause of brain abscess is the hematogenous spread resulting from a systemic infection. This can appear in chronic pyogenic lung disease, endocarditis, intraabdominal abscess, and urinary tract infections.[53] Due to the increase in cases, such as organ transplantation, cancer chemotherapy, and HIV, where immunosuppression has been a major factor, hematogenous (or metastatic) spread is getting more common.[46]

Bacteria are the main reason for most of the brain abscesses and many of them have more than one species. The most common bacteria that cause brain abscesses are *S. aureus* and streptococci (e.g., *Streptococcus milleri* group and viridian group streptococci) because of their extension from the nasopharynx and oropharynx. Enteric Gram-negative bacilli are often recovered in association with an intraabdominal or genitourinary source. *Pseudomonas* spp. can be seen in brain abscesses that arise from otitis media or otitis externa. *Staphylococcus* spp. and aerobic, Gram-negative bacilli are also frequently isolated from brain abscesses related to head trauma or neurosurgical procedures.[54] Brain abscesses caused by anaerobic bacteria usually originate in otorhinolaryngeal infections. Anaerobic streptococci, *Bacteroides, Prevotella, Peptostreptococcus*, and *Fusobacterium* species are the most commonly isolated anaerobic organisms, which often lead to polymicrobial infections.[55,56] Also, *Mycoplasma hominis* and rare *Mycoplasma* species may also be responsible for brain abscesses.[57,58]

Protozoa and helminths can also cause parasitic brain abscesses. CNS toxoplasmosis (*Toxoplasma gondii*), and neurocystcercosis (larval form of *Taenia solium*) are some of the important examples of parasitic infections.[31]

Fungal brain abscesses caused by yeast (e.g., *Candida* spp. and *Cryptococcus* spp.), dimorphic fungi (e.g., *Histoplasma* spp., *Coccidioides* spp., and *Blastomyces* spp.), and molds (e.g., *Aspergillus* spp., *Fusarium*, and *Rhizopus*) are related to the immunocompromised states and poorly controlled diabetes in the case of zygomycosis.[31,59] Branched hyphal–form fungal infections cause cerebral arterial thrombosis and infarction. Sterile infarct can be converted to septic infarct with the formation of the abscess. Despite intensive treatment with amphotericin B (AmB), the mortality rates due to fungal abscesses are as high as 75%–100%. Moreover, the frequency of fungal brain abscess has increased because of the frequent administration of broad-spectrum antimicrobials, immunosuppressive agents, and corticosteroids.[60]

Brain abscesses in immunocompromised or old patients can be caused by a wide spectrum of microorganisms, such as fungi, atypical bacteria, and parasites. Dissemination of cutaneous or pulmonary infection can often cause brain abscesses due to *Nocardia* spp. There are reports showing that *M. tuberculosis* and nontuberculous mycobacteria may cause brain abscesses in patients with HIV. *T. gondii*, *L. monocytogenes*, *Pseudomonas*, *Aspergillus* species, and other opportunistic pathogens can also be found frequently in immunocompromised patients.[61]

2.3.2 Treatment

The treatment of brain abscesses requires a combination of antimicrobials, surgical intervention, and eradication of primary infected foci. Therefore, the treatment of brain abscesses requires a multidisciplinary approach, involving intensivists, neurosurgeons, radiologists, and infectious disease specialists. While empiric antimicrobial therapy should be started, particularly in patients with sepsis or impending herniation, every effort should be made to quickly obtain microbiologic or tissue diagnosis before initiating antimicrobial therapy. As brain abscesses are often polymicrobial, empiric antimicrobial treatment should include Gram-positive, Gram-negative, and anaerobic microorganisms, and should be selected considering predisposing factors. The antimicrobial therapy can be specifically tailored upon the identification of the pathogen organisms and their in vitro susceptibilities. Similar to meningitis, the drugs to be chosen should have broad spectrum activity with the ability to reach optimum levels in the brain parenchyma. An example regimen would cover a third- or fourth-generation cephalosporin, metronidazole, and vancomycin based on predisposing factors. Carbapenems can be used in place of the combination of cephalosporins and metronidazole.[31,47] Causative microorganisms and reduction in the size of the abscess influence the duration of therapy. While at least 6–8 weeks of parenteral therapy is traditionally given for bacterial brain abscesses, cerebral nocardiosis or actinomycosis may require a prolonged course of therapy. Monthly neurological examinations and CT or MRI studies should be conducted to monitor clinical and radiographic changes and to determine the duration of antibiotic treatment. Antibiotic treatment should not be discontinued until the sequestered pus is eliminated, strength of the abscess wall is substantially reduced, and local edema is treated.[25,31]

3 CONSIDERATION OF THE TREATMENT OF BRAIN INFECTIONS

3.1 Lack of Information on Microbial Translocation into the CNS

A relatively small number of microbial pathogens can cause CNS infections in humans. Hence the mechanisms of pathogens entry into the CNS (by crossing the BBB and BCSFB) and subsequent disease manifestations are not completely understood yet.[62–65]

Pathogens can cross the BBB and BCSFB with two different mechanisms: by transcellular and paracellular transversals. Transcellular traversal mechanism is the microbial penetration through barrier cells without any evidence of microorganisms between the cells or intercellular tight junction disruption, whereas in the paracellular mechanism microorganisms can penetrate the barrier cell with and/or without tight junction destruction. The paracellular traversal is also called the Trojan horse mechanism and in this mechanism infected phagocytes carry the microorganism through the BBB. Transcellular traversal of the BBB was demonstrated for several bacterial pathogens (such as *E. coli*), group B *Streptococcus* (such as *S. pneumoniae*, *L. monocytogenes*, and *M. tuberculosis*), fungal pathogens (such as *Candida albicans* and *Cryptococcus neoformans*), and is suggested for the WNV. Paracellular penetration of the BBB has been suggested for the *Trypanosoma* sp. The Trojan horse mechanism has also been suggested for *L. monocytogenes*, *M. tuberculosis*, and HIV.[62,66]

Further studies on the mechanisms of microbial entry into the CNS are still needed. Achieving more knowledge about these entry routes and pathogenesis of CNS infections might enhance the development of new strategies for prevention and therapy through the use of new antibiotic delivery methods.

3.2 BBB and BCSFB

As BBB and BCSFB are explained in more detail in other chapters, the discussion in this chapter is limited to the role of BBB and BCSFB in the antibiotic treatment of CNS infections.

BBB and BCSFB are structural and functional barriers that are formed by brain microvascular endothelial cells. Drug penetration on BBB varies in the presence and absence of the inflammation. The pharmacokinetic and pharmacodynamic characteristics of the antibiotics are critical to this penetration. These characteristics should be considered before choosing the right antibiotic.[67]

The entry of antibiotics into the CSF and extracellular spaces of the brain in the absence of meningeal inflammation depends on: molecular size, lipophilicity, plasma protein binding, and active transport. Antimicrobials with small molecular weight, such as sulfonamides, rifampin, and fluoroquinolones, can penetrate the CSF better compared to other drugs. Hydrophobic drugs (fluoroquinolones, rifamycin, and chloramphenicol) are able to cross the BBB, while hydrophilic antimicrobials (β-lactams and vancomycin) can not. Plasma protein binding is another property of drugs that affects the amount of free drug capable of crossing the BBB. Only a small degree of drug can pass BBB/BCSFB via binding plasma proteins (mainly albumin and globulins) in the presence of an intact barrier. However, in case of an inflammatory response to an infection, protein concentration in the CSF increases. This may cause reduced activity of highly protein bound antibiotics at the site of infection. There are different transport mechanisms in the BBB/BCSFB, which remove toxic compounds from the CNS. The influence of these systems on the antibiotic concentrations in intracranial compartments depends on certain characteristics. Concentrations of the antibiotic in CSF varies with the physicochemical properties and differential affinity of the antibiotic to the transport system.[32] The acidic environment in infected CSF is another factor that influences the activity of specific drugs. The lower pH of infected CSF alters the net electrostatic charge of antibiotics, mainly aminoglycosides. The main explanation is that the aminoglycosides rely on an active charge-dependent transport system to achieve antimicrobial activity against pathogenic bacteria.

In the presence of meningeal inflammation, the BBB/BCSFB becomes more permeable and proinflammatory cytokines can prevent active transport system that leads to an increased rate of antibiotics crossing, irrespective of their physicochemical characteristics.[68]

3.3 Antibiotic Resistance

There is an increasing trend in antibiotic resistance in CNS infection–causing pathogens, which is a serious therapeutic challenge, especially in infections caused by methicillin-resistant *S. aureus*, penicillin-/cefotaxim-/ceftriaxone-resistant pneumococci, multiresistant Gram-negative aerobic bacilli, amoxicillin-resistant *H. influenzae*, and many other organisms, such as *Aspergillus* spp. and *Nocardia asteroides*. This increasing trend has been very strong because these microorganisms have an impact on immunodeficient patients.

Vancomycin or rifampicin should be added in the initial antibiotic treatment protocol, especially for areas where clinical isolates of *S. pneumoniae* have been shown to have intermediate or high levels of resistance to penicillin and third-generation cephalosporins (e.g., cefotaxim and ceftriaxone). Fluoroquinolones might be good alternatives for penicillin-resistant pnuemococci. Although decreased susceptibility to penicillin has been associated with W135 and C serogroups, meningococci is generally susceptible to antibiotics.[69] Considering recent developments in antibiotic resistance, antibiotic therapy should be managed taking into account the culture results, and antibiotic selection should achieve high effectiveness with narrow activity spectrum. More importantly, penicillin G monotherapy for meningococci or pneumococci should be selected only after susceptibility test results have been confirmed.[39]

Antibiotic resistance has been increasingly recognized among CNS pathogens, and continued monitoring of resistant bacteria, as well as development of new antibiotics should be encouraged. Studies are needed to identify novel targets for the prevention and therapy of bacterial meningitis in the era of increasing resistance to conventional antibiotics.

4 NANOTECHNOLOGICAL APPROACH AND RECENT TRENDS IN BRAIN INFECTION TREATMENTS

This section provides information about the use of nanotechnology, for antibiotics that are directed to the brain, in overcoming obstacles for the treatments of brain infections.

Several studies investigated newly developed antimicrobial therapeutics that target brain infections through nanotechnology applications. Within these studies, the most investigated antibiotic-loaded nanoparticle-based systems are liposomes, micelles, polymeric nanoparticles, solid lipid nanocarriers (SLNs), and cell-mediated nanoparticles. These delivery systems have advantages over conventional drugs, such as protection of the drug from chemical and biological degradation in the blood circulation, controllable drug release rate at the site of infection, and surface modification alterations for ligand targeting.[70–73]

4.1 Polymeric Nanoparticles

Among the variety of potential nanoparticles that can be used as delivery vehicles for the treatment of CNS diseases, one of the most promising nanoparticle formulations are those made of poly(lactic-*co*-glycolic acid) (PLGA). FDA has approved PLGA for human use since 1989. However, despite their advantages, the use of drug delivery systems with PLGA nanoparticles is limited in CNS diseases due to their incapability of passing the BBB. Further modifications need to be performed to overcome these limitations, such as taking into account their negative charge that limits cellular uptake and their relatively short blood circulation time.[73,74] In recent years, a variety of approaches have been developed to deliver PLGA nanoparticles to the brain.[75] Cell-penetrating peptides (CPPs) are one of the most promising candidate biomolecules targeting the underlying transport mechanisms of drug delivery into the brain.[76–78] CPPs are positively charged at physiological pH and are biomolecules with different amino acid sequences and sizes. During the transition through the BBB, CPPs are carrying agents for many therapeutic molecules, such as proteins, peptides, nucleic acids, small molecules, and nanoparticles used in the treatment of CNS diseases. Transactivator of transcription (TAT), Angiopep, penetratin, rabies virus glycoprotein, and synthetic peptide family (e.g., SynB) are among the first CPPs used for this purpose.[79] An important study performed by Qin et al.,[79] working on transfer of TAT-mediated clinically applicable liposomes to the brain for the treatment of brain infections and tracking of nanoparticles in vivo. They reported that even though not all of the formulations were selectively targeted to the brain, TAT-modified liposome (TAT-LIP) accumulated in the brain within 24 h after administration via tail vein injection. Similarly, in other studies also, a stable drug both with improved CNS penetration and reduced side effects was achieved using conjugated nanocarrier–TAT peptide systems.[80,81] Liu et al.[82] showed that CPPs were also exploited for improving the transport of small molecules across the BBB. This study is particularly important, as it reveals that ciprofloxacin, an antibiotic commonly used in brain infections, can actually be delivered across the BBB to the brain when TAT-modified micelles were used to improve its delivery.[82] Similarly, a higher CSF concentration of ritonavir was achieved through the conjugation of its nanoformulation with TAT peptide.[83]

Furthermore, conjugation of drugs to members of the SynB family of peptides, which are cationic CPPs, increased their brain uptake and their in vivo activity in the brain via an adsorptive-mediated transcytosis mechanism.[84] Rousselle et al.[85] concluded that the brain penetration of a variety of poor brain-penetrating drugs, such as doxorubicin, benzylpenicillin, paclitaxel, and dalargin, was significantly increased when the drugs were conjugated to SynB1 or SynB3 and injected intravenously into mice. Although only limited knowledge about the BBB transport mechanism of CPPs exists, an adsorptive-mediated translocation mechanism is proposed for SynB vectors.[86] Another study by Adenot et al.,[84] using both in situ brain perfusion and in vitro BBB model, showed that the conjugation of SynB3 with benzylpenicillin enhanced its brain penetration without affecting tight junction integrity.

In addition to TAT and SynB peptide family, other CPPs have also been successfully employed for drug delivery to the brain.[77] For example, Angiopep-2 conjugated to AmB-loaded dendrimers can be a potential delivery system for CNS yeast infections.[87]

It is reported that antibiotic treatment used for the prevention of infections after surgical operations of the CNS is unsuccessful because of the weak vascularization in infected tissues.[54,88] Vancomycin is a common IV-administered antibiotic used against Gram-negative bacterial infections. However, side effects, such as nephrotoxicity and autotoxicity, limit the use of this drug in high doses. In a study by Tseng et al., the researchers treated experimental postoperation CNS infection with antibiotic-loaded PLGA nanofibers.[89,90]

4.2 Polymeric Micelles

Difficulties in transporting antibiotics through the BBB have also been overcome by polymeric micelles (PMs) prepared from cholesterol-conjugated PEG, anchored with transcript or activator TAT peptide. Liu et al.[91] evidenced that TAT-PEG-*b*-Col micelles have sustained antibacterial activity against *Bacillus subtilis* and *E. coli*, and that these nanoparticles reside in the nucleus of neurons by crossing the BBB. This study therefore highlighted the applicability of these micelles for developing nanodelivery systems to treat brain infections. Besides, there are studies that used PMs in the treatment of viral brain infections. These studies proved that by administering the efavirenz-loaded Pluronic block copolymer [poly(ethylene oxide)–poly(propylene oxide)] by intranasal route, PMs could facilitate the passage of the drug through the BBB due to the inhibition of the efflux transporters on brain microvascular endothelial cells.[92,93]

4.3 Solid Lipid Nanoparticles

Another drug delivery vehicle used in the research for infectious brain diseases are SLNs. An important study confirmed the potential ability of SLNs to overcome the P-gp efflux pump and pass through the BBB, when loaded with an antibiotic.[94] Furthermore, Bargoni et al.[95] reported that aminoglycosides, which are one of the antibacterial therapeutic agents, have low permeability across the BBB when administered via the parenteral route. By using SLNs as a drug delivery vehicle, tobramycin (an aminoglycoside antibiotic) was able to pass through the BBB as showed in tissue distribution studies.[95]

SLNs have been evaluated in advanced antiretroviral drug pharmacokinetics and CNS transport. Kuo and Su demonstrated that stavudine, delavirdine, and saquinavir encapsulated in SLNs delivered easily through the artificial BBB in comparison to unencapsulated drugs.[96]

More recently, Kuo and Ko found that saquinavir-encapsulated SLNs prepared by the addition of insulin-like peptidomimetic monoclonal antibody (83-14 mAb) increased the penetration of drugs across the BBB.[97] Although these in vitro study results seem promising, the utility of targeting SLNs for advanced CNS antiretroviral therapy needs to be supported by further in vivo studies.

Dendrimers, a type of SLNs, have been evaluated for CNS delivery of antiretroviral drugs as well. In a recent study an evaluation of in vitro antiviral activity was carried out for lamivudine-loaded polyamidoamine dendrimers. Compared to control group, cellular lamivudine uptake was increased and viral p24 levels were reduced.[98,99] Although these results are encouraging, dendrimers are still at a relatively early development stage for their use in brain infections.

4.4 Cationic Antimicrobial Peptides

Cationic antimicrobial peptides have received increased attention due to their broad-spectrum activities and ability to combat multidrug-resistant microbes.[100]

Liu et al.[101] revealed in their study that self-assembled cationic peptide nanoparticles were effective against neural tissue infection. These nanoparticles were incorporated with a cationic peptide. Compared with conventional antibiotics, it achieved better results, such as broader antimicrobial activity and more efficient killing of bacteria and fungi. In the same study, it was further revealed that nanoparticles crossing the BBB and targeting brain infections were as effective as vancomycin alone.[101]

Other studies, using the same cationic peptide nanoparticle with TAT–cholesterol–G_3R_6 formulation in *C. neoformans and C. albicans* meningitis rabbit models, showed that these nanoparticles went through the BBB and were active against pathogens in the brain tissue.[102,103]

4.5 Liposomes

To improve pharmacokinetics of hydrophobic and hydrophilic agents, liposomal formulations have been used. As demonstrated by Jin et al. in their study, liposomal formulations consisting of a zidovudine prodrug (AZT-myristate) increased AZT in the brain more than an equivalent dose of free AZT.[104] Among the alternatives explored in this study, the liposomal formulations targeted to transferrin and insulin receptors on endothelial cells enhanced delivery of drugs to the brain.[105,106] It was demonstrated in a study by Saiyed et al. that upon application of an external magnetic field, an improved penetration of magnetic azidothymidine 5′-triphosphate liposomal nanoformulations was observed.[107]

Another study in which AmB-loaded magnetic liposomes were administered into the carotid artery of rats, reported that this formulation decreased the toxic effects of AmB and ensured a high concentration in the brain.[108]

4.6 Cell-Mediated Drug Delivery

Another alternative that has been developed for the transport of the drug loaded nanoparticles through biologic barriers, like the BBB, is targeted cell-based delivery. Using targeted systems to deliver drugs to CNS disease sites and to other protected sites, such as lymphoid tissue can utilize the phagocytic and chemotactic characteristics of mononuclear phagocytes (MPs).[109,110]

Sufficient amount of therapeutically active drug must be delivered to the infection site in cell-based drug delivery. In order to achieve the delivery to the site, the drug must be localized in stable and nondegrading subcellular compartments that accelerate drug release for a variety of infectious diseases after penetrating into the cell.[111] Cationic nanoparticles are less likely to be degraded than anionic particles because they decrease acidification of lysosomal compartments.[110]

Immunocytes [including MPs (such as dendritic cells, monocytes, and macrophages), neutrophils, and lymphocytes] are highly mobile; they can cross the BBB, and release their load at infection site or tissue injury. Thus, immune cells can be used as Trojan horses for drug delivery. Targeted drug transport and longer circulation times, as well as reductions in cell and tissue toxicities could be achieved by using these cells as drug transport vehicles while reducing untoward immune reactivity.

Formulations are being prepared for the treatment of CNS HIV infections, through the use of the cell-mediated drug delivery strategy. Researchers have developed the concept of using MPs for the delivery of nanoantiretroviral therapy to increase circulating drug levels at target sites of HIV replication, including the CNS.[112–115]

The efficiency of antifungal agents was mostly diminished by their severe side effects and poor pharmacokinetics. To this end, erythrocytes were suggested as delivery vehicles for AmB. To avoid toxic effects and achieve efficient drug loading, AmB was encapsulated in nanosuspension (AmB-NS) by high-pressure homogenization.[116] Hence, red blood cells served as primary carriers. Results showed that more than 98% of the phagocyte population (granulocytes and peripheral monocytes) accumulated AmB-NS after 4 h of incubation, and subsequently showed a slow AmB release over 10 days without any alteration in cell viability. This inhibited intra- and extracellular fungal activity immediately and permanently. Furthermore, the delivery of AmB by AmB-NS-RBC-leukocyte caused a reduction in AmB dosage lowering the toxic levels. Khan et al. designed chloroquine-loaded liposomes (phosphatidylserine-containing negatively charged) for MP-mediated delivery against *Cryptococcus* infection in the mouse brain.[117] Administration of chloroquine-loaded liposomes accumulated inside macrophage phagolysosomes, which resulted in a remarkable reduction of the fungal load in the brain, thus increasing the antifungal activity of macrophages, even in the case when low doses of this formulation were compared with free drug in high doses.[117] This drug delivery method is also effective for the transport of water-insoluble substances, such as AmB, which underlines the need for further testing.[118]

In order to reduce the inflammatory events contributing to CNS damage and long-term impairments associated with CNS infectious diseases, supplemental therapies have been used commonly in combination with antimicrobials. On the other hand, their use is limited due to high rate of adverse side effects with corticosteroids.[5] A recent study was conducted to investigate the possibility of enhancing the efficacy of the antiplasmodial using experimental mouse model of cerebral malaria with no glucocorticoid-related side effects. In conjunction with artemisone, nanosterically stabilized liposomal formulations of the glucocorticoid β-methasone hemisuccinate were used in the study.[119] Importantly, the liposomal formulation of the glucocorticoid, accumulated the drug in the brains of infected mice excluding healthy mice.

5 CONCLUSIONS

CNS infections continue to be a major cause of mortality and morbidity around the world; the main reasons being delays and shortages in infection management, failure in proper distribution of vaccines, antibiotic resistance development in microorganisms, and lack of information on pathogenesis of the infections. However, promising studies have been conducted intensively to meet these challenges. Novel diagnosis techniques, a recent rise in epidemiological studies, and development of new vaccines and antibiotics can be used in the direct treatment and targeting of the infection site, which will help in achieving higher activity with lower resistance rates.

Abbreviations

AmB Amphotericin B
AmB-NS AmB encapsulated in nanosuspension
AZT Zidovudine prodrug
BBB Blood–brain barrier
BCSFB Blood–cerebrospinal fluid barrier
CNS Central nervous system
CPPs Cell-penetrating peptides
CSF Cerebrospinal fluid
HEV Human enterovirus
HIV Human immunodeficiency virus
HSV Herpes simplex virus
IDSA Infectious Diseases Society of America
IV Intravenous
LCMV Lymphocytic choriomeningitis virus
MPs Mononuclear phagocytes
PLGA Poly(lactic-*co*-glycolic acid)
PMs Polymeric micelles
SLNs Solid lipid nanocarriers

TAT Transactivator of transcription
TAT-LIP TAT-modified liposome

References

1. World Health Organization*World Health Report: Changing History*. Geneva: WHO; 2004.
2. National Institute of Neurological Disorders and Stroke. Available from: http://www.ninds.nih.gov
3. van de Beek D, de Gans J, Spanjaard L, Weisfelt M, Reitsma JB, Vermeulen M. Clinical features and prognostic factors in adults with bacterial meningitis. *N Engl J Med*. 2004;351(18):1849–1859.
4. Hoffman O, Weber RJ. Pathophysiology and treatment of bacterial meningitis. *Ther Adv Neurol Disord*. 2009;2(6):1–7.
5. Borchorst S, Møller K. The role of dexamethasone in the treatment of bacterial meningitis—a systematic review. *Acta Anaesthesiol Scand*. 2012;56(10):1210–1221.
6. Ceyhan M, Ozsurekci Y, Gürler N, Karadag Oncel E, Camcioglu Y, Salman N, Celik M, Emiroglu MK, Akin F, Tezer H, Parlakay AO. Bacterial agents causing meningitis during 2013-2014 in Turkey: a multi-center hospital-based prospective surveillance study. *Human Vaccines Immunother*. 2016;12(11):2940–2945.
7. Zysk G, Brück W, Gerber J, Brück Y, Prange HW, Nau R. Anti-inflammatory treatment influences neuronal apoptotic cell death in the dentate gyrus in experimental pneumococcal meningitis. *J Neuropathol Exp Neurol*. 1996;55(6):722–728.
8. Nau R, Soto A, Brück W. Apoptosis of neurons in the dentate gyrus in humans suffering from bacterial meningitis. *J Neuropathol Exp Neurol*. 1999;58(3):265–274.
9. Edmond K, Clark A, Korczak VS, Sanderson C, Griffiths UK, Rudan I. Global and regional risk of disabling sequelae from bacterial meningitis: a systematic review and meta-analysis. *Lancet Infect Dis*. 2010;10(5):317–328.
10. McGill F, Heyderman RS, Panagiotou S, Tunkel AR, Solomon T. Acute bacterial meningitis in adults. *Lancet*. 2017;388(10063):3036–3047.
11. Tunkel AR, Scheld WM. Pathogenesis and pathophysiology of bacterial meningitis. *Clin Microbiol Rev*. 1993;6(2):118–136.
12. Hoffman O, Weber JR. Review: Pathophysiology and treatment of bacterial meningitis. *Ther Adv Neurol Disord*. 2009;2(6):401–412.
13. Ren B, Li J, Genschmer K, Hollingshead SK, Briles DE. The absence of PspA or presence of antibody to PspA facilitates the complement-dependent phagocytosis of pneumococci in vitro. *Clin Vaccine Immunol*. 2012;19(10):1574–1582.
14. Hyams C, Trzcinski K, Camberlein E, Weinberger DM, Chimalapati S, Noursadeghi M, et al. *Streptococcus pneumoniae* capsular serotype invasiveness correlates with the degree of factor H binding and opsonization with C3b/iC3b. *Infect Immun*. 2013;81(1):354–363.
15. Liechti FD, Grandgirard D, Leib SL. Bacterial meningitis: insights into pathogenesis and evaluation of new treatment options: a perspective from experimental studies. *Fut Microbiol*. 2015;10(7):1195–1213.
16. Van Furth A, Roord JJ, Van Furth R. Roles of proinflammatory and anti-inflammatory cytokines in pathophysiology of bacterial meningitis and effect of adjunctive therapy. *Infect Immun*. 1996;64(12):4883.
17. Spector R. Advances in understanding the pharmacology of agents used to treat bacterial meningitis. *Pharmacology*. 1990;41(3):113–118.

18. Ribes S, Djukic M, Eiffert H. Strategies to increase the activity of microglia as efficient protectors of the brain against infections. *Front Cell Neurosci*. 2014;8:138.
19. Van de Beek D, de Gans J, Tunkel AR, Wijdicks EF. Community-acquired bacterial meningitis in adults. *N Engl J Med*. 2006;354(1):44–53.
20. Okike IO, Ribeiro S, Ramsay ME, Heath PT, Sharland M, Ladhani SN. Trends in bacterial, mycobacterial, and fungal meningitis in England and Wales 2004-11: an observational study. *Lancet Infect Dis*. 2014;14(4):301–307.
21. Weisfelt M, van de Beek D, Spanjaard L, Reitsma JB, de Gans J. Clinical features, complications, and outcome in adults with pneumococcal meningitis: a prospective case series. *Lancet Neurol*. 2006;5(2):123–129.
22. Goldstein EJ, Overturf GD. Indications for the immunological evaluation of patients with meningitis. *Clin Infect Dis*. 2003;36(2):189–194.
23. Stephens DS. Conquering the *Meningococcus*. *FEMS Microbiol Rev*. 2007;31(1):3–14.
24. Brouwer MC, van de Beek D, Heckenberg SG, Spanjaard L, de Gans J. Community-acquired *Listeria monocytogenes* meningitis in adults. *Clin Infect Dis*. 2006;43(10):1233–1238.
25. Roos KL. Chapter 39—Acute bacterial infections of the central nervous system. In: Aminoff MJ, Josephson SA, eds. *Aminoff's Neurology and General Medicine*. Boston: Academic Press; 2014:795–815.
26. Jackson LA, Hilsdon R, Farley MM, Harrison LH, Reingold AL, Plikaytis BD, et al. Risk factors for group B streptococcal disease in adults. *Ann Int Med*. 1995;123(6):415–420.
27. Bisgard KM, Kao A, Leake J, Strebel PM, Perkins BA, Wharton M. *Haemophilus influenzae* invasive disease in the United States, 1994-1995: near disappearance of a vaccine-preventable childhood disease. *Emerg Infect Dis*. 1998;4(2):229.
28. van de Beek D, de Gans J, Spanjaard L, Sela S, Vermeulen M, Dankert J. Group A streptococcal meningitis in adults: report of 41 cases and a review of the literature. *Clin Infect Dis*. 2002;34(9):e32–e36.
29. Bijlsma MW, Brouwer MC, Kasanmoentalib ES, Kloek AT, Lucas MJ, Tanck MW, et al. Community-acquired bacterial meningitis in adults in the Netherlands, 2006-14: a prospective cohort study. *Lancet Infect Dis*. 2016;16(3):339–347.
30. Palabiyikoglu I, Tekeli E, Cokca F, Akan O, Unal N, Erberktas I, et al. Nosocomial meningitis in a university hospital between 1993 and 2002. *J Hospital Infect*. 2006;62(1):94–97.
31. Honda H, Warren DK. Central nervous system infections: meningitis and brain abscess. *Infect Dis Clin N Am*. 2009;23(3):609–623.
32. Nau R, Sörgel F, Eiffert H. Penetration of drugs through the blood-cerebrospinal fluid/blood-brain barrier for treatment of central nervous system infections. *Clin Microbiol Rev*. 2010;23(4):858–883.
33. Sáez-Llorens X, Castano E, García R, Báez C, Perez M, Tejeira F, et al. Prospective randomized comparison of cefepime and cefotaxime for treatment of bacterial meningitis in infants and children. *Antimicrob Agents Chemother*. 1995;39(4):937–940.
34. Zhanel GG, Wiebe R, Dilay L, Thomson K, Rubinstein E, Hoban DJ, et al. Comparative review of the carbapenems. *Drugs*. 2007;67(7):1027–1052.
35. Stucki A, Gerber P, Acosta F, Cottagnoud M, Cottagnoud P. Efficacy of telavancin against penicillin-resistant pneumococci and *Staphylococcus aureus* in a rabbit meningitis model and determination of kinetic parameters. *Antimicrob Agents Chemother*. 2006;50(2):770–773.
36. Tunkel AR, Hartman BJ, Kaplan SL, Kaufman BA, Roos KL, Scheld WM, et al. Practice guidelines for the management of bacterial meningitis. *Clin Infect Dis*. 2004;39(9):1267–1284.

37. Tessier JM, Scheld WM. Principles of antimicrobial therapy. *Handb Clin Neurol.* 2010;96: 17–29.
38. Brouwer MC, Wijdicks EF, van de Beek D. What's new in bacterial meningitis. *Intensive Care Med.* 2016;42:415.
39. Tan YC, Gill AK, Kim KS. Treatment strategies for central nervous system infections: an update. *Exp Opin Pharmacother.* 2015;16(2):187–203.
40. Dinleyici EC, Ceyhan M. The dynamic and changing epidemiology of meningococcal disease at the country-based level: the experience in Turkey. *Exp Rev Vaccines.* 2012;11(5):515–528.
41. Tezer H, Ozkaya-Parlakay A, Kanik-Yuksek S, Gülhan B, Güldemir D. A Syrian patient diagnosed with meningococcal meningitis serogroup B. *Human Vaccines Immunother.* 2014;10(8):2482.
42. Swanson PA, McGavern DB. Viral diseases of the central nervous system. *Curr Opin Virol.* 2015;11:44–54.
43. Granerod J, Ambrose HE, Davies NW, Clewley JP, Walsh AL, Morgan D, et al. Causes of encephalitis and differences in their clinical presentations in England: a multicentre, population-based prospective study. *Lancet Infect Dis.* 2010;10(12):835–844.
44. Glaser CA, Honarmand S, Anderson LJ, Schnurr DP, Forghani B, Cossen CK, et al. Beyond viruses: clinical profiles and etiologies associated with encephalitis. *Clin Infect Dis.* 2006;43(12):1565–1577.
45. Tyler KL. Encephalitis. 7th ed. Mandell, Douglas and Bennett's Principles and Practice of Infectious Diseases. Philadelphia, PA: Elsevier; 2009:[pp.1243–1264].
46. Muzumdar D, Jhawar S, Goel A. Brain abscess: an overview. *Int J Surg.* 2011;9(2):136–144.
47. Patel K, Clifford DB. Bacterial brain abscess. *Neurohospitalist.* 2014;4(4):196–204.
48. Takeshita M, Kawamata T, Izawa M, Hori T. Prodromal signs and clinical factors influencing outcome in patients with intraventricular rupture of purulent brain abscess. *Neurosurgery.* 2001;48(2):310–317.
49. Tseng J-H, Tseng M-Y. Brain abscess in 142 patients: factors influencing outcome and mortality. *Surg Neurol.* 2006;65(6):557–562.
50. Brouwer MC, Coutinho JM, van de Beek D. Clinical characteristics and outcome of brain abscess systematic review and meta-analysis. *Neurology.* 2014;82(9):806–813.
51. Mathisen GE, Johnson JP. Brain abscess. *Clin Infect Dis.* 1997;:763–779.
52. Xiao F, Tseng M-Y, Teng L-J, Tseng H-M, Tsai J-C. Brain abscess: clinical experience and analysis of prognostic factors. *Surg Neurol.* 2005;63(5):442–449.
53. Heilpern KL, Lorber B. Focal intracranial infections. *Infect Dis Clin N Am.* 1996;10(4):879–898.
54. Kaya RA, Türkmenoglu ON, Çolak I, Aydin Y. Brain abscess: analysis of results in a series of 51 patients with a combined surgical and medical approach during an 11-year period. *Neurosurg Focus.* 2008;24(6):E9.
55. Le Moal G, Landron C, Grollier G, Bataille B, Roblot F, Nassans P, et al. Characteristics of brain abscess with isolation of anaerobic bacteria. *Scand J Infect Dis.* 2003;35(5):318–321.
56. Vishwanath S, Shenoy PA, Gupta A, Menon G, Chawla K. Brain abscess with anaerobic Gram-negative bacilli: case series. *J Case Rep.* 2016;6(4):467–474.
57. Raoultm D, Al Masalma M, Armougom F, Scheld WM, Dufour H, Roche P-H, et al. The expansion of the microbiological spectrum of brain abscesses with use of multiple 16S ribosomal DNA sequencing. *Clin Infect Dis.* 2009;48(9):1169–1178.

58. Ørsted I, Gertsen J, Schønheyder H, Jensen J, Nielsen H. *Mycoplasma salivarium* isolated from brain abscesses. *Clin Microbiol Infect.* 2011;17(7):1047–1049.
59. Garcia RR, Min Z, Narasimhan S, Bhanot N. *Fusarium* brain abscess: case report and literature review. *Mycoses*. 2015;58(1):22–26.
60. Erdogan E, Beyzadeoglu M, Arpaci F, Celasun B. Cerebellar aspergillosis: case report and literature review. *Neurosurgery*. 2002;50(4):874–877.
61. Sims L, Lim M, Harsh GR. Review of brain abscesses. *Oper Tech Neurosurg.* 2004;7(4): 176–181.
62. Kim KS. Mechanisms of microbial traversal of the blood–brain barrier. *Nat Rev Microbiol.* 2008;6(8):625–634.
63. Zhang J-R, Tuomanen E. Molecular and cellular mechanisms for microbial entry into the CNS. *J Neurovirol.* 1999;5(6):591–603.
64. Schwerk C, Tenenbaum T, Kim KS, Schroten H. The choroid plexus—a multi-role player during infectious diseases of the CNS. *Front Cell Neurosci.* 2015;9:80.
65. Nassif X, Bourdoulous S, Eugène E, Couraud P-O. How do extracellular pathogens cross the blood–brain barrier?. *Trends Microbiol.* 2002;10(5):227–232.
66. Pulzova L, Bhide MR, Andrej K. Pathogen translocation across the blood-brain barrier. *FEMS Immunol Med Microbiol.* 2009;57(3):203–213.
67. Lutsar I, McCracken Jr GH, Friedland IR. Antibiotic pharmacodynamics in cerebrospinal fluid. *Clin Infect Dis.* 1998;:1117–1127.
68. de Vries HE, Kuiper J, de Boer AG, Van Berkel TJ, Breimer DD. The blood-brain barrier in neuroinflammatory diseases. *Pharmacol Rev.* 1997;49(2):143–156.
69. Brown EM, Fisman DN, Drews SJ, Dolman S, Rawte P, Brown S, et al. Epidemiology of invasive meningococcal disease with decreased susceptibility to penicillin in Ontario, Canada, 2000 to 2006. *Antimicrob Agents Chemother.* 2010;54(3):1016–1021.
70. Chen Y, Liu L. Modern methods for delivery of drugs across the blood–brain barrier. *Adv Drug Deliv Rev.* 2012;64(7):640–665.
71. Taylor E, Webster TJ. Reducing infections through nanotechnology and nanoparticles. *Int J Nanomed.* 2011;6:1463.
72. Gendelman HE, Anantharam V, Bronich T, Ghaisas S, Jin H, Kanthasamy AG, et al. Nanoneuromedicines for degenerative, inflammatory, and infectious nervous system diseases. *Nanomedicine.* 2015;11(3):751–767.
73. Kasinathan N, Jagani HV, Alex AT, Volety SM, Rao JV. Strategies for drug delivery to the central nervous system by systemic route. *Drug Deliv.* 2015;22(3):243–257.
74. Sah H, Thoma LA, Desu HR, Sah E, Wood GC. Concepts and practices used to develop functional PLGA-based nanoparticulate systems. *Int J Nanomed.* 2013;8:747.
75. Cai Q, Wang L, Deng G, Liu J, Chen Q, Chen Z. Systemic delivery to central nervous system by engineered PLGA nanoparticles. *Am J Transl Res.* 2016;8(2):749.
76. Kreuter J. Drug delivery to the central nervous system by polymeric nanoparticles: what do we know?. *Adv Drug Deliv Rev.* 2014;71:2–14.
77. Zou L-l, Ma J-L, Wang T, Yang T-B, Liu C-B. Cell-penetrating peptide-mediated therapeutic molecule delivery into the central nervous system. *Curr Neuropharmacol.* 2013;11(2): 197–208.
78. Sharma G, Lakkadwala S, Modgil A, Singh J. The role of cell-penetrating peptide and transferrin on enhanced delivery of drug to brain. *Int J Mol Sci.* 2016;17(6):806.
79. Qin Y, Chen H, Yuan W, Kuai R, Zhang Q, Xie F, et al. Liposome formulated with TAT-modified cholesterol for enhancing the brain delivery. *Int J Pharm.* 2011;419(1):85–95.

80. Torchilin VP. Cell penetrating peptide-modified pharmaceutical nanocarriers for intracellular drug and gene delivery. *Peptide Sci.* 2008;90(5):604–610.

81. Rapoport M, Lorberboum-Galski H. TAT-based drug delivery system—new directions in protein delivery for new hopes?. *Exp Opin Drug Deliv.* 2009;6(5):453–463.

82. Liu L, Guo K, Lu J, Venkatraman SS, Luo D, Ng KC, et al. Biologically active core/shell nanoparticles self-assembled from cholesterol-terminated PEG–TAT for drug delivery across the blood–brain barrier. *Biomaterials.* 2008;29(10):1509–1517.

83. Borgmann K, Rao KS, Labhasetwar V, Ghorpade A. Efficacy of Tat-conjugated ritonavir-loaded nanoparticles in reducing HIV-1 replication in monocyte-derived macrophages and cytocompatibility with macrophages and human neurons. *AIDS Res Human Retroviruses.* 2011;27(8):853–862.

84. Adenot M, Merida P, Lahana R. Applications of a blood-brain barrier technology platform to predict CNS penetration of various chemotherapeutic agents. 2. Cationic peptide vectors for brain delivery. *Chemotherapy.* 2007;53(1):73–76.

85. Rousselle C, Clair P, Smirnova M, Kolesnikov Y, Pasternak GW, Gac-Breton S, et al. Improved brain uptake and pharmacological activity of dalargin using a peptide-vector-mediated strategy. *J Pharmacol Exp Ther.* 2003;306(1):371–376.

86. Stalmans S, Bracke N, Wynendaele E, Gevaert B, Peremans K, Burvenich C, et al. Cell-penetrating peptides selectively cross the blood-brain barrier in vivo. *PloS One.* 2015;10(10):e0139652.

87. Shao K, Huang R, Li J, Han L, Ye L, Lou J, et al. Angiopep-2 modified PE-PEG based polymeric micelles for amphotericin B delivery targeted to the brain. *J Control Release.* 2010;147(1):118–126.

88. Moorthy RK, Rajshekhar V. Management of brain abscess: an overview. *Neurosurg Focus.* 2008;24(6):E3.

89. Tseng Y-Y, Wang Y-C, Su C-H, Liu S-J. Biodegradable vancomycin-eluting poly [(D,L)-lactide-*co*-glycolide] nanofibres for the treatment of postoperative central nervous system infection. *Sci Rep.* 2015;:5.

90. Tseng Y-Y, Kao Y-C, Liao J-Y, Chen W-A, Liu S-J. Biodegradable drug-eluting poly [lactic-*co*-glycol acid] nanofibers for the sustainable delivery of vancomycin to brain tissue: in vitro and in vivo studies. *ACS Chem Neurosci.* 2013;4(9):1314–1321.

91. Liu L, Venkatraman SS, Yang YY, Guo K, Lu J, He B, et al. Polymeric micelles anchored with TAT for delivery of antibiotics across the blood–brain barrier. *Peptide Sci.* 2008;90(5):617–623.

92. Chiappetta DA, Hocht C, Opezzo JA, Sosnik A. Intranasal administration of antiretroviral-loaded micelles for anatomical targeting to the brain in HIV. *Nanomedicine.* 2013;8(2):223–237.

93. Batrakova E, Lee S, Li S, Venne A, Alakhov V, Kabanov A. Fundamental relationships between the composition of Pluronic block copolymers and their hypersensitization effect in MDR cancer cells. *Pharm Res.* 1999;16(9):1373–1379.

94. Kalhapure RS, Suleman N, Mocktar C, Seedat N, Govender T. Nanoengineered drug delivery systems for enhancing antibiotic therapy. *J Pharm Sci.* 2015;104(3):872–905.

95. Bargoni A, Cavalli R, Zara GP, Fundarò A, Caputo O, Gasco MR. Transmucosal transport of tobramycin incorporated in solid lipid nanoparticles (SLN) after duodenal administration to rats. Part II—tissue distribution. *Pharm Res.* 2001;43(5):497–502.

96. Kuo Y-C, Su F-L. Transport of stavudine, delavirdine, and saquinavir across the blood–brain barrier by polybutylcyanoacrylate, methylmethacrylate-sulfopropylmethacrylate, and solid lipid nanoparticles. *Int J Pharm.* 2007;340(1):143–152.

97. Kuo Y-C, Ko H-F. Targeting delivery of saquinavir to the brain using 83-14 monoclonal antibody-grafted solid lipid nanoparticles. *Biomaterials*. 2013;34(20):4818–4830.

98. Dutta T, Jain NK. Targeting potential and anti-HIV activity of lamivudine loaded mannosylated poly (propyleneimine) dendrimer. *Biochim Biophys Acta*. 2007;1770(4):681–686.

99. Salouti M, Ahangari A. *Nanoparticle Based Drug Delivery Systems for Treatment of Infectious Diseases. Application of Nanotechnology in Drug Delivery*. InTech; 2014:[pp. 155–192].

100. Hancock RE, Sahl H-G. Antimicrobial and host-defense peptides as new anti-infective therapeutic strategies. *Nat Biotechnol*. 2006;24(12):1551–1557.

101. Liu L, Xu K, Wang H, Tan PJ, Fan W, Venkatraman SS, et al. Self-assembled cationic peptide nanoparticles as an efficient antimicrobial agent. *Nat Nanotechnol*. 2009;4(7):457–463.

102. Wang H, Xu K, Liu L, Tan JP, Chen Y, Li Y, et al. The efficacy of self-assembled cationic antimicrobial peptide nanoparticles against *Cryptococcus neoformans* for the treatment of meningitis. *Biomaterials*. 2010;31(10):2874–2881.

103. Xu K, Wang H, Liu L, Xu W, Sheng J, Fan W, et al. Efficacy of CG3R6TAT nanoparticles self-assembled from a novel antimicrobial peptide for the treatment of *Candida albicans* meningitis in rabbits. *Chemotherapy*. 2011;57(5):417–425.

104. Jin S, Bi D, Wang J, Wang Y, Hu H, Deng Y. Pharmacokinetics and tissue distribution of zidovudine in rats following intravenous administration of zidovudine myristate loaded liposomes. *Die Pharm*. 2005;60(11):840–843.

105. Chen H, Qin Y, Zhang Q, Jiang W, Tang L, Liu J, et al. Lactoferrin modified doxorubicin-loaded procationic liposomes for the treatment of gliomas. *Eur J Pharm Sci*. 2011;44(1):164–173.

106. Soni V, Kohli D, Jain S. Transferrin-conjugated liposomal system for improved delivery of 5-fluorouracil to brain. *J Drug Target*. 2008;16(1):73–78.

107. Saiyed ZM, Gandhi NH, Nair M. Magnetic nanoformulation of azidothymidine 5′-triphosphate for targeted delivery across the blood-brain barrier. *Int J Nanomed*. 2010;5(1):157–166.

108. Zhao M, Hu J, Zhang L, Zhang L, Sun Y, Ma N, et al. Study of amphotericin B magnetic liposomes for brain targeting. *Int J Pharm*. 2014;475(1):9–16.

109. Wagner V, Dullaart A, Bock A-K, Zweck A. The emerging nanomedicine landscape. *Nat Biotechnol*. 2006;24(10):1211–1217.

110. Batrakova EV, Gendelman HE, Kabanov AV. Cell-mediated drug delivery. *Exp Opin Drug Deliv*. 2011;8(4):415–433.

111. Duncan R, Richardson SC. Endocytosis and intracellular trafficking as gateways for nanomedicine delivery: opportunities and challenges. *Mol Pharm*. 2012;9(9):2380–2402.

112. Dou H, Destache CJ, Morehead JR, Mosley RL, Boska MD, Kingsley J, et al. Development of a macrophage-based nanoparticle platform for antiretroviral drug delivery. *Blood*. 2006;108(8):2827–2835.

113. Dou H, Grotepas CB, McMillan JM, Destache CJ, Chaubal M, Werling J, et al. Macrophage delivery of nanoformulated antiretroviral drug to the brain in a murine model of neuroAIDS. *J Immunol*. 2009;183(1):661–669.

114. Nowacek AS, McMillan J, Miller R, Anderson A, Rabinow B, Gendelman HE. Nanoformulated antiretroviral drug combinations extend drug release and antiretroviral responses in HIV-1-infected macrophages: implications for neuroAIDS therapeutics. *J Neuroimm Pharmacol*. 2010;5(4):592–601.

115. Nowacek AS, Miller RL, McMillan J, Kanmogne G, Kanmogne M, Mosley RL, et al. NanoART synthesis, characterization, uptake, release and toxicology for human monocyte–macrophage drug delivery. *Nanomedicine*. 2009;4(8):903–917.

116. Staedtke V, Brähler M, Müller A, Georgieva R, Bauer S, Sternberg N, et al. In vitro inhibition of fungal activity by macrophage-mediated sequestration and release of encapsulated amphotericin B nanosupension in red blood cells. *Small*. 2010;6(1):96–103.

117. Khan MA, Jabeen R, Nasti T, Mohammad O. Enhanced anticryptococcal activity of chloroquine in phosphatidylserine-containing liposomes in a murine model. *J Antimicrob Chemother*. 2005;55(2):223–228.

118. Batrakova EV, Kabanov AV. Cell-mediated drug delivery to the brain. *J Drug Deliv Sci Technol*. 2013;23(5):419–433.

119. Waknine-Grinberg JH, Even-Chen S, Avichzer J, Turjeman K, Bentura-Marciano A, Haynes RK, et al. Glucocorticosteroids in nano-sterically stabilized liposomes are efficacious for elimination of the acute symptoms of experimental cerebral malaria. *PloS One*. 2013;8(8):e72722.

Brain Tumors

CHAPTER 17

Brain Tumors

Sibel Bozdağ-Pehlivan, PhD
Hacettepe University, Ankara, Turkey

1 INTRODUCTION

Among all deaths that result from cancer, brain and central nervous system (CNS) tumors are the third on the list.[1] Malignant brain tumors have a high mortality rate, with an incidence rate of approximately 7.2/100,000 per year for all ages.[2,3] The incidence and mortality rates of primary brain tumors and other CNS tumors, registered between 2009 and 2013 in the United States, are represented per age category in Fig. 17.1A–B. Moreover, the incidence of brain tumors is increasing every year and the expected number of brain and spinal cord cancer cases in the United States in 2016 was 23,770, from which nearly 67.5% of the affected individuals (16,050) died in the same year.[4] The survival of a patient diagnosed with a primary brain tumor strongly depends on the age, tumor histology, behavior, and its molecular markers. Besides, the cost of care for diagnosed patients/survivors is increasing dramatically. According to the predictions of researchers, it is estimated that from a projected 13.8 and 18.1 million US cancer survivors in 2010 and 2020, respectively, the associated total cost of care for these survivors reflects an amount of 124.57 and 157.77 billion US dollars (2010 currency) for 2010 and 2020, respectively.[5] From all cancer groups, brain tumors represent the highest initial cost of care, namely over a 100,000 US dollars (2010 currency) per patient per year.[5] The financial impact, as well as the severity of this burden for both patient and caregivers brings along the need for more effective and efficient treatment methods. Although brain tumors are very complex because of their location, which limits handling procedures, many research studies have been carried out to understand the tumor-surviving/resistance mechanisms, as well as to develop drug delivery methods for noninvasive brain targeting to achieve better, more practical, and economical tumor treatments.

CONTENTS

Nanotechnology Methods for Neurological Diseases and Brain Tumors. http://dx.doi.org/10.1016/B978-0-12-803796-6.00017-4

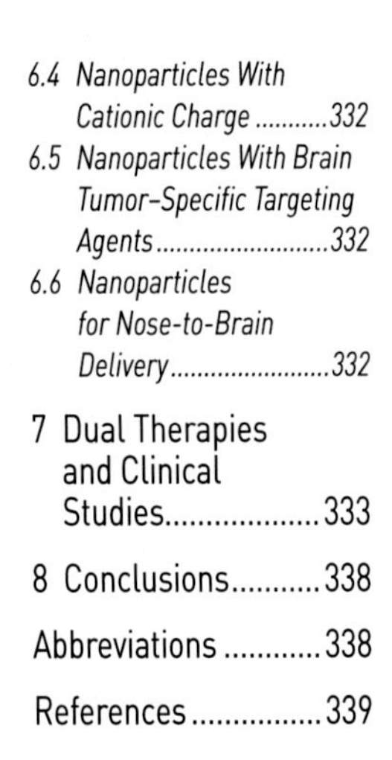

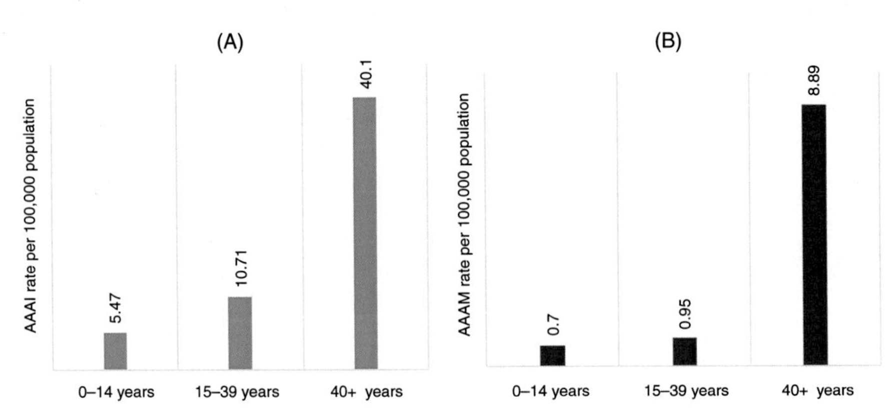

FIGURE 17.1 (A) The average annual age-adjusted incidence (AAAI) (both malignant and nonmalignant tumors) and (B) the average annual age-adjusted mortality (AAAM) rate of primary brain and other central nervous system (CNS) tumors per 100,000 population (age-adjusted to standard population in 2000 in the United States) are represented for three age categories according to data obtained between 2009–13 in the United States. *Data adapted from Ostrom QT, Gittleman H, Xu J, Kromer C, Wolinsky Y, Kruchko C, Barnholtz-Sloan JS. CBTRUS Statistical Report: primary brain and other central nervous system tumors diagnosed in the United States in 2009–2013.* Neurooncol. *2016;18(suppl 5):v1–v75.*[3]

2 TYPES OF BRAIN TUMORS

Brain tumors are caused by the formation of abnormal cells in the brain. There are two major types of brain tumors: malignant or cancerous tumors and benign tumors. To find the most optimal treatment method, it is important to understand the tumor structure and origin.

Typical characteristics of a benign tumor (nonmalignant tumor) are its slow progression behavior and its well-defined border. If removed with surgery, most benign brain tumors do not relapse. Rarely a benign tumor can spread to other locations in the body. In addition to this nature, the location of a benign tumor decides the fate of its treatment; a benign tumor located in a vital area of the brain can cause life-threatening situations similar to a malignant tumor. Meningioma is the most frequently found tumor (53.2%) in all nonmalignant brain and other CNS tumors.[3]

Cancerous tumors can be investigated as primary tumors and secondary tumors. Primary tumors originate in the brain itself, whereas secondary tumors initiate elsewhere and spread to the brain. The cell structure of a malignant brain tumor is significantly different than that of healthy brain cells. For example, gliomas are highly lethal,[6] can contain upto 1×10^6 cells with a weight of approximately 10 g, and may occur along with edema, depending on the localization and the degree of the tumor before it becomes clinically symptomatic. Malignant gliomas consist of glial cells, astrocytes, and oligodendrocytes and cause 13%–22% of all brain tumors.[7] Some extraordinary gliomas may consist of 1×10^{11} cells or even more, with a weight of 100 g, before becoming symptomatic.[8] In all malignant brain and other CNS tumors, glioblastoma is the most frequently found tumor (46.6%).[3]

To diagnose and determine the behavior of the tumor, computed tomography (CT) scans, magnetic resonance imaging (MRI) scans, and cerebrospinal fluid (CSF) samples can be analyzed. These methods can be used to determine the initial tumor size and morphology or to analyze whether the tumor remains after surgery.

According to the World Health Organization (WHO) classification,[9] there are more than 120 types of brain tumors, from which the most common types are listed in Table 17.1. From all these types, malignant gliomas are the most aggressive brain tumors due to their rapid proliferation and infiltration properties. Symptoms of malignant gliomas are increased intracranial pressure, headache, nausea and vomiting (rarely decrease in pulse rate), disorientation, epilepsy, lethargy, apathy, and papilledema.[8]

Brain tumors are further graded into four phases by WHO.[9] An overview of the characteristics of each grade is given in Table 17.2. This grading system allows the understanding of the degree of aggressiveness/malignancy of a tumor. The higher the tumor grade, the more malignant or aggressive the specific tumor

Table 17.1 Most Common Brain and CNS Tumors and Their Associated Tumor Classes

Gliomas • Astrocytoma (e.g., glioblastoma) • Oligoastrocytoma • Oligodendroglioma • Ependymoma • Choroid plexus tumor • Juvenile pilocytic astrocytoma	*Neuronal and mixed neuronal–glial tumors* • Ganglioglioma (e.g. gangliocytoma) • Central neurocytoma • Dysembryoplastic neuroepithelial tumor • Neuroblastoma
Embryonic tumors • Medulloblastoma • Atypical teratoid/rhabdoid tumor	*Meningeal tumors* • Meningioma • Hemangioma • Sarcoma (chondrosarcoma) • Lipoma
Nerve sheath tumors • Neurofibroma • Schwannoma	*Germ cell tumors* • Germinoma • Teratoma
Tumor-like anomalies • Craniopharyngioma • Epidermoid tumor • Dermoid tumor • Colloid cyst	*Other tumors* • Primary CNS lymphoma • Pituitary tumors • Pineal tumors • Metastatic tumors

Table 17.2 Description of the Four Tumor Grades According to the World Health Organization (WHO)

Grade I	Grade II
• Slow progression • Least malignant • Long-term survival	• Slow progression • Abnormal appearance • Spreading to nearby tissue • Recurrence as higher-grade tumor
Grade III	*Grade IV*
• Malignant • Active reproduction of cells • Spreading to nearby tissue • Recurrence often as IV grade	• Most malignant • Cells reproduce rapidly • Very abnormal appearance • Easily spreading to nearby tissue • Increased vascularization • Dead cells at the center

is. On the contrary, a lower tumor grade means a higher chance for recovery. For example III and IV degree gliomas are high-grade gliomas and very difficult to treat. It is also possible for a tumor to consist of areas with different grades simultaneously.[10,11]

2.1 Tumor Cell Kinetics

Knowledge about type and growth kinetics of brain tumor cells leads to the selection of the most convenient chemotherapy. Determination of kinetic parameters, such as cell cycle time, labeling index, duration of DNA synthesis, and the growth fraction decide the choice of drug(s), as well as the treatment schedule.[12]

The difference in the kinetics of normal brain tissue and tumor tissue arises because of the uncontrolled high multiplication of tumor cells. In investigating the cell kinetic parameters for malignant brain tumors, Hoshino et al. found out that glioblastoma has a proliferation rate of 30%–40%, cell cycle time of 2–3 days, and an S phase (replication) time of 4.4–10.5 h.[13,14]

The times during which antineoplastic drugs are active in the cell cycle vary. Antineoplastic drugs, such as vincristine, cytosine arabinoside, and methotrexate, have an effect at a specific point in the cell cycle and kill only those cells. Thus, only a limited proportion of growing tumor cells can be killed by a single dose of such a drug. It has been evaluated that the highest percentage of killed cells achieved in glioblastoma with a single administration of this kind of drug is 30%. Therefore, combination therapy has been recommended to increase treatment success with cell cycle–specific antineoplastic drugs.

Another class of antineoplastic drugs includes cell cycle–nonspecific agents, such as nitrosoureas and alkylating agents, which affect proliferating and nonproliferating cells and are more effective agents compared to cell cycle-specific drugs.[15,16]

3 CONVENTIONAL TREATMENT APPROACHES

Currently, three main methods are used alone or in combination in the treatment of malignant gliomas: surgery, radiation therapy, and chemotherapy.[6,17–21]

3.1 Surgery

The objective of surgical procedures in brain tumors is to confirm the diagnosis, to shrink the tumor to alleviate the symptoms of increased intracranial pressure, and to reduce the tumor mass as a pretreatment. Generally, surgical procedures for all primary malignant tumors result in the removal of 99% of the tumor mass. Consequently, the latter leads to reduced intracranial pressure and symptoms alleviation. However, due to the brain tumor's infiltrating nature and the difficulty in differentiation of malignant tissue from normal tissue in surgery, 1×10^8 to 1×10^9 tumor cells can remain after surgery. As a result, surgical intervention, which increases the average survival time of patients up to about 17 weeks only, is therefore not sufficient as a single treatment.[6] Hence, even in combination with radiotherapy, there is a risk of secondary brain tumors.[22] Eventually, patients undergoing surgery for brain tumors have a higher chance of a longer stay at the hospital, high medical costs, and higher rates of other complications.[23]

3.2 Radiation Therapy

In addition to the surgery procedure, radiation therapy is usually an effective support in the treatment of high-grade tumors, which can increase the average survival time of patients with gliomas up to about 37.5 weeks.[18] A daily amount of radiation, with a maximum dose of 1.8–2.0 Gy/day, is associated with the incidence of complications. The total applied radiation dose depends on the type, localization, and size of the tumor. For gliomas this total applied dose is usually 45–60 Gy. Besides, increasing this total radiation dose increases the risk of brain necrosis.[19] Furthermore, radiation is associated with increased side effects or postradiation tumors of the CNS, such as induced gliosarcoma,[24] glioma,[25,26] and meningioma.[27]

3.3 Chemotherapy

Chemotherapy is a commonly used conventional treatment approach to use drugs, such as carmustine, procarbazine, vincristine, and temozolomide, for the treatment of brain tumors.[20] *Intravenous* chemotherapy is the most preferred treatment method due to the absence of any surgical procedure. However, many of the cytotoxic drugs used via this route are not able to achieve effective concentrations at the tumor site due to the blood–brain barrier (BBB) or used at a high concentration in the blood, which is toxic to other tissues. One of the strategies for providing efficacious concentrations of the drug in this area is by increasing the administered dose, which unfortunately can cause severe systemic toxicity.[21]

Malignant brain tumors usually develop due to control failure in the local tumor environment. Hence, local drug application has been developed to prevent this condition. *Intratumoral* chemotherapy, is the administration of the drug directly into the tumor or into the cavity formed after removal of the tumor. The advantages of *intratumoral* chemotherapy are:

1. Increased *intratumoral* drug concentration and increased tumor–drug contact time.
2. Ensured bypass of the BBB.
3. Drug delivery to nonvascular areas in the tumor.
4. Reduction in side effects due to a reduction in the total administered dose of drug.

Nonetheless, the main drawback of the *intratumoral* method is that it is an invasive method. Additionally, due to the nonorganized structure of the tumor, the drug distribution within the tumor may not be in proportion.[21]

Another route for chemotherapy is via *intracarotid* administration. The main purpose of *intracarotid* administration is to ensure that the maximum amount of drug is carried to the tumor site, while a minimum amount of drug is passed into the systemic circulation. Hereby, both the intratumoral drug concentration and tumor–drug contact time are increased.[21] However, the drawbacks of this method are the occurrence of neurotoxicity and loss of vision due to the proximity of the application area (carotid artery) to the ophthalmic artery.[21]

4 BLOOD–BRAIN TUMOR BARRIER

The inefficient and/or ineffective ability of chemotherapeutic agents to treat brain tumors can be explained by limited drug delivery across the BBB, low drug uptake at the targeted tissue, as well as the involvement of tumor resistance. In fact, the BBB is the main obstacle for targeted therapies. In a healthy condition, the BBB consists of mainly capillary endothelial cells (ECs) interconnected through tight junctions.[28] The differences between the BBB and peripheral endothelium are the absence of fenestration, the presence of intercellular tight junctions, the lack of pinocytic vesicles, and the presence of high amount of mitochondria. The BBB allows the diffusion of lipophilic molecules, O_2, and CO_2, but restricts hydrophilic molecules. Moreover, nutrients, such as hormones and growth factors, are transported across the ECs by specific membrane transporters.[29] Macromolecules on the other hand, are transported across the BBB via specific receptors expressed on the surface of ECs, which are insulin, transferrin, and low-density lipoprotein receptor–related proteins.[30–33] For a drug to pass the BBB it should: (1) be of low molecular weight, (2) lipid soluble, (3) nonionized at physiological pH, and (4) have low affinity to serum proteins. Moreover, it has been reported that the local cerebral blood flow rate also influences the penetration of drugs through the BBB.[8]

In the presence of brain tumors, partial destruction of the capillaries results in edema and causes a more permeable BBB.[34] However, as the brain tumor grows, the BBB is disrupted and brain metastasis occurs due to angiogenesis by tumor cells, which leads to more blood supply.[35] If the tumor becomes larger than 0.2 mm^3 and the BBB is disrupted, then this existing barrier is called the blood–brain tumor barrier (BBTB) (Fig. 17.2).[36] As the tumor grows, it needs more energy, which leads to hypoxia and subsequently increased angiogenesis. After considerable significant changes in the permeability and morphology of the BBTB, the gaps developed within the capillaries and between ECs allow the passage of molecules in the range of 48 nm to 1 μM.[37]

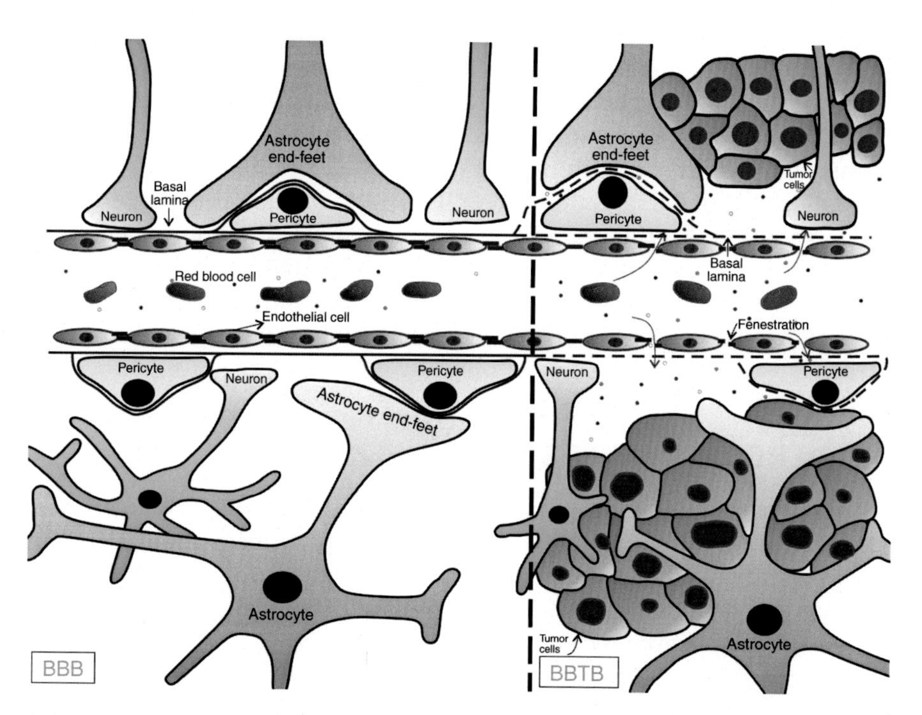

FIGURE 17.2 Schematic representation of the normal blood–brain barrier (BBB) versus blood–brain tumor barrier (BBTB) establishment.

After tumor growth reaches a size larger than 0.2 mm^3 and the basal lamina is disrupted, as well as less interaction is present between neurovascular unit cells, the molecules from the blood can easily enter into the brain, creating a protumor environment.

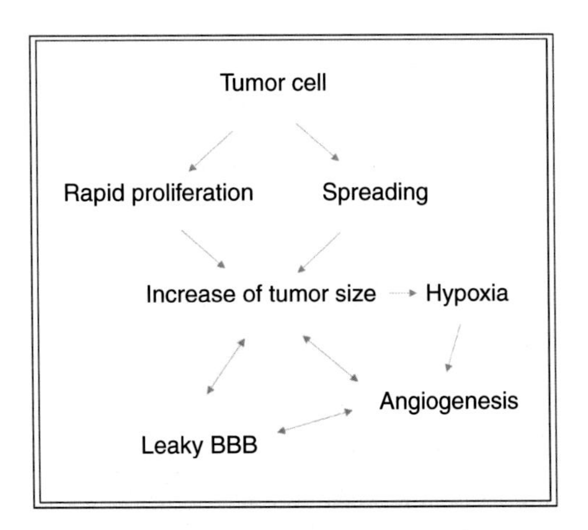

FIGURE 17.3 Mechanism of the formation of the BBTB and its characteristic loop.

The latter might be beneficial in providing drug-loaded delivery systems access through the BBB. However, due to the leaky behavior of the BBB, the drug influx rate will counteract the efflux rate, and therefore the drug might not retain its effective therapeutic concentration in the brain. Therefore, the BBTB remains an obstacle for brain-targeted drug delivery due to the low amount of drug at the target site; even the leaky BBB remains nonpermeable to many hydrophilic molecules, and therefore other approaches are necessary to overcome this hurdle. To this end, it can be assumed that the BBB integrity has a major impact on the brain tumor progression, as well as the opposite, the tumor growth and its invasion pattern affects the BBB integrity (Fig. 17.3), and both mechanisms influence one another reciprocally, making treatment very difficult.

5 OVERCOMING THE BBB

Until now, many methods have been used to overcome the BBB. Nanotechnology-based techniques for brain tumors are thoroughly discussed in the next section. Three of other methods are discussed in this section: modification of drugs, reversible opening of the BBB for passage of molecules, and the use of implants as drug delivery systems (Fig. 17.4).

5.1 Drug Modification

Lipophilic low–molecular weight drugs are able to easily penetrate via transcellular diffusion through the BBB. Therefore, it is essential to prepare lipophilic analogs for drugs that can not pass through the BBB.[38–42] For example, chlorambucil is normally ionized at physiological pH, has high affinity for plasma

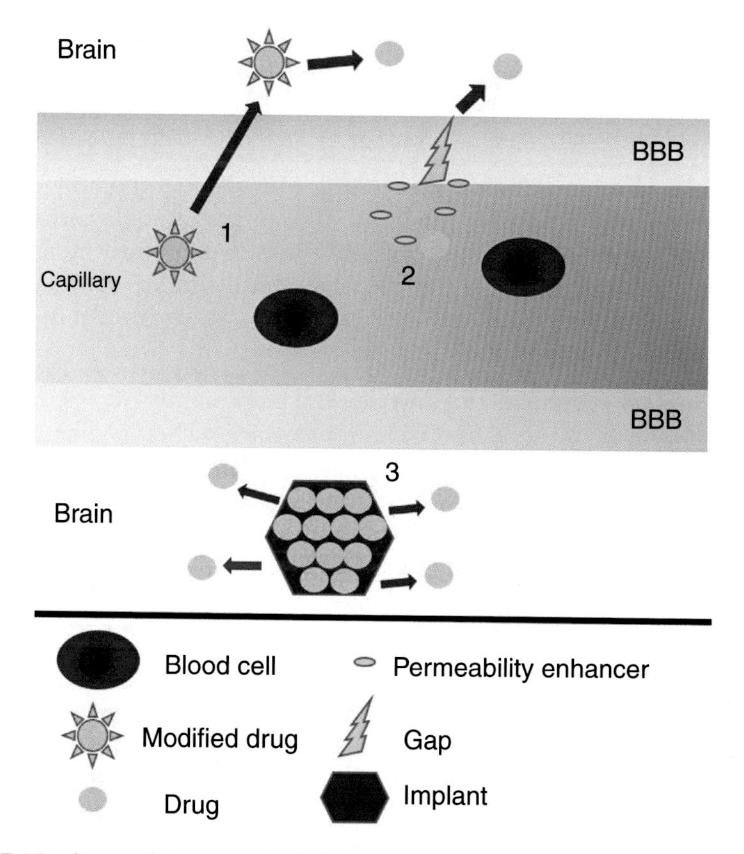

FIGURE 17.4 Several BBB-overcoming techniques.
(1) Modified drug (e.g., prodrug) crosses through the BBB and is then transformed to the drug. (2) With the use of permeability enhancers, a gap between endothelial cells (ECs) is achieved and the drug of interest can pass through the BBB. (3) Sustainable controlled release from polymeric drug carrier implants in the brain is achieved.

proteins, and therefore has minimum transition ability through the BBB. However, Greig et al.[8] were able to mask the carboxylic acid groups of chlorambucil by lipophilic ester forms. After application of chlorambucil, with and without ester forms, to rat brains, it was determined that ester-masked chlorambucil was present in higher concentrations in the brain.[8]

Among the applied approaches in drug modification, prodrugs can also used to overcome the BBB. For this approach, binding of a chemical agent to the drug can result in a lipophilic drug conjugate. This prodrug can reach the brain and can be hydrolyzed by reduction–oxidation or esterases within the target tissues in the brain. In this way, the active ingredient in the prodrug conjugate becomes free and can reach its effective concentration locally.[42]

5.2 Reversible Opening of the BBB

The BBB can be reversibly opened for a short time to allow passage of otherwise nonpermeable molecules. This can be achieved by either osmotic or biochemical approach.

The osmotic effect of hypertonic solutions is used to temporarily open the BBB to allow drug transport to the brain. The application of hypertonic solutions of substances, such as mannitol, arabinose, and dehydrocholate, causes wrinkling of brain ECs due to water loss and results in the opening of tight junctions of the BBB. This method can be used for the treatment of different brain tumors.[43]

Leukotrienes are substances that increase the permeability of peripheral capillaries. Normal brain capillaries contain enzymes, such as γ-glutamyl transpeptidase (which is not found in peripheral capillaries), which can inactivate C4–D4 and E4–F4 leukotrienes. However in the presence of brain tumors, these capillaries lose their features and after infusion of leukotriene C4, an increase in BBB permeability is reported.[44–46]

5.3 Implantation of Drug Delivery Systems

Various drug delivery systems, such as implants and microspheres, for the treatment of brain tumors have been studied by numerous researchers.[47–55] These systems increase local drug concentration and efficacy and decrease the systemic toxicity of the drug. Furthermore, due to their size (from 1 to 1000 μm), microparticles in suspensions can easily be implanted into functional areas of the brain by stereotaxic instruments, without damaging the surrounding tissues. With this technique, difficult surgery procedures for large implants are therefore avoided.

Menei et al. implanted poly(lactic glycolide acid) (PLGA) microspheres, containing 5-fluorouracil (5-FU), in the wall of the cavity formed after tumor removal in eight glioblastoma-diagnosed patients.[48] Patients were subsequently subjected to a conventional fractionated radiotherapy for 6 weeks (60 Gy). Neurological or systemic toxicity was not observed in any of the patients treated with microspheres containing 70 mg (group I) or 132 mg (group II) 5-FU. In the study, the active substance could be detected at sustained concentrations in the CSF for at least 30 days. Furthermore, in the last follow-up examination it was reported that for all patients in the latest control, the overall median survival time was 98 weeks, starting from the moment of implantation. Several years later, the same microsphere formulation with 132-mg 5-FU dose was implanted by stereotaxy into the tumors of 10 patients with inoperable malignant gliomas, and the results were again promising, with an average median survival time of 40 weeks, out of which 2 patients had 71 and 89 weeks, respectively.[49]

Bis-chloroethylnitrosourea (BCNU) also called carmustine or Gliadel, which is effectively used in the treatment of brain tumors, is a lipophilic drug with low molecular weight. Due to its physicochemical properties, BCNU can pass through the BBB when *intravenously* administered. However, the short half-life of BCNU in biological fluids (about 12 min) and its systemic toxicity at significant levels limit its usage in brain tumor treatment.[56] For this reason, many researchers aimed to overcome this problem with the use of polymeric drug delivery systems and even in combination with anticancer therapies.[50–57]

Brem et al. prepared implants containing BCNU by polyanhydride copolymer poly[bis(*p*-carboxyphenoxy)propane]anhydride (PCPP) and sebacic acid (SA) (PCPP–SA) (20:80) copolymer and performed Phase I–II studies in 21 patients with recurrent malignant glioma. After removal of the tumor, the prepared implants were placed in the formed cavity, and drug release was observed for the subsequent 3 weeks. In the study, three groups were divided into low-, medium-, and high-dose groups, and the obtained overall median survival times were determined to be 65, 64, and 32 weeks, respectively. Furthermore, it was reported that 38% of the patients lived more than 1 year after the implantation and no toxic effects were observed.[58]

Moreover, randomized placebo-controlled Phase III study using BCNU-containing biodegradable polymers showed significant improvements in the survival time of patients.[59] This local chemotherapeutic approach, which is a direct application into the resection cavity, immediately acts against residual tumor cells. Although, many clinical studies were effective and increased survival times of patients, side effects, such as significant brain edema during BCNU exposure, also occurred, which is currently under investigation.[60]

6 NANOTECHNOLOGY-BASED DRUG DELIVERY SYSTEMS

Many nanotechnology-based drug delivery systems (NBDDS) are currently under investigation as potential candidates for drug delivery to malignant brain tumors. One of these approaches is the development of polymeric drug carrier systems.[61] These systems can be either: (1) systemic administration (noninvasive) (colloidal systems, e.g., nanoparticles, liposomes, micelles, and dendrimers) or (2) local administration [by injection or implantation (invasive) into the tumor or into the cavity after tumor removal as explained before (implants, microspheres, and colloidal systems, e.g., nanoparticles, liposomes, micelles, and dendrimers)] (see Chapter 3).[62]

The application of NBDDS provides benefits in overcoming the BBB.[63] For example, Sharma et al. prepared poly(ethylene glycol) (PEG)-incorporated doxorubicin-loaded liposome formulations. Doxorubicin is an anticancer drug that has a low passage through the BBB. In their study, Sharma et al.[64] applied free doxorubicin, doxorubicin-containing liposomes, and saline (control) *intravenously* to brain tumor–bearing rats. While, free doxorubicin-treated rats showed no effective increase in live span as compared to the control group, for the doxorubicin-encapsulated liposomes–treated group, the average median survival time increased by around 29%.[64]

Among the other NBDDS, nanoparticles have been intensively investigated for the diagnosis, research, imaging, and treatment of brain tumors.[65] Nanoparticles can be designed for systemic and local administration for the treatment of brain tumors.[62,66] Due to their particle size ranging from 1–1000 nm,[67] they provide better penetration of therapeutic and diagnostic agents into target tissues, and also provide controlled long-term release of active substances.[68] After reaching the target site, drug release can be achieved by desorption of the nanoparticle, diffusion through the polymeric matrix, or via both ways.[69] The main purpose of this system is to provide effective release of therapeutic agents to the target site through the control of their particle size, surface properties, and release kinetics. Nanoparticles, due to their polymeric structure, are more stable in biological fluids and during storage as compared to liposomes,[70] and have many other advantages as a drug delivery system.[71] However, nanoparticles can interact with plasma proteins after *intravenous* administration. This process is called "opsonization." Nanoparticles with hydrophobic surface properties are defined as "foreign" by plasma components, and are therefore quickly removed from the circulation. However, very small nanoparticles with hydrophilic surface properties can escape from elimination and may remain longer in the blood circulation.[72] Therefore, nanoparticles need to be optimized for transition through the BBB as drug carriers for brain tumors and for longer retention times in the blood or brain tissue. Many techniques have been used widely for nanoparticles and several of the frequently used ones are discussed here.

6.1 Nanoparticles With Surfactants

To improve drug transition through the BBB, Kreuter et al.[73] prepared dalargin-containing poly(butyl cyanoacrylate)–adsorbed nanoparticles coated with polysorbate 80. In this study it was found that free dalargin, *intravenously* applied to mice, was not able to pass though the BBB. On the other hand, an analgesic effect was observed 45 min after application of the nanoparticle formulation.[73] The mechanism by which nanoparticles coated with polysorbate 80 pass through the BBB is not yet fully understood, but it has been reported that the detergent characteristic of polysorbate 80 opens tight junctions. Similarly, it has been found in many other studies that drugs that failed to pass the BBB were adsorbed or loaded to polysorbate 80–, poloxamer 188–, and poloxamer 407–coated nanoparticles to increase their transition.[4,74–78]

In another study, Wohlfart et al.[79] prepared doxorubicin-loaded poly(isohexyl cyanoacrylate) (PIHCA) nanoparticle formulations coated with polysorbate 80, and administered them *intravenously* to brain tumor (glioblastoma 101/8)–bearing rats. In this study, doxorubicin-loaded polysorbate 80–coated PIHCA nanoparticles showed a 10-fold higher reduction in tumor volume compared to free doxorubicin and empty polysorbate 80–coated PIHCA nanoparticle formulations.[79]

6.2 Nanoparticles With PEG

PEG is used as a hydrophilic polymer to coat nanoparticles or to copolymerize the used polymer to achieve a longer retention time of nanoparticles in the bloodstream. PEGylation has protective properties toward nanosized drug carriers, namely, it avoids the opsonization and the removal process from the blood. This protective behavior of PEGylation has been extensively studied by researchers.[80–85] It has been shown that PEG surface coating prevents the interaction of nanoparticles with other structures (such as blood components) by steric hindrance[86] and can prolong their retention time in the bloodstream.[87]

In a study by Brigg et al.,[88] PEG-coated hexadecylcyanoacrylate nanoparticles were prepared and *intravenous*ly administered to 9L gliosarcoma–bearing rats. Compared to the uncoated nanoparticle formulation, PEG-coated nanoparticles remained longer in the circulation, and a threefold accumulation was found in the tumor tissue.[88]

Cole et al.[89] aimed to prevent rapid elimination during the imaging of glioblastomas, and therefore coupled starch-coated iron oxide nanoparticles with PEG chains on their surfaces and applied them *intravenous*ly to 9L glioma–bearing rats to determine their deposition rates in the brain. In this study, compared to nanoparticles without PEG, it was found that the amount of PEG-coupled nanoparticles was 15-fold higher in the brain.[89] Furthermore, it was found that PEGylation of nanoparticles does not only extend the retention time, but also increases cellular uptake.[90,91]

6.3 Nanoparticles With Lipid/Lipid-PEG

Another approach to increase the transition of nanoparticles through the BBB is by lipid or lipid-PEG coating. Huang et al.[92] prepared iron oxide nanoparticles coated with cationic lipids [1,2-dioleoyl-3-(trimethylammonium) propane or DOTAP] and PEG-modified cationic lipids (PEG-2000–1,2-distearyl-3-*sn*-phosphatidylcholine or PEG–DSPE) mixture (3:1 molar ratio), and investigated their uptake by human cervical cancer (HeLa) cells, human prostatic adenocarcinoma (PC-3) cells, mouse neuroblastoma (Neuro-2a) cells, and mouse colorectal adenocarcinoma cells. The uptake of the coated nanoparticles into the cells was found to be higher compared to the control group.[92]

6.4 Nanoparticles With Cationic Charge

Providing cationic charge to the nanoparticles is another way to increase the transition of nanoparticles through the BBB. This process includes the preparation of biocompatible nanoparticles using cationic polymers or conjugation or coating nanoparticles with cationic materials.[93,94]

Chertok et al.[94] prepared cationic iron oxide nanoparticles with heparin–PEG adsorbed on the surface for the treatment of brain tumors. The formulations were *intravenously* administered to rats to evaluate the transition through the BBB. The study showed that, heparin–PEG–adsorbed nanoparticles retained longer in the circulation and had a twofold more accumulation ratio at the tumor site compared to the control group.[94]

6.5 Nanoparticles With Brain Tumor–Specific Targeting Agents

Targeting tumor tissue without causing damage to the healthy tissue can be achieved by incorporating specific targeting agents onto the surface of nanoparticles. For example, these nanoparticles can be targeted to the tumor receptor, prepared with biomimetic properties, antibody linked, and incorporated with immunological and genetic material.[62]

Chang et al.[95] conducted a study where PLGA nanoparticles were prepared and coated with albumin or transferrin proteins and administered *intravenously* to healthy or F98 glioma–bearing mice and rats. In this study it was found that in both healthy and tumor-bearing animals, the nanoparticles remained for a long time in the plasma. Moreover, due to the high expression of transferrin receptors in glioma cells, the transferrin-coated nanoparticles were found to be present in higher amounts in tumor-bearing animals.[95]

Guo et al.[96] prepared paclitaxel-loaded PEG–PLGA nanoparticles conjugated with AS1411, a DNA aptamer expressed at high levels in glioma cells, on its surface and administered the formulation *intravenously* to C6 glioma–bearing rats. In this study, compared to the control group, it was reported that AS1411-conjugated PEG–PLGA nanoparticles remained longer in the bloodstream, and a high amount was present at the tumor site, resulting in a decreased tumor volume and a significant increase in the survival time of the rats.[96]

6.6 Nanoparticles for Nose-to-Brain Delivery

An alternative route for noninvasive delivery of drugs to the brain is the nose-to-brain route. Nanoparticles for this route have been prepared with different polymers and investigated in many studies.[97] These studies have shown that using nanoparticles in the nose-to-brain route increased drug transport to the brain. For example, when coadministered with morphine, the 60-nm sized maltodextrin nanoparticles have been reported to prolong the antinociceptive activity time.[98]

Further addition of PEG to the nanoparticle surface has been reported to increase diffusion through the mucus and therefore the transition into the brain.[97]

In a yet unpublished study of Sekerdag et al.,[99] farnesylthiosalicylic acid (FTA)–loaded hybrid (lipid/DOTAP–PEG–PLGA) nanoparticles were prepared and a single dose of 500 μM FTA-containing nanoparticle shot was *intravenously* or *intranasally* applied to glioblastoma (RG2)–bearing rats.[99] Tumor volumes before and after treatment were detected by MRI with the use of a handmade coil for rats (Fig. 17.5). Although a very small amount of drug dose used in this study (compared to oral FTA doses of 600 mg in humans[100]), both pathways caused a significant decrease in the tumor volume after 5 days, with a 31.0% tumor reduction for *intranasally* applied FTA-loaded hybrid nanoparticles.

7 DUAL THERAPIES AND CLINICAL STUDIES

In fact, most of the NBDDS, mentioned in the previous section, are used for killing tumor cells only. However, a combination of methods for both transport drugs to the targeted tumor cells and simultaneous inhibition the local angiogenesis is of great interest currently. As discussed earlier, the tumor growth, which is reinforced by a leaky BBB, leads to increase in tumor size, hypoxia, and angiogenesis. Therefore, a combination treatment for anticancer cells and antiangiogenesis can be an ideal option, by cutting off the blood supply to these tumor cells, thereby leading to starvation of the tumor cells.

In the study of Lv et al., PEG–PLGA nanoparticles were loaded with paclitaxel and conjugated with Pep-1 and CGKRK PEG–PLGA. Pep-1 is a ligand for interleukin 13 receptor a2, which is overexpressed on glioma cells and CGKRK is a ligand for heparin sulfate, which is overexpressed on endothelial cells of tumor vasculature. With this formulation, the nanoparticle could cross the BBTB by endocytosis (CGKRK binding) through tumor vasculature, could reach the target receptor (Pep-1 binding), and release its payload paclitaxel to kill tumor cells. Results of the in vivo study showed an average median survival time of 61 days for the formulation compared to Taxol (22 days), and further showed targeted accumulation and enhanced antiglioma efficacy with almost negligible toxicity.[101]

These promising results require further investigation of dual therapy combinations for the prevention of angiogenesis and tumor growth in brain tumor treatments. Furthermore, not only dual therapies, but multitherapies also, such as surgery with the addition of radiotherapy, multidrug combinations, and antiangiogenesis could be investigated for optimal tumor eradication.

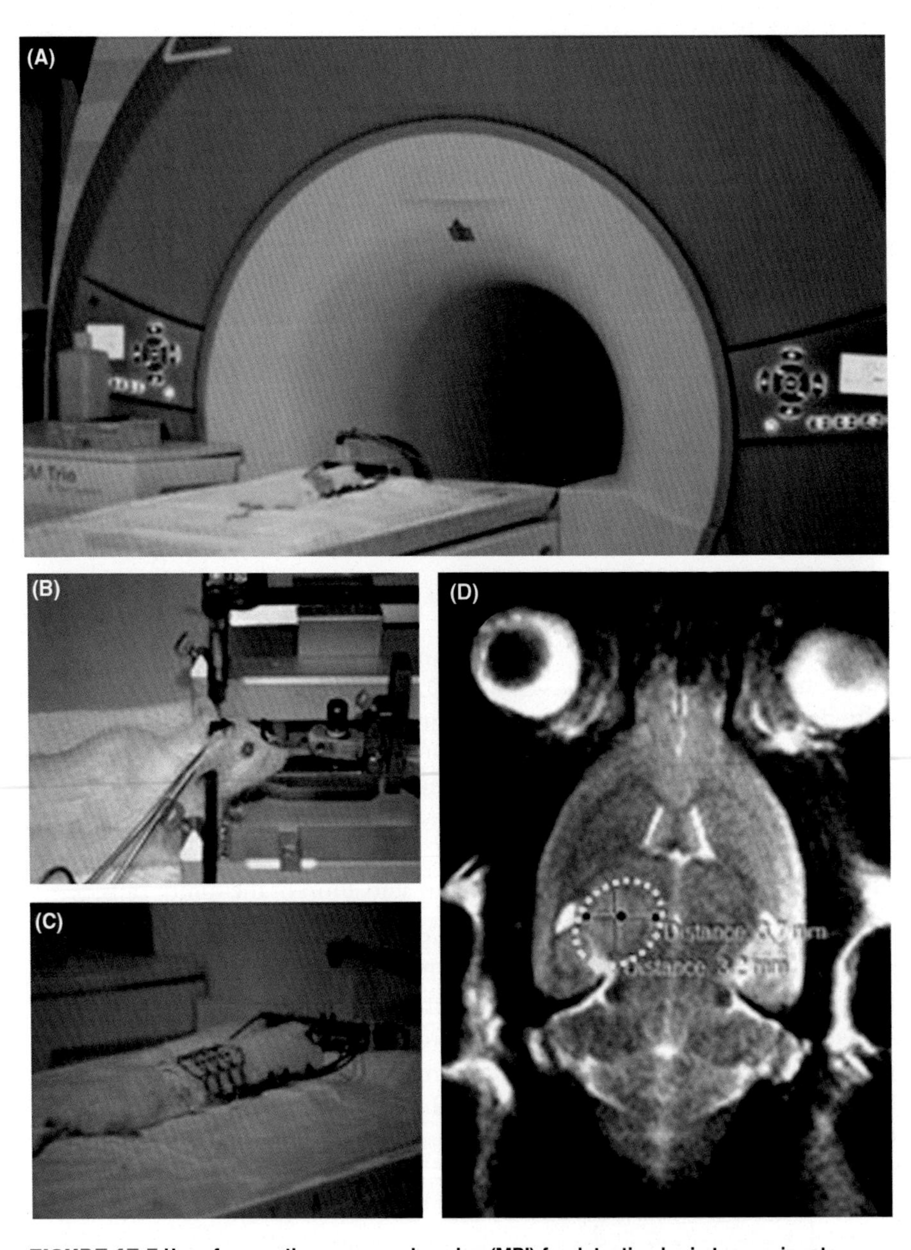

FIGURE 17.5 Use of magnetic resonance imaging (MRI) for detecting brain tumors in rats.
(A) MRI device, (B) glioblastoma inoculation in rat brain with the use of stereotaxy, (C) a special handmade coil for MRI detection of rat brain, and (D) an MRI scan of a rat brain bearing brain tumor *(dotted yellow circle).*

Nowadays, several NBDDS are already in clinical use and many other potential candidates are in the developmental stages. However, only a few of these NBDDS, such as liposomes, albumin nanoparticles, and polymeric micelles, have been approved by the FDA for tumor treatment. Table 17.3 provides a list of examples of NBDDS used or planned in approved clinical trials for brain tumor treatment. For example, one of the marketed NBDDS in the treatment of brain tumors is PEGylated liposomal doxorubicin (Doxil/Caelyx).

An interesting clinical trial, which has not started yet, is intended to investigate Phase I stage of doxorubicin-loaded epidermal growth factor receptor (EGFR) targeted by bacterially derived nanocells (EDVs) in glioblastoma and astrocytoma grade IV patients.[108] The cell particles, around 400 nm, are developed using bispecific antibodies that target specific cancer cells at the tumor vasculature only, and thereby do not leak through endothelial cells in the healthy areas in the brain. Once EDV is taken up by tumor cells, it will be degraded, and its payload, doxorubicin, will be released to kill the tumor cells.

The use of nanotechnology for radiotherapy is one of the many potential effective treatments. Namely, radiation therapy causes many side effects due to high-energy X-rays, which affect not only to tumor cells, but healthy tissues also. For this matter, tumor cells can be made more sensitive for radiation by the use of AGuIX particles. These polysiloxane gadolinium chelate–based nanoparticles can be administered for radiosensitization of tumor cells and radiotherapy can be applied. This potential method could reduce radiation sessions and be beneficial for patients.[104]

Furthermore, chemotherapy can be combined with gene therapy to achieve better effects in brain tumor treatment. In the majority of human tumors, p53 suppressor dysfunction leads to disruption/dysregulation of cell cycle control, apoptosis, senescence, and DNA and genomic integrity, as well as angiogenesis. The p53 mutation is also involved in chemotherapy and radiotherapy resistance. In one of the clinical studies, researchers intended to test the restoration of the p53 wildtype function. SGT-53 is a complex cationic liposome incorporated with the human wildtype p53 DNA sequence in a plasmid backbone. This complex can pass through the BBB and target tumor cells. However, the cytotoxicity inducing effect of restored p53 is best achieved when an apoptosis-initiating agent is applied. Therefore in this Phase II clinical trial, this vehicle will be administered to patients along with chemotherapy using the standard of care for glioblastoma, which is temozolomide.[102] This seems to be a very promising direction in brain tumor treatment strategies.

Table 17.3 Examples of Nanotechnology-Based Drug Delivery Systems (NBDDS) Used (or Currently in Progress) in Clinical Trials for Brain Tumors

Formulation	Drugs/Agents[a]	Generic/Other Names	Approaches	Tumor Types	Status
Cationic liposome	Human wildtype p53 (Temozolomide)	SGT-53	Dual therapy: gene therapy + chemotherapy	Glioblastoma	Phase II[102]
Liposome	Doxorubicin (Temozolomide)	Caelyx	Multitherapy: radiotherapy + dual chemotherapy	Glioblastoma	Phase II[103]
Nanoparticle	Gadolinium	AGuIX	Radiosensitizing tumor cells for radiotherapy	Brain metastases	Phase I[104]
Nanoparticle	Iron oxide	Nano-therm	Hyperthermia	Glioblastoma	Europe[105]
Liposome	Cytarabine (Temozolomide)	DepoCyt/Ara-C	Dual therapy: antigliomagenesis + chemotherapy	Glioblastoma multiforme Glioma Astrocytoma Brain tumor	Phase I/II[106]
Liposome	Doxorubicin	MYOCET	Reduce toxicity of drug in children	Malignant glioma	Phase I[107]
Nanocell	Doxorubicin	EnGeneIC Dream Vector	Bispesific antibody targeting	Glioblastoma Astrocytoma, grade IV	Phase I[108]
Nanoliposome	Irinotecan	NL CPT-11	Reduce toxicity of drug	Glioblastoma Gliosarcoma Anaplastic astrocytoma Anaplastic oligodendroglioma	Phase I[109]
Nanoliposome	Irinotecan	MM-398	Chemotherapy with CED	High-grade glioma	Phase I[110]
Glutathione PEGylated liposome	Doxorubicin (Trastuzumab)	2B3-101	Dual chemotherapy for metastases	Brain metastases Lung cancer Breast cancer Melanoma Malignant glioma	Phase I/IIa[111]
Nanoliposome	Rhenium	186RNL	Radiotherapy	Glioblastoma Astrocytoma	Phase I/II[112]

Liposome	Cytarabine (Bevacizumab) (Thalidomide) (Celecoxib) (Fenofibric acid) (Etoposide) (Cyclophosphamide) (Etoposide phosphate)	DepoCyt/Ara-C	Multidrug + antiangiogenic	Medulloblastoma	Phase II[113]
Liposome	Vincristine sulfate	Marqibo	Multitherapy: radiotherapy + multidrug chemotherapy	Childhood infratentorial ependymoma Childhood supratentorial ependymoma Newly diagnosed childhood ependymoma	Phase III[114]

CED, Convection-enhanced delivery; PEG, poly(ethylene glycol).
[a]*Drugs mentioned in brackets are chemotherapeutic agents that are not incorporated in the nanocarrier formulation, but are/will be administered along in the study.*

8 CONCLUSIONS

The complexity and heterogeneity of the brain tumor structure, cell kinetics, as well as the presence of the BBTB make it difficult to treat/cure brain tumors. Novel brain-targeted strategies are therefore needed in the clinical and scientific field of neurooncology. The main challenge is to cross or bypass the BBB, which prevents many drugs from reaching the brain. If NBDDS are designed with targeting agents or injected into the desired regions, they provide continuous and high amount of drugs in the targeted tissues. However, they must be designed in a way that leads to increase in drug uptake and transport across the BBB. This situation led to an intensive research on these drug delivery systems in the last few years, especially for the treatment of malignant gliomas. Furthermore, advanced nanotechnology-based formulations are being developed for dual and multidrug therapies for brain tumor treatment. For the latter, tumor cell eradication, as well as diminishing tumor neovasculature, seems to be very promising currently. Further, combining several targets in one vehicle might be a relief for patients who are taking a combination of medicines and therapies in both short and long terms. All the same, nanotechnology only requires deciphered formulations and combinations for an optimal end result. Therefore, further directions and possibilities within this area, as well as between disciplines, will be of great value to achieve better clinical outcomes.

Abbreviations

AAAI Average annual adjusted incidence
AAAM Average annual adjusted mortality
BBB Blood–brain barrier
BBTB Blood–brain tumor barrier
BCNU Bis-chloroethylnitrosourea
CNS Central nervous system
CSF Cerebrospinal fluid
CT Computed tomography
DOTAP 1,2-Dioleoyl-3-(trimethylammonium) propane
ECs Endothelial cells
EDV Bacterially derived nanocell
EGFR Epidermal growth factor receptor
5-FU 5-Fluorouracil
FTA Farnesylthiosalicylic acid
MRI Magnetic resonance imaging
NBDDS Nanotechnology-based drug delivery systems
PCPP Polyanhydride copolymer poly[bis(*p*-carboxyphenoxy)propane]anhydride
PEG Poly(ethylene glycol)
PEG–DSPE Poly(ethylene glycol) 2000–1,2-distearyl-3-*sn*-phosphatidylcholine
PIHCA Poly(isohexyl cyanoacrylate)
PLGA Poly(lactic glycolide acid)
SA Sebacic acid
WHO World Health Organization

References

1. Ostrom QT, Gittleman H, de Blank PM, Finlay JL, Gurney JG, McKean-Cowdin R, et al. American Brain Tumor Association adolescent and young adult primary brain and central nervous system tumors diagnosed in the United States in 2008–2012. *Neurooncol.* 2016;18(suppl 1):i1–i50.
2. Ostrom JB-S. Current state of our knowledge on brain tumor epidemiology. *Curr Neurol Neurosci Rep.* 2011;11(3):329–335.
3. Ostrom QT, Gittleman H, Xu J, Kromer C, Wolinsky Y, Kruchko C, et al. CBTRUS Statistical Report: primary brain and other central nervous system tumors diagnosed in the United States in 2009-2013. *Neurooncol.* 2016;18(suppl 5):v1–v75.
4. Wang CX, Hou LB, Jiang L, Yan ZT, Wang YL, Chen ZL. Antitumor effects of polysorbate-80 coated gemcitabine polybutylcyanoacrylate nanoparticles in vitro and its pharmacodynamics in vivo on C6 glioma cells of a brain tumor model. *Brain Res.* 2009;19(1261):91–99.
5. Mariotto AB, Robin Yabroff K, Shao Y, Feuer EJ, Brown ML. Projections of the cost of cancer care in the United States: 2010-2020. *J Natl Cancer Inst.* 2011;103(2):117–128.
6. Walker MD, Byar DP. Randomized comparisions of radiotheraphy and nitrosoureas for the treatment of malignant glioma after surgery. *N Eng J Med.* 1980;303:1323–1329.
7. Menei MB-C P, Croué A, Guy G, Benoit JP. Effect of stereotactic implantation of biodegradable 5-fluorouracil-loaded microspheres in healthy and C6 glioma-bearing rats. *Neurosurgery.* 1996;39(1):117–123.
8. Greig NH, Rapoport SI. Delivery of vital drugs to the brain for the treatment of brain tumors. *J Control Release.* 1990;11:61–78.
9. Louis DN, Perry A, Reifenberger G, von Deimling A, Figarella-Branger D, Cavenee WK, et al. The 2016 World Health Organization Classification of Tumors of the Central Nervous System: a summary. *Acta Neuropathol.* 2016;131(6):803–820.
10. Fuhrman SA, Lasky LC, Limas C. Prognostic significance of morphologic parameters in renal cell carcinoma. *Am J Surg Pathol.* 1982;6(7):655–664.
11. Kenmochi K, Sugihara S, Kojiro M. Relationship of histologic grade of hepatocellular carcinoma (HCC) to tumor size, and demonstration of tumor cells of multiple different grades in single small HCC. *Liver Int.* 1987;7(1):18–26.
12. Poplack DG, Horowitz ME, Bleyer WA. Pharmacology of antineoplastic agents in cerebrospinal fluid. *Neurobiology of Cerebrospinal Fluid 1*. Berlin: Springer; 1980:[pp. 561–578].
13. Hoshino T, Kobayashi S, Townsend JJ, Wilson CB. A cell kinetic study on medulloblastomas. *Cancer.* 1985;55(8):1711–1713.
14. Hoshino T, Wilson CB. Cell kinetic analyses of human malignant brain tumors (gliomas). *Cancer.* 1979;44(3):956–962.
15. Nabors LB, Surboeck B, Grisold W. Complications from pharmacotherapy. *Handb Clin Neurol.* 2016;134:235.
16. Malhotra V, Perry MC. Classical chemotherapy: mechanisms, toxicities and the therapeutc window. *Cancer Biol Ther.* 2003;2(suppl 1):1–3.
17. Bhojani MS, Rehemtulla A, Ross BD. Targeted imaging and therapy of brain cancer using theranostic nanoparticles. *Mol Pharm.* 2010;7(6):1921–1929.
18. Paoletti P, Adinolfi D, Butti G, Pezotta G. Therapy for central nervous system malignant tumors. *Brain Tumors: Biopathology and Therapy, Advances in the Biosciences*. Oxford: Pergamon; 1986:[pp. 223–235].
19. Kaye A. *Essential Neurosurgery*. Singapore: Churchill Livingstone Inc.; 1997.
20. Kim L, Glantz M. Chemotherapeutic options for primary brain tumors. *Curr Treat Opt Oncol.* 2006;7(6):467–478.

21. Berger MS. *The Gliomas*. Philadelphia, PA: WB Saunders Company; 1999.
22. Brada M, Ford D, Ashley S, Bliss JM, Crowley S, Mason M, et al. Risk of second brain tumour after conservative surgery and radiotherapy for pituitary adenoma. *Br Med J*. 1992;304(6838):1343.
23. De la Garza-Ramos R, Kerezoudis P, Tamargo RJ, Brem H, Huang J, Bydon M. Surgical complications following malignant brain tumor surgery: an analysis of 2002–2011 data. *Clin Neurol Neurosurg*. 2016;140:6–10.
24. Kaschten B, Flandroy P, Reznik M, Hainaut H, Stevenaert A. Radiation-induced gliosarcoma: case report and review of the literature. *J Neurosurg*. 1995;83(1):154–162.
25. Liwnicz BH, Berger TS, Liwnicz RG, Aron BS. Radiation-associated gliomas: a report of four cases and analysis of postradiation tumors of the central nervous system. *Neurosurgery*. 1985;17(3):436–445.
26. Simmons NE, Laws Jr ER. Glioma occurrence after sellar irradiation: case report and review. *Neurosurgery*. 1998;42(1):172–178.
27. Sadetzki S, Flint-Richter P, Ben-Tal T, Nass D. Radiation-induced meningioma: a descriptive study of 253 cases. *J Neurosurg*. 2002;97(5):1078–1082.
28. Reese T. Fine structural localization of a blood-brain barrier to exogenous peroxidase. *J Cell Biol*. 1967;34:207–217.
29. Pardridge WM. Drug transport across the blood–brain barrier. *J Cereb Blood Flow Metab*. 2012;32(11):1959–1972.
30. Descamps L, Dehouck M-P, Torpier G, Cecchelli R. Receptor-mediated transcytosis of transferrin through blood-brain barrier endothelial cells. In: Couraud P-O, Scherman D, eds. *Biology and Physiology of the Blood-Brain Barrier: Transport, Cellular Interactions, and Brain Pathologies*. Boston, MA: Springer; 1996:51–54.
31. Pardridge WM, Kang Y-S, Buciak JL, Yang J. Human insulin receptor monoclonal antibody undergoes high affinity binding to human brain capillaries in vitro and rapid transcytosis through the blood–brain barrier in vivo in the primate. *Pharm Res*. 1995;12(6):807–816.
32. Demeule M, Currie J-C, Bertrand Y, Ché C, Nguyen T, Régina A, et al. Involvement of the low-density lipoprotein receptor-related protein in the transcytosis of the brain delivery vector Angiopep-2. *J Neurochem*. 2008;106(4):1534–1544.
33. Benchenane K, Berezowski V, Ali C, Fernández-Monreal M, López-Atalaya JP, Brillault J, et al. Tissue-type plasminogen activator crosses the intact blood-brain barrier by low-density lipoprotein receptor–related protein-mediated transcytosis. *Circulation*. 2005;111(17):2241.
34. Thompson G, Stivaros SM, Jackson A. Imaging of brain tumors: perfusion/permeability. *Neuroimaging Clin N Am*. 2010;20(3).
35. Valiente M, Obenauf AC, Jin X, Chen Q, Zhang XHF, Lee DJ, et al. Serpins promote cancer cell survival and vascular cooption in brain metastasis. *Cell*. 2014;156(5):1002–1016.
36. Pinto MP, Arce M, Yameen B, Vilos C. Targeted brain delivery nanoparticles for malignant gliomas. *Nanomedicine*. 2016;12(1):59–72.
37. Schlageter KE, Molnar P, Lapin GD, Groothuis DR. Microvessel organization and structure in experimental brain tumors: microvessel populations with distinctive structural and functional properties. *Microvasc Res*. 1999;58(3):312–328.
38. Kalyanasundaram S. Intracranial drug delivery systems. *STP Pharm Sci*. 1997;7:62–70.
39. Scherrmann J-M. Drug delivery to brain via the blood–brain barrier. *Vasc Pharm*. 2002;38(6):349–354.
40. Egleton RD, Mitchell SA, Huber JD, Janders J, Stropova D, Polt R, et al. Improved bioavailability to the brain of glycosylated met-enkephalin analogs. *Brain Res*. 2000;881(1):37–46.
41. Adessi C, Soto C. Converting a peptide into a drug: strategies to improve stability and bioavailability. *Curr Med Chem*. 2002;9(9):963–978.

42. Pavan B, Dalpiaz A, Ciliberti N, Biondi C, Manfredini S, Vertuani S. Progress in drug delivery to the central nervous system by the prodrug approach. *Molecules*. 2008;13(5): 1035–1065.

43. Neuwelt EA, Dahlborg SA, Crossen J, Ramsey F, Roman-Goldstain S, Braziel R, Dana B. Primary CNS lymphoma treated with osmotic blood-brain barrier disruption: prolong survival and preservation of cognitive function. *J Clin Oncol*. 1991;9:1580–1590.

44. Black KL, Ikezaki K. Selective opening of the blood-tumor barrier by intracarotid infusion of leukotrien C4. *Acta Neurochir Suppl*. 1990;51:140–141.

45. Hashizume K, Black KL. Increased endothelial vesicular transport correlates with increased blood-tumor barrier permeability induced by bradykinin and leukotriene C4. *J Neuropathol Exp Neurol*. 2002;61(8):725–735.

46. Baba T, Black KL, Ikezaki K, Chen K, Becker DP. Intracarotid infusion of leukotriene C4 selectively increases blood-brain barrier permeability after focal ischemia in rats. *J Cereb Blood Flow Metab*. 1991;11(4):638–643.

47. Menei P, Venier-Julienne M, Benoit J. Drug delivery into the brain using implantable polymeric systems. *STP Pharm Sci*. 1997;7(1):53–61.

48. Menei P, Venier MC, Gamelin E, Saint-André JP, Hayek G, Jadaud E, et al. Local and sustained delivery of 5-fluorouracil from biodegradable microspheres for the radiosensitization of glioblastoma. *Cancer*. 1999;86(2):325–330.

49. Menei P, Jadaud E, Faisant N, Boisdron-Celle M, Michalak S, Fournier D, et al. Stereotaxic implantation of 5-fluorouracil-releasing microspheres in malignant glioma. *Cancer*. 2004;100(2):405–410.

50. McGirt MJ, Than KD, Weingart JD, Chaichana KL, Attenello FJ, Olivi A, et al. Gliadel (BCNU) wafer plus concomitant temozolomide therapy after primary resection of glioblastoma multiforme. *J Neurosurg*. 2009;110(3):583–588.

51. Brem H, Ewend MG, Piantadosi S, Greenhoot J, Burger PC, Sisti M. The safety of interstitial chemotherapy with BCNU-loaded polymer followed by radiation therapy in the treatment of newly diagnosed malignant gliomas: phase I trial. *J Neurooncol*. 1995;26(2): 111–123.

52. Brem H, Gabikian P. Biodegradable polymer implants to treat brain tumors. *J Control Release*. 2001;74(1–3):63–67.

53. Brem H, Piantadosi S, Burger PC, Walker M, Selker R, Vick NA, et al. Placebo-controlled trial of safety and efficacy of intraoperative controlled delivery by biodegradable polymers of chemotherapy for recurrent gliomas. *Lancet*. 1995;345(8956):1008–1012.

54. Della Puppa A, Lombardi G, Rossetto M, Rustemi O, Berti F, Cecchin D, et al. Outcome of patients affected by newly diagnosed glioblastoma undergoing surgery assisted by 5-aminolevulinic acid guided resection followed by BCNU wafers implantation: a 3-year follow-up. *J Neurooncol*. 2016;131:1–10.

55. Sipos EP, Tyler B, Piantadosi S, Burger PC, Brem H. Optimizing interstitial delivery of BCNU from controlled release polymers for the treatment of brain tumors. *Cancer Chemother Pharmacol*. 1997;39(5):383–389.

56. Chasin M, Hollenbeck G, Brem H, Grossman S, Colvin M, Langer R. Interstitial drug therapy for brain tumors: a case study. *Drug Dev Ind Pharm*. 1990;16(18):2579–2594.

57. Grossman SA, Reinhard C, Colvin OM, Chasin M, Brundrett R, Tamargo RJ, et al. The intracerebral distribution of BCNU delivered by surgically implanted biodegradable polymers. *J Neurosurg*. 1992;76(4):640–647.

58. Brem H, Mahaley Jr MS, Vick NA, Black KL, Schold Jr SC, Burger PC, et al. Interstitial chemotherapy with drug polymer implants for the treatment of recurrent gliomas. *J Neurosurg*. 1991;74(3):441–446.

59. Giese A, Kucinski T, Knopp U, Goldbrunner R, Hamel W, Mehdorn HM, et al. Pattern of recurrence following local chemotherapy with biodegradable carmustine (BCNU) implants in patients with glioblastoma. *J Neurooncol*. 2004;66(3):351–360.

60. Murai S, Ichikawa T, Kurozumi K, Shimazu Y, Oka T, Otani Y, et al. Quantitative analysis of brain edema in patients with malignant glioma treated with BCNU wafers. *J Clin Neurosci*. 2016;33:148–153.

61. Benoit JP, Venier-Julienne MC, Menei P. Development of microspheres for neurological disorders: from basics to clinical applications. *J Control Release*. 2000;65:285–296.

62. Haque S, Sahni JK, Ali J, Baboota S. Nanostructure-based drug delivery systems for brain targeting. *Drug Dev Ind Pharm*. 2011;38:387–411.

63. Shibata S, Mori K. Liposomes as carriers of cisplatin in to the central nervous system, experiments with 9L gliomas in rats. *Neurol Med Chir*. 1990;30:242–245.

64. Sharma US, Chau RI, Straubinger RM. Liposome-mediated therapy of intracranial brain tumors in a rat model. *Pharm Res*. 1997;14:992–998.

65. Caruso G, Alafaci C, Raudino G, Cafarella D, Lucerna S, Salpietro FM, Tomasello F. Could nanoparticle systems have a role in the treatment of cerebral gliomas? *Nanomedicine*. 2011;7(6):744–752.

66. Invernici G, Alessandri G, Navone SE, Canzi L, Tavian D, Redaelli C, Acerbi F, Parati EA. Nanotechnology advances in brain tumors: the state of the art. *Recent Pat Anticancer Drug Discov*. 2011;6(1):58–69.

67. Soppimath KS, Kulkarni AR, Rudzinski AE. Biodegradable polymeric nanoparticles as drug delivery devices. *J Control Release*. 2001;70:1–20.

68. Vila A, Tobio M, Calvo P, Alonso MJ. Design of biodegradable particles for protein delivery. *J Control Release*. 2002;78(1–3):15–24.

69. Lockman PR, Khan MA, Allen DD. Nanoparticle technology for drug delivery across the blood-brain barrier. *Drug Dev Ind Pharm*. 2002;28(1):1–13.

70. Pinto-Alphandary H, Couvreur P. Targeted delivery of antibiotics using liposomes and nanoparticles: research and applications. *Int J Antimicrob Agents*. 2000;13(3):155–168.

71. Schlossauer H. Comparative anatomy, physiology and in vitro models of the blood-brain and blood-retina barrier. *Curr Med Chem*. 2002;2:175–186.

72. Garcia E, Gil S, Couvreur P. Colloidal carriers and blood-brain barrier (BBB) translocation: a way to deliver drugs to the brain? *Int J Pharm*. 2005;298(2):274–292.

73. Kreuter J, Kharkevich DA, Ivanov AA. Passage of peptides through the blood-brain barrier with colloidal polymer particles (nanoparticles). *Brain Res*. 1995;674:171–174.

74. Petri B, Khalansky A, Hekmatara T, Müller R, Uhl R, Kreuter J, Gelperina S. Chemotherapy of brain tumour using doxorubicin bound to surfactant-coated poly(butyl cyanoacrylate) nanoparticles: revisiting the role of surfactants. *J Control Release*. 2007;117(1):51–58.

75. Gelperina S, Khalansky A, Vanchugova L, Shipulo E, Abbasova K, Berdiev R, Wohlfart S, Chepurnova N, Kreuter J. Drug delivery to the brain using surfactant-coated poly(lactide-co-glycolide) nanoparticles: influence of the formulation parameters. *Eur J Pharm Biopharm*. 2010;74(2):157–163.

76. Kulkarni SA. Effects of surface modification on delivery efficiency of biodegradable nanoparticles across the blood-brain barrier. *Nanomedicine*. 2011;6(2):377–394.

77. Gajbhiye V. The treatment of glioblastoma xenografts by surfactant conjugated dendritic nanoconjugates. *Biomaterials*. 2011;32(26):6213–6225.

78. Tian XH, Wei F, Feng W, Huang ZC, Wang P, Ren L, Diao Y. Enhanced brain targeting of temozolomide in polysorbate-80 coated polybutylcyanoacrylate nanoparticles. *Int J Nanomed*. 2011;6:445–452.

79. Wohlfart S, Gelperina S, Maksimenko O, Bernreuther C, Glatzel M, Kreuter J. Efficient chemotherapy of rat glioblastoma using doxorubicin-loaded PLGA nanoparticles with different stabilizers. *PLoS One*. 2011;6(5):e19121.

80. Harris JM, Chess RB. Effect of pegylation on pharmaceuticals. *Nat Rev Drug Discov*. 2003;2(3):214–221.

81. Veronese FM, Pasut G. PEGylation, successful approach to drug delivery. *Drug Discov Today*. 2005;10(21):1451–1458.

82. Jokerst JV, Lobovkina T, Zare RN, Gambhir SS. Nanoparticle PEGylation for imaging and therapy. *Nanomedicine*. 2011;6(4):715–728.

83. Mishra S, Webster P, Davis ME. PEGylation significantly affects cellular uptake and intracellular trafficking of non-viral gene delivery particles. *Eur J Cell Biol*. 2004;83(3):97–111.

84. El-Hammadi MM, Delgado ÁV, Melguizo C, Prados JC, Arias JL, Folic acid-decorated, PEGylated PLGA nanoparticles for improving the antitumour activity of 5-fluorouracil. *Int J Pharm*. 2017;516(1):61–70.

85. Suk JS, Xu Q, Kim N, Hanes J, Ensign LM. PEGylation as a strategy for improving nanoparticle-based drug and gene delivery. *Adv Drug Deliv Rev*. 2016;99:28–51.

86. Torchilin V. Polymer-coated long-circulating microparticulate pharmaceuticals. *J Microencapsul*. 1998;15(1):1–19.

87. Moghimi SM, Murray JC. Long-circulating and target-specific nanoparticles: theory to practice. *Pharmacol Rev*. 2001;53(2):283–318.

88. Brigger I, Aubert G, Chacun H, Terrier-Lacombe MJ, Couvreur P, Vassal G. Poly(ethylene glycol)-coated hexadecylcyanoacrylate nanospheres display a combined effect for brain tumor targeting. *J Pharmacol Exp Ther*. 2002;303:928–936.

89. Cole AJ, Wang J, Galbán CJ, Yang VC. Magnetic brain tumor targeting and biodistribution of long-circulating PEG-modified, cross-linked starch-coated iron oxide nanoparticles. *Biomaterials*. 2011;32(26):6291–6301.

90. Dmitri B, Yi Shao D, Shalaby R, Hong K, Nielsen UB, Marks JD, Benz CC, Park JW. Antibody targeting of long-circulating lipidic nanoparticles does not increase tumor localization but does increase internalization in animal models. *Cancer Res*. 2006;66:6732–6740.

91. Kaul G. Long-circulating poly(ethylene glycol)-modified gelatin nanoparticles for intracellular delivery. *Pharm Res*. 2002;19(7):1061–1067.

92. Huang HC, Chang K, Chen CY, Lin CW, Chen JH, Mou CY, Chang ZF, Chang FH. Formulation of novel lipid-coated magnetic nanoparticles as the probe for in vivo imaging. *J Biomed Sci*. 2009;16:86.

93. Agarwal A, Tiwari S, Jain S, Agrawal GP. Cationic ligand appended nanoconstructs: a prospective strategy for brain targeting. *Int J Pharm*. 2011;421(1):189–201.

94. Chertok B, Moffat BA, Yang VC. Substantiating in vivo magnetic brain tumor targeting of cationic iron oxide nanocarriers via adsorptive surface masking. *Biomaterials*. 2009;30(35):6780–6787.

95. Chang J, Passirani C, Morille M, Benoit JP, Betbeder D, Garcion E. Transferrin adsorption onto PLGA nanoparticles governs their interaction with biological systems from blood circulation to brain cancer cells. *Pharm Res*. 2011;29:1495–1505.

96. Guo J, Su L, Xia H, Gu G, Pang Z, Jiang X, Yao L, Chen J, Chen H. Aptamer-functionalized PEG-PLGA nanoparticles for enhanced anti-glioma drug delivery. *Biomaterials*. 2011;32(31):8010–8020.

97. van Woensel M, Rosière R, Amighi K, Mathieu V, Lefranc F, Van Gool SW, Vleeschouwer S. Formulations for intranasal delivery of pharmacological agents to combat brain disease: a new opportunity to tackle GBM? *Cancers*. 2013;5:1020–1048.

98. Betbeder D, Latapie JP, de Nadai J, Etienne A, Zajac JM, Frances B. Biovector nanoparticles improve antinociceptive efficacy of nasal morphine. *Pharm Res*. 2000;17:743–748.

99. Sekerdag E, Bozdag Pehlivan S, Kara A, Ozturk N, Kaffashi A, Vural I, Yavuz B, Oguz KK, Soylemezoglu F, Gursoy-Ozdemir Y, Mut M. Potential nose-to-brain delivery of hybrid nanoparticles in the treatment of glioblastoma induced female wistar rats. *First International Gazi Pharma Symposium Series (GPSS)*. Antalya, Turkey; 2015.

100. Tsimberidou AM, Rudek MA, Hong D, Ng CS, Blair J, Goldsweig H, et al. Phase 1 first-in-human clinical study of S-*trans*, *trans*-farnesylthiosalicylic acid (salirasib) in patients with solid tumors. *Cancer Chemother Pharmacol*. 2009;65(2):235.

101. Lv L, Jiang Y, Liu X, Wang B, Lv W, Zhao Y, et al. Enhanced antiglioblastoma efficacy of neovasculature and glioma cells dual targeted nanoparticles. *Mol Pharm*. 2016;13(10): 3506–3517.

102. Phase II study of combined temozolomide and SGT-53 for treatment of recurrent glioblastoma. Available from: https://clinicaltrials.gov/show/NCT02340156

103. PEGgylated liposomal doxorubicine and prolonged temozolomide in addition to radiotherapy in newly diagnosed glioblastoma. Available from: https://clinicaltrials.gov/show/NCT00944801

104. Radiosensitization of multiple brain metastases using AGuIX gadolinium based nanoparticles (NANO-RAD). Available from: https://clinicaltrials.gov/show/NCT02820454

105. Verma J, Lal S, Van Noorden CJF. Nanoparticles for hyperthermic therapy: synthesis strategies and applications in glioblastoma. *Int J Nanomed*. 2014;9:2863–2877.

106. A study of intraventricular liposomal encapsulated Ara-C (DepoCyt) in patients with recurrent glioblastoma. Available from: https://clinicaltrials.gov/show/NCT01044966

107. Myocet® in children with relapsed or refractory non-brainstem malignant glioma (MYOCET). Available from: https://clinicaltrials.gov/show/NCT02861222

108. A Phase 1 study to evaluate the safety, tolerability, and immunogenicity of EGFR (Vectibix® sequence)-targeted EnGeneIC dream vectors containing doxorubicin (EGFR(V)-EDV-Dox) in subjects with recurrent glioblastoma multiforme (GBM). Available from: https://clinicaltrials.gov/show/NCT02766699

109. A Phase I trial of nanoliposomal CPT-11 (NL CPT-11) in patients with recurrent high-grade gliomas. Available from: https://clinicaltrials.gov/show/NCT00734682

110. Study of convection-enhanced, image-assisted delivery of liposomal-irinotecan in recurrent high grade glioma. Available from: https://clinicaltrials.gov/show/NCT02022644

111. An open-label, Phase I/IIa, dose escalating study of 2B3-101 in patients with solid tumors and brain metastases or recurrent malignant glioma. Available from: https://clinicaltrials.gov/show/NCT01386580

112. Maximum tolerated dose, safety, and efficacy of rhenium nanoliposomes in recurrent glioblastoma. Available from: https://clinicaltrials.gov/show/NCT01906385

113. Metronomic and targeted anti-angiogenesis therapy for children with recurrent/progressive medulloblastoma (MEMMAT). Available from: https://clinicaltrials.gov/show/NCT01356290

114. Maintenance chemotherapy or observation following induction chemotherapy and radiation therapy in treating younger patients with newly diagnosed ependymoma. Available from: https://clinicaltrials.gov/show/NCT01096368

Future Outlook

Neurological diseases are one of the world's leading causes of morbidity and mortality. In spite of novel techniques that provide enormous data and new knowledge obtained from experimental studies, brain and central nervous system disorders are difficult to treat due to a limited number of brain-penetrating drugs. One of the major challenges to overcome in brain drug delivery is the blood–brain barrier. Efficient transfer of materials to the brain tissue must be achieved, so that potent treatment strategies for neurological diseases and brain tumors can be established.

Although the exact mechanisms of most neurological diseases are not yet fully understood, it is necessary to establish new treatment targets, as well as extensive collaborations between several disciplines and specializations. The lack of proper understanding of the mechanisms and ways involved in disease treatment poses a constant hurdle in research. Moreover, experimental design is a concern that may significantly hinder accurate research results that could otherwise have been promising and could have had the potential to be translated to the clinical level. Therefore, scientists from the fields of pharmacy to neurology, to biology and nanotechnology, as well as to clinicians, such as neurosurgeons, neurologists, radiologists, and those from several other disciplines, altogether have to find ways for close collaborations to unravel disease patterns and to open up new insights, so that new treatment approaches can emerge.

A great part of today's research is very successful in finding promising tools, therapies, and drug candidates, as well as new simulation models to prevent, treat, unravel, and/or diagnose neurological diseases and brain tumors. However, most of these developments have been studied in vivo, in vitro, in situ, or via simulation programs and thereby still lack validation in humans, which only a few successful studies have accomplished at the clinical level.

On the other hand, nanotechnology is a rapidly growing area of science, but its application to the medical research area, especially to neurological diseases, is in its infancy. Nonetheless, nanotechnology will be an important diagnostic tool for neuroimaging and an easy way of drug transport to the brain. Researchers with basic multidisciplinary knowledge and the ability to think "out of the box" might be valuable in this aspect. With a more comprehensive overview of the disease mechanisms and promising targeted drug delivery techniques, it might be possible to reach successful treatment approaches for patients with complex neurological diseases and/or brain tumors.

In the light of all this information, we hope that this book might trigger more multidisciplinary collaborations for emerging therapies in the field of nanotechnology and neuroscience.

Yasemin Gürsoy-Özdemir
Sibel Bozdağ-Pehlivan
Emine Sekerdag

Index

A

B

I

J

K

L

M

N

O

P

Q

R

S

Printed in the United States
By Bookmasters